Susan Bryant
RR#1, Rochester
In.
219-223-4437

Arithmetic
and
Algebra

THIRD EDITION

Arithmetic and Algebra

Daniel D. Benice
Montgomery College

Prentice-Hall, Inc., Englewood Cliffs, New Jersey   07632

Library of Congress Cataloging in Publication Data

BENICE, DANIEL D.
   Arithmetric and algebra.

   Includes index.
     1. Arthmetric—1961–   . 2. Algebra.
I. Title.
QA107.B46 1985   513′.12   84-13402
ISBN 0-13-046111-3

Editorial/production supervision
   and interior design: *Kathleen M. Lafferty*
Editorial assistant: *Susan Pintner*
Manufacturing buyer: *Anthony Caruso*
Art production: *Toni Sterling and Jill S. Packer*

Printed in the United States of America

10 9 8 7 6 5 4 3 2

ISBN 0-13-046111-3 01

PRENTICE-HALL INTERNATIONAL, INC., *London*
PRENTICE-HALL OF AUSTRALIA PTY. LIMITED, *Sydney*
EDITORA PRENTICE-HALL DO BRASIL, LTDA., *Rio de Janeiro*
PRENTICE-HALL CANADA INC., *Toronto*
PRENTICE-HALL OF INDIA PRIVATE LIMITED, *New Delhi*
PRENTICE-HALL OF JAPAN, INC., *Tokyo*
PRENTICE-HALL OF SOUTHEAST ASIA PTE. LTD., *Singapore*
WHITEHALL BOOKS LIMITED, *Wellington, New Zealand*

To Sylvia

Contents

PREFACE *xi*

PART I

1 *REVIEW OF BASIC OPERATIONS* *3*

2 *ORDER OF OPERATIONS* *27*

3 *EVALUATION OF GEOMETRIC FORMULAS* *32*

4 *SQUARE ROOTS* *39*

5 *PROPERTIES OF ARITHMETIC* *45*

6 *ARITHMETIC OF FRACTIONS* *51*

7   *FRACTIONS, DECIMALS, AND PERCENT*   *78*

8   *UNITS OF MEASURE*   *90*

9   *METRIC SYSTEM*   *97*

PART 2

10   *SIGNED NUMBERS*   *109*

11   *ALGEBRAIC EXPRESSIONS*   *140*

12   *SOLVING LINEAR EQUATIONS*   *159*

13   *WORD PROBLEMS*   *179*

14   *GRAPHING STRAIGHT LINES*   *196*

15   *SYSTEMS OF LINEAR EQUATIONS*   *211*

16   *INEQUALITIES*   *223*

17   *FACTORING*   *234*

PART 3

18   *FRACTIONS CONTAINING VARIABLES*   *257*

19   *PROPERTIES OF EXPONENTS*   *282*

20   *FORMULA AND EQUATION MANIPULATION*   *305*

21　*RADICALS AND ROOTS*　*314*

22　*MORE QUADRATIC EQUATIONS*　*330*

23　*SLOPES AND INTERCEPTS (OPTIONAL)*　*349*

24　*SETS AND FUNCTIONS*　*362*

APPENDICES

A　*PRETESTS*　*387*

B　*ANSWERS TO PRETESTS*　*396*

C　*RATIO AND PROPORTION*　*401*

D　*RIGHT TRIANGLE TRIGONOMETRY*　*411*

E　*NUMBER SYSTEMS*　*418*

F　*ANSWERS TO SELECTED EXERCISES*　*430*

INDEX　*471*

Preface

Goals of This Book

The third edition of *Arithmetic and Algebra* is a considerably revised version, improved by suggestions from teachers and students throughout the country who have used the text in their classes. The spirit of the text remains unchanged.

1. The presentation of arithmetic leads gently into a study of standard elementary algebra topics. The text is suitable for courses in elementary algebra, arithmetic and algebra, pre-algebra, or arithmetic.

2. Theory and discussions have been kept brief and are followed by numerous examples.

3. Because I have found that an extra step or example can make the difference between understanding and confusion, you will be led step by step through basic concepts.

4. Many exercises are included to ensure mastery of the material.

5. Answers to more than half of the exercises are provided.

6. Pretests (both questions and answers) are available in the appendices.

7. Important ideas are boxed for easy reference.

New in This Edition

Here are a few of the changes that have been made.

1. Chapter 17 (Factoring) offers expanded coverage and draws all factoring together into one chapter.

2. Chapter 11 (Algebraic Expressions) provides an expanded and reorganized introduction to algebraic expressions, including addition, subtraction, multiplication, short division, and long division. Also, the chapter offers an early introduction to some basic properties of exponents.

3. The arithmetic portion of the text is now more complete, including expansion of number structures (1.1), rounding (1.6), geometry (Chapter 3), fractions (Chapter 6), percent (7.4), improper fractions and mixed numbers (6.7), and reading decimal numbers (7.1).

4. Chapter 12 (Solving Linear Equations) has been expanded and reorganized to offer a slower-paced approach.

5. A simpler approach to complex fractions is provided (18.6).

6. Sections are numbered, example numbers are simpler, more results and suggestions are boxed for easy reference, more "notes" are given on things to watch out for.

7. More step-by-step procedures are presented, and the procedures are usually boxed (see, for example, 12.6 and 20.1).

8. The number of exercises has been greatly increased. Each set of exercises has been examined and refined, reorganized and expanded to better suit the needs of both the students and the instructor. In addition, review exercises are given at the end of each chapter.

Instructor's Manual

An Instructor's Manual is available. It contains answers to those exercises not answered in the text. The manual also contains course outlines and teaching suggestions.

Test Bank

A separate book of chapter tests is also available to instructors.

Reviewers

I am grateful for the assistance of the following reviewers of *Arithmetic and Algebra* who have helped me to improve the text.

Third Edition

Dorys J. Barban, *Montgomery College*
Marty Campbell, *College of the Virgin Islands, St. Croix*
Gerhart Moore, *Southeastern Louisiana University*
Patricia D. Roecklein, *Montgomery College*
Robin Symonds, *Indiana University at Kokomo*

Second Edition

James W. Gorham, Jr., *Harris-Stowe College*
Anita Kitchens, *Appalachian State University*
Janice McFatter, *Gulf Coast Community College*
Carla B. Oviatt, *Montgomery College*
Dudley R. Pitt, *Northwestern State College of Louisiana*
Patricia H. Rubenstein, *Montgomery College*
Ronald E. Ruemmler, *Middlesex County College*

First Edition

Margaret G. Aldrich, *Montgomery College*
Robert Alwin, *St. Petersburg Junior College*
George Schultz, *St. Petersburg Junior College*

Daniel D. Benice
Rockville, Maryland

PART I

CHAPTER **1**

Review of Basic Operations

In order to succeed in your study of algebra, you must have a solid foundation in arithmetic. Why? Because most of the properties and manipulations used in algebra are the same as those used in arithmetic. Algebra is a generalization and extension of arithmetic. You will use a great deal of arithmetic when you study algebra. So if your arithmetic background is weak, you will not be able to master algebra.

The assumption is made that arithmetic is not totally foreign to you. In other words, we assume that you have seen fractions and decimals before, and that you have added, subtracted, multiplied, and divided whole numbers before. On the other hand, the assumption is made that many things have been forgotten, confused, or never really learned.

1.1 NUMBER STRUCTURE

We begin our study of arithmetic with a look at the structure of numbers and the basic operations used to combine numbers. Much of our attention will be on the *whole numbers*—that is, 0, 1, 2, 3, 4, 5, 6, 7,   **whole number**

3

8, 9, 10, 11, 12, 13, 14, 15, and so on. At the very beginning we will be
digit   using the numbers called *digits*: 0, 1, 2, 3, 4, 5, 6, 7, 8, 9.

When you read the number 572, you don't think five-seven-two but rather five hundred seventy-two. In other words, each digit (5, 7, and 2) has an additional meaning that corresponds to where it appears in the number, that is, its *place value*.

$$572 = 500 + 70 + 2$$
$$752 = 700 + 50 + 2$$
$$275 = 200 + 70 + 5$$

A digit placed in the rightmost column of a number contributes that many *ones* to the value of the number. In *572* there are *2* ones.

A digit placed in the next column to the left contributes that many *tens* to the number. In *572* there are *7* tens.

A digit placed in the next column over contributes that many *hundreds* to the number. In *572* there are *5* hundreds.

Thus

$$572 = (5 \times 100) + (7 \times 10) + (2 \times 1)$$
$$894 = (8 \times 100) + (9 \times 10) + (4 \times 1)$$

Similarly,

$$5360 = (5 \times 1000) + (3 \times 100) + (6 \times 10) + (0 \times 1)$$
$$47,815 = (4 \times 10,000) + (7 \times 1000) + (8 \times 100) +$$
$$(1 \times 10) + (5 \times 1)$$

Here is a chart showing place values.

billions	hundred millions	ten millions	millions	hundred thousands	ten thousands	thousands	hundreds	tens	ones
↓	↓	↓	↓	↓	↓	↓	↓	↓	↓
3 ,	2	1	5 ,	7	8	4 ,	9	1	6

The *word equivalent* of any whole number can be read from the number. Here are a few examples.

894 = eight hundred ninety-four

47,815 = forty-seven thousand eight hundred fifteen

3,050,907 = three million fifty thousand nine hundred seven

Notice the use of commas to separate at every three digits, counting from the right.

EXERCISES 1.1

Answers to starred exercises are given in the back of the book.

1. Write each number as the sum of ones, tens, hundreds, and so on as in the example below.

   Example: $1563 = (1 \times 1000) + (5 \times 100) + (6 \times 10) + (3 \times 1)$

*(a) 143	(b) 159	*(c) 50
*(d) 1603	*(e) 777	(f) 490
*(g) 600	*(h) 19,734	*(i) 34,851
(j) 65,127	*(k) 648,713	(l) 952,006
*(m) 4,980,617	(n) 1,187,543	

2. Write the word equivalent of each whole number.

*(a) 754	(b) 879	*(c) 1341
(d) 8694	*(e) 54,617	(f) 71,955
*(g) 563,432	(h) 782,996	*(i) 1,897,400
(j) 7,734,015	*(k) 78,800,167	(l) 44,608,500
*(m) 654,456,546	(n) 780,078,009	*(o) 8,098,764,123
(p) 9,017,658,432		

*3. Which of the following numbers are digits?

(a) 9	(b) 0	(c) 15	(d) 3	(e) 2
(f) 10	(g) 11	(h) 167	(i) 23	(j) 32

*4. Which of the following numbers are whole numbers?

(a) 6	(b) 16	(c) 160	(d) $4\frac{1}{2}$
(e) $\frac{1}{4}$	(f) 6.4	(g) .78	(h) 10,000,000

1.2 ADDITION

Hopefully you will not need the basic addition table presented next. But if you don't already know it, memorize it. The entry 8 is circled to demonstrate how to read the table: $3 + 5 = 8$.

+	0	1	2	3	4	↓5	6	7	8	9
0	0	1	2	3	4	5	6	7	8	9
1	1	2	3	4	5	6	7	8	9	10
2	2	3	4	5	6	7	8	9	10	11
→3	3	4	5	6	7	⑧	9	10	11	12
4	4	5	6	7	8	9	10	11	12	13
5	5	6	7	8	9	10	11	12	13	14
6	6	7	8	9	10	11	12	13	14	15
7	7	8	9	10	11	12	13	14	15	16
8	8	9	10	11	12	13	14	15	16	17
9	9	10	11	12	13	14	15	16	17	18

You should be able to add 59 and 25 and get 84 as the sum.

$$\begin{array}{r} 59 \\ 25 \\ \hline 84 \end{array}$$

Your reasoning may be as follows.

1. $9 + 5 = 14$.
2. Put down *4* and carry 1.
3. $5 + 2 = 7$ plus 1 (carried) gives *8*.

But do you *understand* what you are doing? Does the procedure "put down 4 and carry 1" make sense to you or is it something that you memorized in grade school without any real understanding? Let's look at the problem differently. The number 59 has 5 tens and 9 ones; 25 contains 2 tens and 5 ones. Thus

$$\begin{array}{cc} & \text{tens} \quad \text{ones} \\ \begin{array}{r} 59 \\ 25 \\ \hline \end{array} \longrightarrow & \begin{array}{cc} 5 & 9 \\ 2 & 5 \\ \hline \end{array} \end{array}$$

Add the ones to get

$$\begin{array}{cc} \text{tens} & \text{ones} \\ 5 & 9 \\ 2 & 5 \\ \hline & 14 \end{array}$$

Since 14 ones are the same as 1 ten and 4 ones, we can rewrite the problem as

$$
\begin{array}{cc}
\text{tens} & \text{ones} \\
\mathbf{1} & \\
5 & 9 \\
2 & 5 \\
\hline
& 4
\end{array}
$$

where the boldface 1 indicates the "1 ten" part of the 14 ones. This is the "carry." The 4 is the "4 ones" part of the 14 ones. The final sum is obtained by adding the 5, 2, and carried 1. Thus

$$
\begin{array}{r}
\mathbf{1} \\
59 \\
25 \\
\hline
84
\end{array}
$$

The process is similar for addition of numbers having more than two digits; for example,

$$
\begin{array}{r}
395 \\
280 \\
155 \\
\hline
\end{array}
\rightarrow
\begin{array}{r}
\mathbf{1} \\
395 \\
280 \\
155 \\
\hline
0
\end{array}
\rightarrow
\begin{array}{r}
\mathbf{2\,1} \\
395 \\
280 \\
155 \\
\hline
30
\end{array}
\rightarrow
\begin{array}{r}
\mathbf{2\,1} \\
395 \\
280 \\
155 \\
\hline
830 \checkmark
\end{array}
$$

Addition may involve decimal numbers as well as whole numbers. We will postpone until Chapter 6 any attempt to explain the meaning of decimal numbers. But here we will present the mechanics of adding decimal numbers, since we know that you have used decimals before.

When the numbers being added have decimals, arrange the numbers so that all the decimal points are aligned. (This will align all the place values, so you will be adding tenths to tenths, and so on.) Then add as in previous examples. To add 342.6, 17.93, and 4.65, arrange the numbers and add as

$$
\begin{array}{r}
342.6 \\
17.93 \\
4.65 \\
\hline
365.18 \checkmark
\end{array}
$$

A whole number, such as 173, can be written as the decimal number 173. or 173.0. Thus to add the numbers 154.7, 173, and 23.02, set up the problem as

$$
\begin{array}{r}
154.7 \\
173.0 \\
23.02 \\
\hline
\end{array}
$$

and then add. You may prefer to supply enough zeros after the last decimal digit to make the numbers themselves align on the right side, as

$$154.70$$
$$173.00$$
$$\underline{23.02}$$

EXERCISES 1.2

*1. Perform each basic addition.

(a) 5	(b) 0	(c) 6	(d) 7	(e) 8	(f) 5
4	6	2	6	0	9

(g) 9	(h) 7	(i) 1	(j) 9	(k) 8	(l) 6
9	8	8	1	3	7

(m) 6	(n) 8	(o) 3	(p) 5	(q) 6	(r) 7
4	8	7	8	6	7

2. Perform each addition.

*(a) 57	*(b) 36	*(c) 49	(d) 58
12	53	26	37

(e) 67	*(f) 99	(g) 428	*(h) 785
79	54	351	517

*(i) 670	(j) 906	*(k) 6520	(l) 4503
920	809	1349	2917

*(m) 9743	*(n) 94	*(o) 691	(p) 157
1059	37	473	845
	15	120	419

(q) 452	*(r) 9503	(s) 9913	*(t) 1508
107	7914	4527	1974
59	1506	1602	6461
			3915

*(u) 5218	(v) 4397	*(w) 59406	*(x) 78532
4729	1562	18733	94164
1001	2468	29464	60878
7362	1619		

(y) 46521
94374
16857

3. Perform each addition.

 *(a) 76.2
   81.9

 *(b) 1.89
   7.68

 (c) 54.107
   77.099

 *(d) 145.09
    23.992

 *(e) 876.9
   752.8

 (f) 78.981
   47.097

 *(g) 86.74
   17.91
   23.85

 (h) 185.09
   329.19
   762.29

 (i) 5692.13
   9987.98

 *(j) 7888.99
   2111.01

1.3 SUBTRACTION

Consider the subtraction problem

$$5947$$
$$\underline{1362}$$

Just as in addition, you proceed from column to column: first subtract the digits in the ones column, then in the tens column, and so forth. To begin, 2 from 7 is 5.

$$5947$$
$$\underline{1362}$$
$$5$$

Next, try to subtract 6 from 4. But wait! 6 is larger than 4; so the subtraction cannot be performed at this point. Instead "borrow" 1 from the column to the left of the 4. And since numbers placed in that column are worth ten times as much, the borrowed 1 is like a borrowed 10. Thus 10 is added to the 4, and the 9 is reduced to an 8. The effect is

$$\overset{10}{5847}$$
$$\underline{1362}$$
$$5$$

As you can see, we didn't *borrow* anything. We changed the form of 94 to 84 + 10; that is, 90 + 4 became 80 + 14.

 Now subtract 6 from 14(10 + 4).

$$\overset{10}{5847}$$
$$\underline{1362}$$
$$85$$

Next, subtract 3 from 8, and subtract 1 from 5.

$$\begin{array}{r} {\scriptstyle 10} \\ 5847 \\ \underline{1362} \\ 4585 \ \checkmark \end{array}$$

We can check our result by an addition process. If $5947 - 1362 = 4585$, then $1362 + 4585 = 5947$.

$$Check: \quad \begin{array}{r} 1362 \\ \underline{4585} \\ 5947 \ \checkmark \end{array}$$

Note, however, that we usually do this addition mentally by looking at the completed subtraction problem.

$$\left.\begin{array}{r} 5947 \\ 1362 \\ \overline{4585} \end{array}\right\} \quad \text{Add these two mentally to get 5947.}$$

As another example, consider the subtraction below.

$$\begin{array}{r} 4783 \\ 2459 \\ \hline \end{array} \longrightarrow \begin{array}{r} {\scriptstyle 10} \\ 4773 \\ 2459 \\ \hline 4 \end{array} \longrightarrow \begin{array}{r} {\scriptstyle 10} \\ 4773 \\ 2459 \\ \hline 24 \end{array} \longrightarrow \begin{array}{r} {\scriptstyle 10} \\ 4773 \\ 2459 \\ \hline 324 \end{array} \longrightarrow \begin{array}{r} {\scriptstyle 10} \\ 4773 \\ 2459 \\ \hline 2324 \ \checkmark \end{array}$$

$$Check: \quad \left.\begin{array}{r} 4783 \\ 2459 \\ \overline{2324} \end{array}\right\} \quad 2459 + 2324 = 4783$$

Subtraction may involve decimal numbers. As in addition, align the numbers according to their decimal points. Then subtract as before. Here are two examples.

$$\begin{array}{r} 65.37 \\ \underline{19.42} \\ 45.95 \end{array}$$

$$\begin{array}{r} 139.07 \\ \underline{85.4} \\ \end{array} \longrightarrow \begin{array}{r} 139.07 \\ \underline{85.40} \\ 53.67 \ \checkmark \end{array}$$

EXERCISES 1.3

*1. Perform each basic subtraction.

(a) 7   (b) 9   (c) 6   (d) 8   (e) 16   (f) 18
    2   8   0   6   9   8
  ___   ___   ___   ___   ___   ___

(g) 23   (h) 24   (i) 35   (j) 26   (k) 37   (l) 26
     9   7   8   17   18   18
  ___   ___   ___   ___   ___   ___

2. Subtract and check.

*(a)   53   (b) 74   *(c) 94   (d) 253
       29   38   65   134
     ___   ___   ___   ___

*(e) 540   (f) 619   *(g) 2538   (h) 1108
      265   387   1939   794
     ___   ___   ___   ___

*(i) 9420   *(j) 16046   *(k) 59213   (l) 35984
      3428   9477   49896   26415
     ___   ___   ___   ___

*(m) 94603   (n) 84716   *(o) 666666
     82583   75908   99999
     ___   ___   ___

3. Subtract and check.

*(a) 29.7   *(b) 87.6   *(c) 92.37   (d) 132.1
      15.2   23.7   45.43   85.1
     ___   ___   ___   ___

*(e) 8.765   (f) 56.90   *(g) 65.00   (h) 19.3
      1.766   47.82   17.32   8.07
     ___   ___   ___   ___

*(i) 212.66   (j) 654.783
     132.76   453.974
     ___   ___

1.4 MULTIPLICATION

Multiplication can be viewed as a shortcut for repeated addition. For instance, $5 \times 7 =$ five times seven = five sevens (added) $= 7 + 7 + 7 + 7 + 7$. You should have learned multiplication tables like the one shown here. If you do not know this table, memorize it. The entry 42 is circled to demonstrate how to read the table: $6 \times 7 = 42$.

×	0	1	2	3	4	5	6	↓7	8	9
0	0	0	0	0	0	0	0	0	0	0
1	0	1	2	3	4	5	6	7	8	9
2	0	2	4	6	8	10	12	14	16	18
3	0	3	6	9	12	15	18	21	24	27
4	0	4	8	12	16	20	24	28	32	36
5	0	5	10	15	20	25	30	35	40	45
→6	0	6	12	18	24	30	36	(42)	48	54
7	0	7	14	21	28	35	42	49	56	63
8	0	8	16	24	32	40	48	56	64	72
9	0	9	18	27	36	45	54	63	72	81

You should know how to apply the multiplication tables to more complicated problems, such as

$$
\begin{array}{r}
27 \\
45 \\
\hline
\end{array}
$$

giving

$$
\begin{array}{r}
27 \\
45 \\
\hline
135 \\
108 \\
\hline
1215
\end{array}
$$

Some understanding of this mechanical operation can be gained by noting the following.

$$45 \times 27 = 5 \times 27 + 40 \times 27 \qquad \text{(Consider 45 as 40 and 5.)}$$

and

$$
\begin{array}{l}
5 \times 27 = 27 + 27 + 27 + 27 + 27 = 135 \\
40 \times 27 = \underbrace{27 + 27 + \quad \cdots \quad + 27}_{40 \quad 27\text{'s}} = 1080 \\
 1215 \checkmark
\end{array}
$$

The sum of *5* 27's and *40* 27's is 1215. This is the same as *45* 27's or 45 *times* 27.

> If the multiplication involves numbers with decimal points, then the result will have after its decimal point the total number of digits found after the decimal points of the numbers being multiplied.

For example,

$$
\begin{array}{r}
14.2 \\
21.3 \\
\hline
42\ 6 \\
142 \\
284 \\
\hline
302.46
\end{array}
$$

14.2 ⟵ *one* digit after decimal
21.3 ⟵ *one* digit after decimal
302.46 ⟵ *two* (*one + one*) digits after decimal

and

$$
\begin{array}{r}
1.212 \\
.5 \\
\hline
.6060
\end{array}
$$

1.212 ⟵ *three* digits after decimal
.5 ⟵ *one* digit after decimal
.6060 ⟵ *four* (*three + one*) digits after decimal

Although we will not explain here why this process works, note that in the last example the result should be half (.5) of 1.212, or .606. And indeed it is the result that we obtained by using the rule for decimals. An explanation of why this works is given after multiplication of fractions is presented in Chapter 6.

> *Multiplication of any number by zero produces zero as a result.*

$$
\begin{aligned}
0 \times 3 &= 0 \\
0 \times 56 &= 0 \\
0 \times 0 &= 0 \\
0 \times \tfrac{1}{2} &= 0 \\
0 \times 7.4 &= 0
\end{aligned}
$$

If a multiplication involves zero, it can be handled as

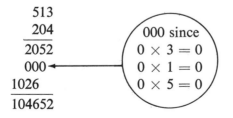

Usually, however, we do not put in the whole row of zeros (as done above); instead we simply use one zero and place the next row of the product on the same line, as

$$
\begin{array}{r}
513 \\
204 \\
\hline
2052 \\
10260 \\
\hline
104652
\end{array}
$$

Using a zero serves to indent the next row of digits (1026). It has the same effect as using the entire row of zeros.

EXERCISES 1.4

*1. Perform each basic multiplication.

(a) 8	(b) 9	(c) 7	(d) 8	(e) 0	(f) 4
9	9	5	1	7	7

(g) 3	(h) 1	(i) 4	(j) 6
6	0	3	2

2. Multiply.

*(a) 193	*(b) 165	*(c) 69	*(d) 75
9	8	52	158

(e) 287	(f) 473	*(g) 4906	(h) 6754
194	509	743	963

*(i) 6578	(j) 1692	*(k) 5600	*(l) 6754
3495	4876	4732	4000

(m) 6700	*(n) 87591	(o) 7000000
9000	60200	7000000

3. Multiply.

*(a) 19.32	*(b) 9.874	(c) 85.14	(d) 78.39
25.49	7.62	.603	56.10

*(e) 6.1697	(f) 842.93	*(g) 549.017	(h) 56.001
42.392	618.01	8581.36	10.070

*(i) 805.001	(j) 8888.8
1.00080	9999.9

1.5 DIVISION

Just as multiplication can be viewed as a shortcut for repeated addition, so can division be viewed as a shortcut for repeated subtraction. For instance, if you are dividing 4 into 12, you are answering the question "How many fours are there in 12?" This question can be answered by subtracting fours from 12 until the result is zero and counting the number of fours subtracted. Thus

$$
\begin{array}{ll}
12 & \text{original number} \\
\underline{4} & \longleftarrow \quad \text{subtract the first 4} \\
8 & \text{result} \\
\underline{4} & \longleftarrow \quad \text{subtract the second 4} \\
4 & \text{result} \\
\underline{4} & \longleftarrow \quad \text{subtract the } \textit{third} \text{ 4} \\
0 & \text{result}
\end{array}
$$

Since subtracting three 4's from 12 left zero as a result, there are three 4's in 12. So 4 goes into 12 *three* times. This can be expressed by using such notation as

$$
4 \overline{)\,12}^{\,3}
$$

$$
12 \div 4 = 3
$$

$$
12/4 = 3
$$

$$
\frac{12}{4} = 3
$$

Again (as in multiplication) we learn tables so that we can divide directly and quickly and avoid the actual subtraction and counting process.

In more complicated division problems, tables cannot be used *directly* to determine the *final result*. Several applications are needed. For example, in

$$
5 \overline{)\,170}
$$

first divide 5 into 17 and get 3. Multiply 3 by 5 to get 15 and subtract the 15 from 17. There is a remainder of 2 at this point.

The 15 is from 3×5.

You then "bring down" the next digit (0) and divide 5 into the newly formed number (20). This result is 4. The final result is 34.

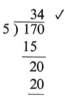

But this is a shortcut. Can you see what is really going on? If you look carefully at

$$5 \overline{)170}$$

you can see that $5 \times 30 = 150$, which is smaller than 170, whereas $5 \times 40 = 200$, which is larger than 170. So divide 5 into 170 and get 30.

$$
\begin{array}{r}
30 \\
5 \overline{)170} \\
150 \\
\hline
20
\end{array}
$$

Now 20 of the 170 is left. Next, divide 5 into that 20 to see how many more fives there are in 170. Since $20 \div 5 = 4$,

$$
\begin{array}{r}
30 + 4 = 34 \ \checkmark \\
5 \overline{)170} \\
150 \\
\hline
20 \\
20 \\
\hline
0
\end{array}
$$

So you can see that division (just as multiplication) can be done without shortcuts; however, meaningful shortcuts are available—and the shortcuts should be used. Consider a more advanced example.

$$34 \overline{)75265}$$

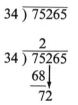

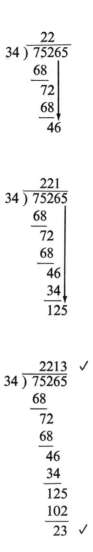

$$
\begin{array}{r}
22 \\
34\,)\,\overline{75265} \\
68 \\
\overline{72} \\
68 \\
\overline{46}
\end{array}
$$

$$
\begin{array}{r}
221 \\
34\,)\,\overline{75265} \\
68 \\
\overline{72} \\
68 \\
\overline{46} \\
34 \\
\overline{125}
\end{array}
$$

$$
\begin{array}{r}
2213\ \checkmark \\
34\,)\,\overline{75265} \\
68 \\
\overline{72} \\
68 \\
\overline{46} \\
34 \\
\overline{125} \\
102 \\
\overline{23}\ \checkmark
\end{array}
$$

The result is not exact, as before. There is a *remainder* of 23. The answer   **remainder**
can be written as 2213 R23 (R for *R*emainder) or as $2213\frac{23}{34}$, where the
fraction $\frac{23}{34}$ means that the process could be continued by dividing 34
into 23. But how can we divide 34 into 23, since 34 is larger than 23?
Continue the computation by placing a decimal point after 75265 and
zeros after the decimal point. This does not change the value of the number 75265.

```
                            2213.
                    34 ) 75265.0000
                         68
                         ──
                          72
                          68
                          ──
                           46
                           34
                           ──
                          125
                          102
                          ───
                           23
```

Let us continue as before.

```
              2213.6                         2213.67
      34 ) 75265.0000                 34 ) 75265.0000
           68         │                     68         │
           ──         │                     ──         │
            72        │                      72        │
            68        │                      68        │
            ──        │                      ──        │
             46       │      ⟶               46       │
             34       │                      34       │
             ──       │                      ──       │
            125│                            125        │
            102↓                            102        │
            ───                             ───        │
             230                             230       │
             204                             204↓
             ───                             ───
              26                             260
                                             238
                                             ───
                                              22
```

We can continue until the result comes out exact (with no remainder) or until we have enough digits after the decimal point. The result of division does not always come out exact; for example, $\frac{1}{3} = .333333333. \ldots$ It is a nonterminating decimal.

Division involving numbers with decimal points is similar except that the decimal points of the numbers involved are shifted, as explained in the next example. The reason that this shifting process works is given after multiplication of fractions is explained in Chapter 6.

If you have a problem like

$$7.42 \overline{)\ 269.3516}$$

then shift the decimal two places to the right in both numbers. Doing so will make the divisor a whole number, 742.

$$7.\underset{\curvearrowright}{42}\,) \, 269.3\underset{\curvearrowright}{5}16$$

So we get

$$742 \,) \, \overline{26935.16}$$

instead of

$$7.42 \,) \, \overline{269.3516}$$

We use multiplication to *check* that the result produced by division is correct. That is,

$$\begin{array}{r} 34 \\ 5 \,) \, \overline{170} \end{array} \quad \text{if} \quad 34 \times 5 = 170$$

When the division does not produce an exact result, the check is not complete until the remainder is added to the product. To illustrate, consider the division

$$34 \,) \, \overline{75063}$$

The division:

$$\begin{array}{r} 2207 \\ 34 \,) \, \overline{75063} \\ \underline{68} \\ 70 \\ \underline{68} \\ 26 \\ \underline{0} \\ 263 \\ \underline{238} \\ 25 \; R \end{array}$$

The check:

$$(2207 \times 34) + 25 = 75038 + 25$$
$$= 75063 \; \checkmark$$

EXERCISES 1.5

*1. Perform each basic division. Then check, using multiplication.

(a) $5 \,) \, \overline{20}$ (b) $6 \,) \, \overline{54}$ (c) $9 \,) \, \overline{63}$

(d) $8 \,) \, \overline{24}$ (e) $4 \,) \, \overline{28}$ (f) $3 \,) \, \overline{18}$

(g) $1 \,) \, \overline{8}$ (h) $2 \,) \, \overline{18}$ (i) $7 \,) \, \overline{35}$

(j) $6 \,) \, \overline{42}$ (k) $5 \,) \, \overline{45}$ (l) $3 \,) \, \overline{27}$

2. Divide and check.
   *(a) 4) 2844   *(b) 2) 16846   *(c) 3) 2763
    (d) 9) 1827   (e) 8) 64016   *(f) 7) 5915
   *(g) 21) 1974   (h) 15) 5730

3. Divide. Obtain the remainder in each case. Then check.
   *(a) 9) 78   (b) 7) 59   *(c) 12) 1531
    (d) 15) 9409   *(e) 23) 5827   (f) 31) 14762
   *(g) 84) 19526   (h) 64) 37682

4. Divide. Continue to two decimal places.
   *(a) 27) 8416   *(b) 45) 7894   (c) 40) 15433
    (d) 102) 56781   (e) 153) 67854   *(f) 78) 99999
   *(g) 205) 19864   (h) 234) 78695

5. Divide. Continue until your answer contains five digits.
   *(a) 7.9) 259.204   (b) 2.1) 54923   *(c) .62) 78.65
    (d) 10.3) 5.73029 *(e) 2.83) 67435   (f) 8.79) .015934
   *(g) .06) 987   (h) .009) 743.64

1.6 ROUNDING

Suppose that you paid $10 for six house plants and you want to know how much you paid per plant. Naturally you would divide the $10 by 6 in order to obtain the cost per plant. The division:

$$\frac{\$\ 1.666666}{6\ )\ \$10.000000}$$

The division could be continued indefinitely. Actually, it has been carried too far already. Certainly $1.666666 is a strange form for an amount of money. Since dollars and cents are the standard way in which we express money, two digits after the decimal point are sufficient. But should the result be simply $1.66?

When *rounding* a number, the last digit retained should reflect the size of the digits being dropped. Specifically, if a number is being rounded to two decimal places (that is, two digits after the decimal point), then the third digit after the decimal should be examined. If it is 5 or greater (5, 6, 7, 8, or 9), the second digit should be increased by 1. If it is 4 or smaller (0, 1, 2, 3, or 4), the second digit should not be changed. All digits beyond the second digit are dropped. Here are some examples. More will be said about tenths and hundredths in Section 7.1.

Number	Rounded to Two Decimal Places (*nearest hundredth*)
$1.666666	$1.67
75.287	75.29
1.492	1.49
849.1254	849.13
16.2964	16.30
735.1337	735.13

The procedure for rounding to one decimal place is similar, of course.

Number	Rounded to One Decimal Place (*nearest tenth*)
9.27	9.3
9.24	9.2
18.35	18.4
1.791	1.8
2.96	3.0

Here are some examples of numbers rounded to the nearest whole number.

Number	Rounded to Nearest Whole Number
81.6	82
81.2	81
9.51	10
9.49	9
86.499	86
89.96	90

When rounding to the nearest ten, hundred, thousand, etc. (that is, to the *left* of the decimal point), digits being "dropped" to the left of the decimal point are replaced by zeros used as placeholders.

Number	Nearest Ten	Comment
863	860	since 3, leave 6 as is
7249	7250	since 9, add 1 to the 4

Number	Nearest Hundred	Comment
863	900	since 6, add 1 to the 8
7249	7200	since 4, leave 2 as is

EXERCISES 1.6

1. Round each number to two decimal places (nearest hundredth).
   *(a) 73.769   *(b) 15.222   (c) 4.128991   *(d) 98.1351
   (e) 1.87499   *(f) 765.919   *(g) 56.895   (h) 159.995

2. Round each number to one decimal place (nearest tenth).
   *(a) 17.83   *(b) 99.68   *(c) 187.15   *(d) 40.049
   (e) 76.555   *(f) 456.789   (g) 29.008   *(h) 45.986

3. Round each number to the nearest whole number.
   *(a) 19.8   *(b) 23.43   (c) 67.9102   (d) 56.5
   (e) 456.7   *(f) 9.999   *(g) 34.4999   (h) 298.18

4. Round each number to three decimal places.
   *(a) 58.6439   *(b) 46.1631   (c) 1.6765   *(d) 98.0017
   (e) 1009.9944   (f) 76.7996   *(g) 9.9998   (h) 1.78632
   *(i) 5.76509   (j) 384.0008

5. Round each number to the nearest ten.
   *(a) 86   *(b) 147   *(c) 541   (d) 894
   *(e) 5675   (f) 8365   *(g) 90   (h) 50

6. Round each number to the nearest hundred.
   *(a) 1770   *(b) 1850   *(c) 5449   (d) 1784
   *(e) 1990   (f) 3956   (g) 2849   *(h) 9979

1.7 DIVISION INVOLVING ZERO

Why does $20 \div 5 = 4$? It checks! That is, $20 \div 5 = 4$, since $4 \times 5 = 20$.

$$5\overline{)20} \quad \text{since} \quad 4 \times 5 = 20$$

Similarly,

$$6\overline{)18} \quad \text{since} \quad 3 \times 6 = 18$$

and

$$3\overline{)21} \quad \text{since} \quad 7 \times 3 = 21$$

Then what about

$$5\overline{)0}$$

The result is 0:

$$5 \overline{)0}^{\,0} \quad \text{since} \quad 0 \times 5 = 0; \text{ that is, 0 checks!}$$

Similarly,

$$3 \overline{)0}^{\,0} \quad \text{since} \quad 0 \times 3 = 0$$

$$29 \overline{)0}^{\,0} \quad \text{since} \quad 0 \times 29 = 0$$

So *if we divide a number into* 0, *the result is* 0.

Another interesting case is division of a nonzero number by zero; for example,

$$0 \overline{)3}$$

The result of this division is the number that when multiplied by zero produces 3. In other words, we want some number (call it □) such that $0 \times \square = 3$. But zero times any number is zero; so there is no number that checks. Consequently, we *cannot divide by zero*; it is an *undefined operation*, an *impossible operation*.

But what about the strange-looking case

$$0 \overline{)0}$$

It is obvious again, isn't it?

$$\text{Is} \quad 0 \overline{)0}^{\,0} \quad \text{since} \quad 0 \times 0 = 0?$$

You should also see that *any number* will work in the check.

$$0 \overline{)0}^{\,5} \quad \text{since} \quad 5 \times 0 = 0$$

$$0 \overline{)\ 0}^{\,17} \quad \text{since} \quad 17 \times 0 = 0$$

$$0 \overline{)\ 0}^{\,29\frac{17}{31}} \quad \text{since} \quad 29\tfrac{17}{31} \times 0 = 0$$

In other words, we *cannot determine* what *the* result is when we divide 0 by 0. We call such a "result" *indeterminate* and make no attempt to **indeterminate** guess an answer in such cases; instead we will say that "the answer cannot be determined" or is "indeterminate." Such division is undefined, just as any other division by 0 is undefined. However, the distinction as "indeterminate" is important in more advanced mathematics.

EXERCISES 1.7

1. Each division problem below involves *zero*. Indicate whichever of the following is the case:

   i. Result is 0.
  ii. Operation is undefined (indeterminate).
 iii. Operation is undefined (impossible).

  *(a) 16) 0   (b) 3) 0   *(c) 0) 0   (d) 0) 6
  *(e) 0) 17   (f) 5) 0   *(g) 0) 0   (h) 9) 0
  *(i) 0) 13   (j) 8) 0   *(k) 0) 0   (l) 49) 0

1.8 TERMINOLOGY

The following examples introduce the terminology often used to identify the different numbers used in addition, subtraction, multiplication, and division.

Addition:
    17 ← addend
    35 ← addend
    52 ← sum

Subtraction:
    37 ← minuend
    12 ← subtrahend
    25 ← difference

Multiplication:
    16 ← multiplicand
     4 ← multiplier
    64 ← product

Division:
       5 ← quotient
   4) 23 ← dividend
      20
       3 ← remainder
divisor

1.9 THE MEANING OF EXPONENT

Look at the examples in the table below to see how *exponents* can be used as a shorthand notation for multiplication. Here we are using a dot to indicate multiplication, instead of using $\times$. Thus 4×5 can be written as $4 \cdot 5$. Still other ways of writing it include (4)(5), 4(5), (4)5, and variations of them. We use the word *factor* to mean any of the numbers that are multiplied together to form a product; for example, 3 and 7 are factors of 21.

Multiplication Form	Exponent Form	Base	Exponent
$5 \cdot 5$	5^2	5	2
$3 \cdot 3 \cdot 3 \cdot 3 \cdot 3 \cdot 3 \cdot 3$	3^7	3	7
$4 \cdot 4 \cdot 4$	4^3	4	3
$2 \cdot 2$	2^2	2	2
$x \cdot x \cdot x \cdot x \cdot x$	x^5	x	5

factor

The number that is repeated in the multiplication is called the *base*. The number of times that it appears as a factor of the multiplication is called the *exponent* or *power*. Thus in 5^7 (or $5 \cdot 5 \cdot 5 \cdot 5 \cdot 5 \cdot 5 \cdot 5$), 5 is the base and 7 is the exponent. The base and exponent are given for each example in the chart.

base

exponent, power

Example 1. *Write* $7 \cdot 7 \cdot 7 \cdot 7 \cdot 7 \cdot 7 \cdot 7 \cdot 7$, *using exponents.*

The base is 7. It appears 8 times as a factor; so the exponent is 8. Thus $7 \cdot 7 \cdot 7 \cdot 7 \cdot 7 \cdot 7 \cdot 7 \cdot 7 = 7^8$.

Example 2. *Write* 5^6 *without using exponents.*

$5^6 = 5 \cdot 5 \cdot 5 \cdot 5 \cdot 5 \cdot 5$, since the base 5 must appear 6 (exponent) times as a factor.

If a number is raised to the second power, we say that it is *squared*. Thus 5^2 can be read "5 squared" as well as "5 to the second power."

square

If a number is raised to the third power, we say it is *cubed*. Therefore 5^3 can be read "5 cubed" as well as "5 to the third power."

cube

EXERCISES 1.9

1. Express each by using exponents—for example, $3 \cdot 3 \cdot 3 \cdot 3 \cdot 3 = 3^5$.
   *(a) $6 \cdot 6$
   (b) $3 \cdot 3$
   *(c) $2 \cdot 2 \cdot 2$
   *(d) $9 \cdot 9 \cdot 9 \cdot 9 \cdot 9 \cdot 9 \cdot 9$
   (e) $4 \cdot 4 \cdot 4 \cdot 4 \cdot 4$
   *(f) $3 \cdot 3 \cdot 3 \cdot 3 \cdot 3 \cdot 3 \cdot 3 \cdot 3 \cdot 3 \cdot 3$
   (g) $12 \cdot 12 \cdot 12 \cdot 12 \cdot 12 \cdot 12$
   *(h) $m \cdot m$
   *(i) $x \cdot x \cdot x \cdot x \cdot x$
   *(j) $(x + y)(x + y)(x + y)$

2. Express as a product of factors, for example, $3^5 = 3 \cdot 3 \cdot 3 \cdot 3 \cdot 3$.
   *(a) 8^2 *(b) 2^3 (c) 5^4 *(d) 6^8
   (e) 3^7 (f) 10^{13} *(g) x^2 (h) y^5
   *(i) $(a + b)^4$

3. Compute the value; for example, $3^4 = 3 \cdot 3 \cdot 3 \cdot 3 = 81$.
   *(a) 2^3 (b) 4^2 *(c) 5^4 (d) 2^5
   (e) 9^2 *(f) 7^3 *(g) 1^8 (h) 15^2
   *(i) 3^5 (j) 6^3 *(k) 0^3 (l) 4^4

Chapter 1. REVIEW EXERCISES

1. Perform each addition.

   *(a) 768   (b) 7699   *(c) 67.89
          924   9901   82.04
          659   5643

   (d) 679.06   *(e) 56789
        321.94   12345
        756.71   77777
                     76804

2. Perform each subtraction.

   *(a) 8976   (b) 7644   *(c) 96666
         4563   6545   77676

   *(d) 45.987   (e) 516.765
        32.899   498.886

3. Perform each multiplication.

   *(a) 638   (b) 1007   *(c) 8739
         542   588   4702

   *(d) 56.78   (e) 459.12
        10.80   88.46

4. Perform each division if possible. Continue to two decimal places.
   *(a) $45 \overline{)\, 7389}$ *(b) $67 \overline{)\, 0}$ *(c) $0 \overline{)\, 892}$
   *(d) $67.1 \overline{)\, 987.3}$ (e) $8.92 \overline{)\, 999.99}$

5. Round off each number to one decimal place.
   *(a) 5.98   (b) 87.04   *(c) 9.91
   (d) 19.99   *(e) .85216

*6. Find the value of each expression by actual computation.
   (a) 2^4 (b) 0^2 (c) 5^3
   (d) 12^2 (e) 1.3^2

Order
of Operations

2.1 ORDER OF OPERATIONS

In Chapter 1 you reviewed the basic arithmetic operations of addition, subtraction, multiplication, division, and exponentiation. Hopefully you will have no problem carrying out such operations in the future. Unfortunately, however, when a *series* of such operations is encountered, the order in which they must be carried out may not be clear. For example, in the calculation $2 + 3 \cdot 4$, should you first add the 2 and 3 or should you multiply the 3 and 4? Certainly if parentheses were used, there would be no doubt. Thus $(2 + 3) \cdot 4$ specifies that the 2 and 3 should be added and that sum should then be multiplied by 4. The result is 20. On the other hand, $2 + (3 \cdot 4)$ specifies that the 3 and 4 should be multiplied and that result should be added to 2. The result is 14.

But what do you do when there are no parentheses to guide you? There are rules for the order in which operations must be carried out when parentheses do not specify. For instance, multiplication is done before addition. So in the case of $2 + 3 \cdot 4$, the result is 14 rather than 20.

Here is the entire set of rules for determining the order of operations *when there are no parentheses to guide you.*

> When there are no parentheses,
>
> *First:* Apply all exponents.
>
> *Second:* Perform all multiplications and divisions, proceeding from left to right.
>
> *Third:* Perform all additions and subtractions, proceeding from left to right.

In the examples that follow, the rules for order of arithmetic operations (given above) will be applied. In each case the expression is simplified down to a one-number final answer. The directions, however, may be given in a variety of ways, including "Simplify the expression," "Simplify," "Evaluate the expression," "Determine the value of," "Calculate," and "Evaluate."

Example 1. *Simplify the expression* $4 + 9 \cdot 2$.

$$4 + 9 \cdot 2 = 4 + 18 \qquad \text{Do multiplication before addition.}$$
$$= 22 \ \checkmark$$

Example 2. *Simplify* $2 \cdot 3^2$.

$$2 \cdot 3^2 = 2 \cdot 9 \qquad \text{Apply the exponent before multiplying.}$$
$$= 18 \ \checkmark$$

Example 3. *Evaluate the expression* $6 \cdot 5 \div 2 \cdot 3$.

$$6 \cdot 5 \div 2 \cdot 3 = 30 \div 2 \cdot 3 \qquad \left\{ \begin{array}{l} \text{since multiplication and division} \\ \text{are performed from left to right} \end{array} \right.$$

$$= 15 \cdot 3 \qquad \left\{ \begin{array}{l} \text{since multiplication and division} \\ \text{are performed from left to right} \end{array} \right.$$

$$= 45 \ \checkmark$$

Example 4. *Evaluate* $15 + \dfrac{5}{5}$.

$$15 + \frac{5}{5} = 15 + 1 \qquad \text{Do division before addition.}$$
$$= 16 \ \checkmark$$

Example 5. *Determine the value of* $5 \cdot 2^2 + 7$.

$$5 \cdot 2^2 + 7 = 5 \cdot 4 + 7 \qquad \text{Do exponentiation before multiplication.}$$
$$= 20 + 7 \qquad \text{Do multiplication before addition.}$$
$$= 27 \ \checkmark$$

Example 6. *Calculate* $10 + 8 \div 2 + 1$.

$$10 + 8 \div 2 + 1 = 10 + 4 + 1 \qquad \text{Do division before addition.}$$
$$= 15 \checkmark$$

Example 7. *Simplify* $2 + 3^2 + 8 + \dfrac{12}{2}$.

$$2 + 3^2 + 8 + \frac{12}{2} = 2 + 9 + 8 + \frac{12}{2} \qquad \text{Apply the exponent first.}$$
$$= 2 + 9 + 8 + 6 \qquad \text{Do division before addition.}$$
$$= 25 \checkmark$$

Parentheses and other grouping symbols can be used to change the order of operations when desired. *If parentheses are used, the expression inside them is treated as if it were one number. That is, its value is computed before it is combined with other quantities.*

> When parentheses are present, simplify
> all expressions inside parentheses.
> *Then* apply the order of operations rules.

Example 8. *Simplify* $(5 + 2)6$.

$$(5 + 2)6 = (7)6 \qquad \begin{cases} \text{because parentheses indicate that} \\ \text{the addition is done first} \end{cases}$$
$$= 42 \checkmark$$

Example 9. *Evaluate* $(9 + 3) \div 4 + 2$.

$$(9 + 3) \div 4 + 2 = 12 \div 4 + 2 \qquad \begin{cases} \text{First, evaluate the expression} \\ \text{inside parentheses.} \end{cases}$$
$$= 3 + 2 \qquad \text{Do division before addition.}$$
$$= 5 \checkmark$$

In the next example the notation / is used for division, as is often the case in computer programming.

Example 10. *Simplify* $(9 + 3)/(4 + 2)$.

$$(9 + 3)/(4 + 2) = 12/6 \qquad \begin{cases} \text{after first determining the values} \\ \text{of the quantities in parentheses} \end{cases}$$
$$= 2 \checkmark$$

Very often division, such as that used in Example 10, is not written on one line. In other words, $(9 + 3)/(4 + 2)$ would appear on two lines, as

$$\frac{9 + 3}{4 + 2}$$

In this latter form, it is obvious that the 9 and 3 should be added and that the 4 and 2 should be added before any division takes place. The long bar of the division serves the same purpose as the parentheses in the one-line form. Thus

$$\frac{9 + 3}{4 + 2} = \frac{12}{6} = 2$$

If the expression of Example 9 is written in this form, it becomes

$$\frac{9 + 3}{4} + 2 = \frac{12}{4} + 2 = 3 + 2 = 5$$

Example 11. *Simplify* $(7 + 2)^2$.

$$(7 + 2)^2 = (9)^2 \qquad \text{Parentheses specify addition first.}$$
$$= 81 \quad \checkmark$$

Example 12. *Evaluate* $(16 + 2^2) \div 2 \cdot 2$.

$$(16 + 2^2) \div 2 \cdot 2 = (16 + 4) \div 2 \cdot 2 \qquad \left\{ \begin{array}{l} \text{since exponentiation precedes} \\ \text{addition in computing the} \\ \text{value of } (16 + 2^2) \end{array} \right.$$

$$= 20 \div 2 \cdot 2 \qquad \left\{ \begin{array}{l} \text{after finally determining the} \\ \text{value of the quantity in} \\ \text{parentheses} \end{array} \right.$$

$$= 10 \cdot 2 \qquad \left\{ \begin{array}{l} \text{since we proceed from left} \\ \text{to right with a series of} \\ \text{multiplication and division} \end{array} \right.$$

$$= 20 \quad \checkmark$$

Example 13. *Simplify* $(16 + 2^2) \div (2 \cdot 2)$.

$$(16 + 2^2) \div (2 \cdot 2) = (16 + 4) \div (2 \cdot 2) \qquad \text{just as in Example 12}$$
$$= 20 \div (2 \cdot 2) \qquad \text{just as in Example 12}$$
$$= 20 \div 4 \qquad \left\{ \begin{array}{l} \text{Parentheses overrule} \\ \text{the left-to-right order.} \end{array} \right.$$
$$= 5 \quad \checkmark$$

Be sure to compare Examples 12 and 13.

EXERCISES 2.1

Answers to starred exercises are given in the back of the book.

*1. Simplify each expression.

(a) $8 + 2 \cdot 6$

(b) $15 + 3 \cdot 4$

(c) $6 + 10 \div 2$

(d) $12 + 8 \div 4$

(e) $9 \div 3 + 6$

(f) $12 \div 6 \cdot 2$

(g) $20 \div 4 \cdot 5$

(h) $5 + 4 \div 2 + 1$

(i) $5 \cdot 9 + 2 \cdot 8$

(j) $5 \cdot 2^3 + 1$

(k) $2 + 3^2$

(l) $8 \cdot 10 \div 5 \cdot 2$

(m) $7 + 6 \cdot 4 - 3$

(n) $2 + 3 \cdot 7 - 4$

(o) $20 + \dfrac{4}{2} + 8$

(p) $16 + 10/1 + 1$

2. Evaluate each expression.

*(a) $7 + 3 - 4 - 2 + 1$

(b) $10 - 7 + 4 - 5 + 1 - 2$

*(c) $5 \cdot 2 + 4 + 3 \cdot 2^4 + 4$

(d) $3 + 4^2 + 5 \cdot 3^2 + 8 \cdot 2$

*(e) $6 \cdot 3 + 4 \cdot 9 \div 6 \cdot 6 + 3$

(f) $5 - 3 + 4 \cdot 6 - 2 \cdot 3/2$

*(g) $18 \cdot 2 \div 3 \cdot 6 \div 2$

(h) $20 \cdot 4 \div 2 \cdot 2 \div 2$

*3. Determine the value of each expression.

(a) $8 + 2(5 + 3)$

(b) $(4 + 1)^2$

(c) $(6 + 4) \div 2 + 8$

(d) $(6 + 4) \div (2 + 8)$

(e) $(6 + 4) \div (2 + 3)$

(f) $(4 + 2)^2/6$

(g) $(21 + 3^2) \div 5 \cdot 6$

(h) $(21 + 3^2) \div (5 \cdot 6)$

(i) $1^2 + 2^2 + 3^2 + 4^2 \div 8$

(j) $12 + 18 \div 3 + 6$

(k) $\dfrac{8 + 8}{3 + 1} + 8 - 3$

(l) $3 \cdot 4^2 + 7 - \dfrac{2 \cdot 3^2}{6}$

(m) $6 - 4 \div 2 + 5$

(n) $19 + 5^2 - 4 \cdot 2 + 5(7 + 3^4)$

Chapter 2. REVIEW EXERCISES

*1. Simplify each expression.

(a) $9 + 3 \cdot 5$

(b) $15 \div 3 \cdot 5$

(c) $5 + 2^2$

(d) $9 - 4 \cdot 2$

*2. Evaluate each expression.

(a) $5 \cdot 4 \div 2 \cdot 2 + 5$

(b) $5 \cdot 3^2 + 4 \cdot 3$

(c) $9 + 3(4 + 2)$

(d) $8 - 2 \div 2 + 1$

(e) $14 + \dfrac{8}{2} - 6$

(f) $9 + 3^2 - 5 \cdot 2$

*3. Determine the value of each expression.

(a) $(6 + 2) \div 4 + 4$

(b) $(6 + 2) \div (4 + 4)$

(c) $12 + 12 \div 6 + 6$

(d) $12 + 12 \div (6 + 6)$

(e) $(12 + 12) \div 6 + 6$

(f) $(12 + 12) \div (6 + 6)$

(g) $4^2 + 6^2 \div 4$

(h) $(6 - 2)^2$

(i) $5 \cdot 6 + 3$

(j) $19 - 16 \div 4 + 12$

(k) $15 - 4 + 11 - 8 + 1 - 9$

(l) $5 \cdot 2 + 6(3 + 4^2)$

Evaluation of Geometric Formulas

3.1 EVALUATION OF GEOMETRIC FORMULAS

Here we apply our knowledge of arithmetic to the evaluation of geometric formulas. In each case, an area, volume, or perimeter is determined by substituting numerical values for the letters given in a formula. Such units as inches and centimeters are omitted but are discussed in detail in Chapters 8 and 9.

The formulas use information about a geometric figure in order to determine its *perimeter* or *area* or *volume*.

> perimeter = distance around
> area = amount enclosed (2 dimensions)
> volume = space enclosed (3 dimensions)

Example 1. *Determine the area of the rectangle whose length is* 12 *and width is* 10.

The formula for area is $A = l \cdot w$ (usually written without the dot for multiplication, as $A = lw$), where

$$A = \text{area}$$
$$l = \text{length}$$
$$w = \text{width}$$

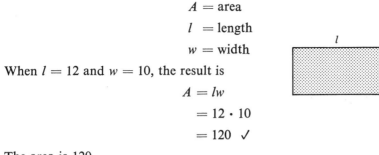

When $l = 12$ and $w = 10$, the result is

$$A = lw$$
$$= 12 \cdot 10$$
$$= 120 \checkmark$$

The area is 120.

Example 2. *Determine the volume of the box whose length is 5, width is 3, and height is 2.*

The formula for volume is $V = lwh$ (note that lwh means $l \cdot w \cdot h$), where

$$V = \text{volume}$$
$$l = \text{length}$$
$$w = \text{width}$$
$$h = \text{height}$$

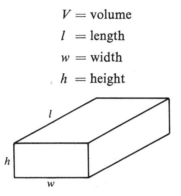

When $l = 5$, $w = 3$, and $h = 2$, the result is

$$V = lwh$$
$$= 5 \cdot 3 \cdot 2$$
$$= 30 \checkmark$$

The volume is 30.

Example 3. *Determine the area of a trapezoid whose bases are 6 and 10 and whose height is 5.*

The formula for area is $A = \frac{1}{2}(b_1 + b_2)h$, where

$$A = \text{area}$$
$$b_1 = \text{one base}$$
$$b_2 = \text{the other base}$$
$$h = \text{height}$$

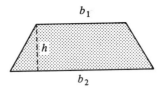

When $b_1 = 6$, $b_2 = 10$, and $h = 5$, the result is

$$A = \tfrac{1}{2}(6 + 10)5$$
$$= \tfrac{1}{2}(16)5$$
$$= 8 \cdot 5$$
$$= 40 \ \checkmark$$

The area is 40.

Example 4. *Determine the circumference (distance around) of a circle whose radius is 92.*

The formula for circumference is $C = 2\pi r$, where

$$C = \text{circumference}$$
$$\pi = \text{approximately } 3.14$$
$$r = \text{radius}$$

When $r = 92$, the result is

$$C = 2\pi r$$
$$= 2 \cdot (3.14)(92)$$
$$= 577.76 \ \checkmark$$

The approximate circumference is 577.76.

Example 5. *Determine the perimeter of the triangle having sides of lengths 13, 17, and 21.*

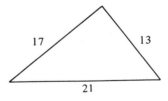

The perimeter (P) of a triangle with sides of lengths a, b, and c is simply $P = a + b + c$, since perimeter means distance around. Here

$$P = a + b + c$$
$$= 13 + 17 + 21$$
$$= 51 \ \checkmark$$

Example 6. *Determine (a) the area and (b) the perimeter of the figure shown below.*

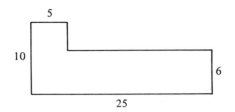

(a) One way to determine the area is to draw the dotted line shown below and consider the area to be the sum of the two rectangular areas, where $A = lw$ in each case,

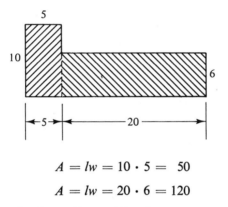

$$A = lw = 10 \cdot 5 = 50$$
$$A = lw = 20 \cdot 6 = 120$$

The total area is $50 + 120$, or 170. ✓

(b) The perimeter is the distance around. Examining the figure having vertical pieces of lengths 10 and 6, you can reason that the other vertical piece is 4 (from 10 minus 6). Also, from the reasoning in part (a), the unlabeled horizontal piece is of length 20.

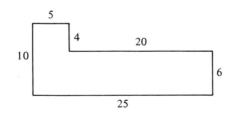

Since perimeter is the distance around,

$$P = 10 + 25 + 6 + 20 + 4 + 5$$
$$= 70 \checkmark$$

EXERCISES 3.1

Answers to starred exercises are given in the back of the book.

1. Use the formulas of the examples to determine what is requested. Use 3.14 as an approximation for π.
    *(a) The area of a rectangle having length 37 and width 28.
    (b) The area of a rectangle having length 26.4 and width 15.2.
    *(c) The volume of a box having length 7, width 6, and height 10.
    (d) The volume of a box having length 12.3, width 9.1, and height 10.
    *(e) The area of a trapezoid with bases 7 and 9 and height 15.
    (f) The area of a trapezoid with bases 14 and 18 and height 13.1.
    *(g) The circumference of a circle with radius 14.
    (h) The circumference of a circle with radius 10.
    *(i) The circumference of a circle with radius 39.7.
    *(j) The perimeter of a triangle with sides 16, 20, and 23.
    (k) The perimeter of a triangle with sides 15, 19, and 27.

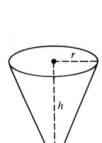

2. The area of a triangle is $A = \frac{1}{2}bh$, where b is the base of the triangle and h is the height.
    *(a) Find the area when $b = 6$ and $h = 10$.
    *(b) Find the area when $b = 300$ and $h = 12$.
    (c) Find the area when $b = 9$ and $h = 5$.

*3. Evaluate each formula, using the values given. Use 3.14 as an approximation for π.
    (a) $P = 2l + 2w$; $l = 6$, $w = 5.4$. This is the perimeter (distance around) of a rectangle having length l and width w.
    (b) $A = \pi r^2$; $r = 7$. This is the area of a circle having radius r.
    (c) $S = 4\pi r^2$; $r = 9$. This is the area of the surface of a sphere having radius r.
    (d) $V = \dfrac{4\pi r^3}{3}$; $r = 5$. This is the volume of a sphere having radius r.
    (e) $V = \frac{1}{3}\pi r^2 h$; $r = 8$, $h = 15$. This is the volume of a cone having radius r and height h.

*4. Determine the area and perimeter of each figure.

(a)

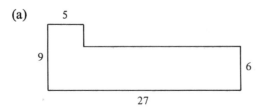

(b)
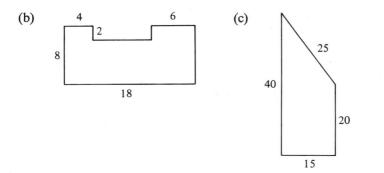

(c)

(d)
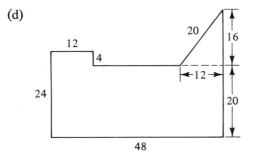

(e)
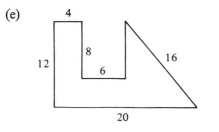

(f)
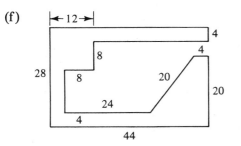

Chapter 3.   REVIEW EXERCISES

*1. Determine each area or perimeter as requested. Use 3.14 as an approximation for π.
   (a) The area of a rectangle with length 17 and width 12.
   (b) The perimeter of a triangle with sides 23, 26, and 36.
   (c) The circumference of a circle with radius 20.
   (d) The area of a triangle with base 12 and height 30.
   (e) The area of a circle with radius 10.

CHAPTER 4

Square Roots

4.1 FUNDAMENTALS

Up to this point we have used the arithmetic operations of addition, subtraction, multiplication, division, and exponentiation. Obtaining square roots is another operation that is important in both arithmetic and algebra.

We begin by using examples to introduce the concept of square root. The square root of 25 is 5, since $5 \times 5 = 25$. The square root of 81 is 9, since $9 \times 9 = 81$. Using the symbol $\sqrt{}$ for square root, we can write

$$\sqrt{25} = 5$$
$$\sqrt{81} = 9$$

A *square root* of a quantity is a number that when squared (that is, **square root** multiplied by itself) produces the original quantity.

$$\sqrt{0} = 0 \quad \text{since} \quad 0 \times 0 = 0$$
$$\sqrt{1} = 1 \quad \text{since} \quad 1 \times 1 = 1$$
$$\sqrt{4} = 2 \quad \text{since} \quad 2 \times 2 = 4$$
$$\sqrt{9} = 3 \quad \text{since} \quad 3 \times 3 = 9$$
$$\sqrt{16} = 4 \quad \text{since} \quad 4^2 = 16$$

$$\sqrt{25} = 5 \quad \text{since} \quad 5^2 = 25$$
$$\sqrt{36} = 6 \quad \text{since} \quad 6^2 = 36$$
$$\sqrt{49} = 7 \quad \text{since} \quad 7^2 = 49$$
$$\sqrt{64} = 8 \quad \text{since} \quad 8^2 = 64$$
$$\sqrt{81} = 9 \quad \text{since} \quad 9^2 = 81$$
$$\sqrt{100} = 10 \quad \text{since} \quad 10^2 = 100$$
$$\sqrt{121} = 11$$
$$\sqrt{144} = 12$$
$$\sqrt{169} = 13$$
$$\sqrt{196} = 14$$
$$\sqrt{225} = 15$$
$$\sqrt{256} = 16$$
$$\sqrt{289} = 17$$
$$\sqrt{324} = 18$$
$$\sqrt{361} = 19$$
$$\sqrt{400} = 20$$

perfect square   The numbers 1, 4, 9, 16, 25, 36, . . . are called *perfect squares*. A whole number is a perfect square if its square root is a whole number.
radical   The symbol $\sqrt{}$ is called a *radical*.

right triangle   One important application of square root is in working with right triangles. A *right triangle* is one that contains a 90° angle (a *right angle*, ∟). Below are examples of right triangles, where the right angle is marked by a small square.

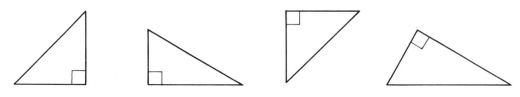

The side opposite the right angle is called the *hypotenuse*. The other two sides are called *legs*.

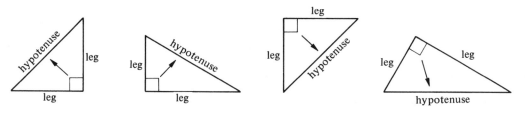

In any right triangle the following relationship is always true.

$$(\text{hypotenuse})^2 = (\text{leg 1})^2 + (\text{leg 2})^2$$

In words, the square of the length of the hypotenuse equals the sum of the squares of the lengths of the legs. This famous theorem was first proved by the Greek mathematician Pythagoras in the sixth century B.C. It is called the *Pythagorean theorem*.

We often denote the length of the hypotenuse by the letter c and the lengths of the legs by the letters a and b. Thus

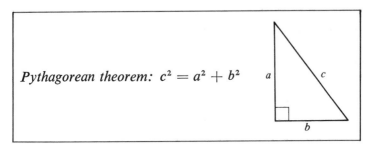

Pythagorean theorem: $c^2 = a^2 + b^2$

Example 1. *Determine the length of the hypothenuse of a right triangle if the legs are 3 inches and 4 inches.*

Since $c^2 = a^2 + b^2$

$\quad c^2 = 4^2 + 3^2 \quad$ using 4 for a and 3 for b

$\quad c^2 = 16 + 9 \quad$ since $4^2 = 16$ and $3^2 = 9$

$\quad c^2 = 25 \quad$ since $16 + 9$ is 25

or $\quad c = \sqrt{25}$

so $\quad c = 5 \checkmark$

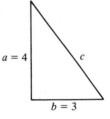

The hypotenuse is 5 inches.

This example illustrates the popular 3–4–5 right triangle. Another combination is 5–12–13, as shown in the next example.

Example 2. *If the hypotenuse of a right triangle is 13 and one of the legs is 12, what is the length of the other leg?*

Since $\quad c^2 = a^2 + b^2$

$\quad 13^2 = 12^2 + b^2$

$\quad 169 = 144 + b^2$

or $\quad b^2 = 25 \quad$ since $144 + \underline{25} = 169$

so $\quad b = \sqrt{25} = 5 \checkmark$

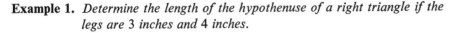

The other leg is 5.

Example 3. *Determine the unknown side, x, of the triangle shown.*

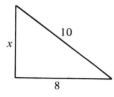

$$10^2 = x^2 + 8^2$$
$$100 = x^2 + 64$$
so $x^2 = 36$ since $\underline{36} + 64 = 100$
or $x = 6$ ✓

Note that this is a 6–8–10 right triangle. The relationship 6–8–10 is the same as 3–4–5. Each length of the 6–8–10 is double that of the 3–4–5.

Example 4. *Find the value of x in the triangle at the left.*

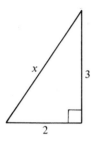

$$x^2 = a^2 + b^2$$
$$x^2 = 3^2 + 2^2$$
$$x^2 = 9 + 4$$
$$x^2 = 13$$
$$x = \sqrt{13} ✓$$

Since 13 is not a perfect square, we cannot obtain a whole number value for $\sqrt{13}$. If desired, a good approximation to the value of $\sqrt{13}$ can be obtained by using a calculator with a square root key or by looking up $\sqrt{13}$ in a table of square roots.

EXERCISES 4.1

Answers to starred exercises are given in the back of the book.

1. Study the square roots of perfect squares given at the beginning of the chapter. When you feel that you know them, evaluate each of the following without looking at that table of square roots.
   (a) $\sqrt{49}$ (b) $\sqrt{81}$ (c) $\sqrt{4}$ (d) $\sqrt{25}$ (e) $\sqrt{400}$
   (f) $\sqrt{121}$ (g) $\sqrt{36}$ (h) $\sqrt{289}$ (i) $\sqrt{0}$ (j) $\sqrt{169}$
   (k) $\sqrt{144}$ (l) $\sqrt{1}$ (m) $\sqrt{324}$ (n) $\sqrt{361}$ (o) $\sqrt{9}$
   (p) $\sqrt{225}$ (q) $\sqrt{16}$ (r) $\sqrt{100}$ (s) $\sqrt{196}$ (t) $\sqrt{256}$

2. Evaluate each square.
   (a) 9^2 (b) 11^2 (c) 19^2 (d) 2^2 (e) 0^2
   (f) 20^2 (g) 13^2 (h) 17^2 (i) 3^2 (j) 18^2
   (k) 12^2 (l) 1^2 (m) 4^2 (n) 10^2 (o) 14^2
   (p) 7^2 (q) 5^2 (r) 16^2 (s) 8^2 (t) 15^2
   (u) 6^2

3. Use the Pythagorean theorem to find the length of the side labeled x in each triangle below.

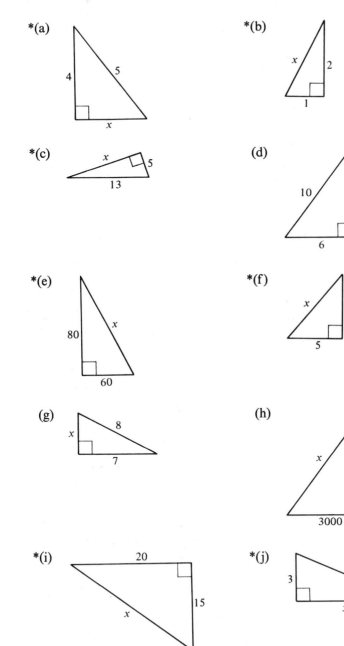

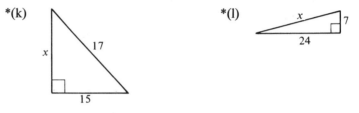

*(k)   *(l)

4. Evaluate each formula, using the values given.

*(a) $t = \sqrt{\dfrac{2s}{a}}$ $s = 64$, $a = 32$

(b) Redo (a) with $s = 256$ and $a = 32$.

*(c) $x = \dfrac{7 + \sqrt{b^2 - 4ac}}{2a}$ $a = 1$, $b = 3$, $c = 2$

(d) Redo (c) with $a = 6$, $b = 7$, and $c = 1$.

*(e) $A = \sqrt{s(s - a)(s - b)(s - c)}$,   where $s = \frac{1}{2}(a + b + c)$, $a = 3$,
$b = 4$, $c = 5$.

(f) Verify that $\sqrt{576} = 24$ by squaring 24.

(g) Redo (e) with $a = 6$, $b = 8$, and $c = 10$.

Chapter 4.   REVIEW EXERCISES

1. Determine the value of each square root *without* looking at the table at the beginning of the chapter. Then check your answers by using that table.

(a) $\sqrt{64}$ (b) $\sqrt{25}$ (c) $\sqrt{144}$ (d) $\sqrt{1}$ (e) $\sqrt{225}$

(f) $\sqrt{0}$ (g) $\sqrt{196}$ (h) $\sqrt{400}$ (i) $\sqrt{100}$

*2. Use the Pythagorean theorem to determine the length of the side labeled x in each right triangle.

(a)   (b)

(c)   (d)

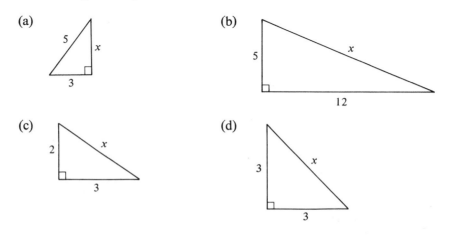

Properties of Arithmetic

5.1 PROPERTIES OF ARITHMETIC

In this section we take a look at some special properties that whole numbers satisfy with respect to the arithmetic operations that we have studied. We can say that addition, subtraction, multiplication, and division are examples of *binary operations*—that is, operations in which *two* numbers are combined to produce a result.

binary operation

The first property considered is the **commutative property** for addition. Here are some examples of the property.

commutative property

$$2 + 3 = 3 + 2$$

This example states that $2 + 3$ and $3 + 2$ are equal; that is, both are equal to 5. Here are the rest of the examples.

$$7 + 9 = 9 + 7$$
$$x + 2 = 2 + x$$
$$p + q = q + p$$
$$5 + (6 + 7) = 5 + (7 + 6)$$

In effect, this commutative property says that if a and b are any two whole numbers, then $(a + b)$ and $(b + a)$ are the same; they are equal.

associative property   Next is the **associative property** for addition. Here are some examples.

$$2 + (3 + 1) = (2 + 3) + 1$$

This example says that $2 + (3 + 1)$ and $(2 + 3) + 1$ are equal; that is, both are equal to 6. Here are the rest of the examples.

$$12 + (6 + 5) = (12 + 6) + 5$$
$$(9 + 2) + 10 = 9 + (2 + 10)$$
$$(x + 5) + 3 = x + (5 + 3)$$
$$a + (d + x) = (a + d) + x$$

This associative property states that when three whole numbers are added, it does not matter with which number (the first or the last) you associate the middle one. Thus in

$$5 + 8 + 2$$

you can associate 8 with 5, as

$$(5 + 8) + 2$$

or associate 8 with 2, as

$$5 + (8 + 2)$$

So $(5 + 8) + 2 = 5 + (8 + 2)$. There are two ways of computing $5 + 8 + 2$.†

The associative property for addition can be written in general terms as $a + (b + c) = (a + b) + c$, where a, b, and c represent any whole numbers.

If you add two whole numbers, you obtain a whole number as the result. This fact should not be particularly surprising. The property has a name—**closure**—and it will be useful later to be able to refer to this property by name. Also included in the meaning of closure is the fact that the sum is *unique*; that is, whenever you add the same two numbers (say 4 and 5), you always get the same result (*9* in the case of 4 and 5).

closure property   Below are examples of the closure property for whole numbers using addition

$$2 + 4 = 6$$
$$3 + 5 = 8$$
$$19 = 7 + 12$$

---

†Recall that addition is a *binary* operation; that is, numbers must be added *two* at a time, not three at a time. Think about it and you will realize that you never add three numbers "at once"; instead you add two of them to get an intermediate result and then immediately add the other number to get the final sum.

In general, $a + b = c$; that is, if a and b are whole numbers, then the sum $(a + b)$ is a whole number; call it c.

Note that the whole numbers using division do not satisfy closure; for example, $3 \div 4 = .75$, and $.75$ is not a whole number.

The closure, associative, and commutative properties are true for *multiplication* as well as for addition. Here are arithmetic examples of each.

$4 \cdot (5 \cdot 7) = (4 \cdot 5) \cdot 7$ associative property for multiplication

$(8 \cdot 6) \cdot 15 = 8 \cdot (6 \cdot 15)$ associative property for multiplication

$4 \cdot 9 = 9 \cdot 4$ commutative property for multiplication

$5 \cdot (9 \cdot 2) = 5 \cdot (2 \cdot 9)$ commutative property for multiplication

$6 \cdot 3 = 18 \cdot$ closure property for multiplication

$10 = 2 \cdot 5$ closure property for multiplication

The last property that we will look at is called the **distributive property**. It involves both multiplication and addition. Here is an example of the distributive property.

distributive property

$$5 \cdot (3 + 4) = 5 \cdot 3 + 5 \cdot 4$$

Notice what has been done.

$$\underline{5} \cdot (3 + 4) = \underline{5} \cdot 3 + \underline{5} \cdot 4$$

We *distribute* 5 to both 3 and 4. That is, *5 times 3* and *5 times 4*, or $5 \cdot 3 + 5 \cdot 4$. Other examples:

$$7 \cdot (2 + 3) = 7 \cdot 2 + 7 \cdot 3$$
$$6(5 + 1) = 6 \cdot 5 + 6 \cdot 1$$
$$5(x + y) = 5 \cdot x + 5 \cdot y$$
$$4 \cdot 2 + 4 \cdot 3 = 4(2 + 3)$$
$$5 \cdot 9 + 5 \cdot 18 = 5(9 + 18)$$
$$5 \cdot a + 5 \cdot b = 5(a + b)$$

Is it obvious that this property is true? Probably not. Let's see an example that might illustrate the point. You can make up additional examples if you are not completely convinced.

$$3(4 + 9) \stackrel{?}{=} 3 \cdot 4 + 3 \cdot 9$$
$$3(13) \stackrel{?}{=} 12 + 27$$
$$39 \stackrel{?}{=} 39$$

The left-hand side shows the work done without using the distributive property; the right-hand side uses the distributed form. Both results are the same, and so $3(4 + 9) = 3 \cdot 4 + 3 \cdot 9$.

Another form of the distributive property is given by

$$(3 + 5)7 = 3 \cdot 7 + 5 \cdot 7$$

Here the multiplier is on the right instead of on the left. It doesn't matter which side the multiplier is on; it is still distributed. Again,

$$(3 + 5)7 = 3 \cdot 7 + 5 \cdot 7$$

We now state both distributive properties in general, for any whole numbers a, b, and c.

$$a(b + c) = a \cdot b + a \cdot c$$

and

$$(b + c)a = b \cdot a + c \cdot a$$

There is also a distributive property that involves multiplication and subtraction. See Exercises 12 and 13.

EXERCISES 5.1

Answers to starred exercises are given in the back of the book.

1. Give two examples of the commutative property for addition and two examples of the commutative property for multiplication.

2. Give two examples of the associative property for addition and two examples of the associative property for multiplication.

3. Give two examples of the closure property for addition and two examples of the closure property for multiplication.

4. Give two examples of the distributive property.

*5. In each case, name the property used.
    (a) $5 + 2 = 2 + 5$ (b) $8 + (6 + 9) = (8 + 6) + 9$
    (c) $8 + (6 + 9) = 8 + (9 + 6)$ (d) $3(2 + 5) = 3 \cdot 2 + 3 \cdot 5$
    (e) $7 \cdot 2 + 7 \cdot 9 = 7(2 + 9)$ (f) $6 + 5 = 11$
    (g) $(8 + 9) + 4 = 8 + (9 + 4)$ (h) $(2 + 3) + 7 = 7 + (2 + 3)$
    (i) $29 = 14 + 15$ (j) $15 \cdot 8 = 8 \cdot 15$
    (k) $(5 + 4)7 = 5 \cdot 7 + 4 \cdot 7$ (l) $9 \cdot 3 + 14 \cdot 3 = (9 + 14)3$

6. For each, name the property used.
    (a) $4(7 + 2) = 4 \cdot 7 + 4 \cdot 2$ (b) $5 \cdot 8 \cdot 9 = 5 \cdot 9 \cdot 8$
    (c) $5(8 \cdot 9) = (5 \cdot 8)9$ (d) $2 + 1 = 3$
    (e) $3(x + y) = 3 \cdot x + 3 \cdot y$ (f) $5(3 + 1) = 5(4)$

(g) $2 + 3 = 3 + 2$

(i) $2 \cdot 3 = 6$

(h) $2 \cdot 3 = 3 \cdot 2$

(j) $9 + (2 + 3) = 9 + 5$

*7. Name the property used in going from each step to the next one.

(a) $5 + (6 + 3) = (5 + 6) + 3$

(b) $\qquad\qquad = (6 + 5) + 3$

(c) $\qquad\qquad = 3 + (6 + 5)$

(d) $\qquad\qquad = 3 + 11$

(e) $\qquad\qquad = 14$

*8. Name the property used in going from each step to the next one.

(a) $7 \cdot 2 + 7 \cdot 3 = 7(2 + 3)$

(b) $\qquad\qquad = 7(3 + 2)$

(c) $\qquad\qquad = 7 \cdot 3 + 7 \cdot 2$

(d) $\qquad\qquad = 3 \cdot 7 + 7 \cdot 2$

(e) $\qquad\qquad = 21 + 7 \cdot 2$

9. Name the property used in going from each step to the next one.

(a) $8(9 + 4) = 8 \cdot 9 + 8 \cdot 4$

(b) $\qquad\qquad = 9 \cdot 8 + 4 \cdot 8$

(c) $\qquad\qquad = (9 + 4)8$

(d) $\qquad\qquad = (13)8$

(e) $\qquad\qquad = 104$

10. Name the property used in going from each step to the next one.

(a) $36 = 4 \cdot 9$

(b) $\quad = 4(6 + 3)$

(c) $\quad = 4 \cdot 6 + 4 \cdot 3$

(d) $\quad = 24 + 12$

(e) $\quad = 12 + 24$

11. State the commutative, associative, and closure properties for *multiplication*, using letters to represent any numbers.

12. A distributive property for multiplication and *subtraction* can be written as $a(b - c) = a \cdot b - a \cdot c$. Verify this property in the cases below by evaluating each side and showing that they are equal.

(a) $5(10 - 2) = 5 \cdot 10 - 5 \cdot 2$

(b) $4(7 - 1) = 4 \cdot 7 - 4 \cdot 1$

(c) $15(8 - 5) = 15 \cdot 8 - 15 \cdot 5$

13. Another distributive property for multiplication and subtraction can be written with the multiplier on the right side, as $(b - c)a = b \cdot a - c \cdot a$. Verify this property in the cases below by evaluating each side and showing that they are equal.

(a) $(6 - 4)7 = 6 \cdot 7 - 4 \cdot 7$

(b) $(12 - 2)5 = 12 \cdot 5 - 2 \cdot 5$

(c) $(8 - 5)9 = 8 \cdot 9 - 5 \cdot 9$

Chapter 5. REVIEW EXERCISES

*1. In each case, name the property used.

(a) $6(7 \cdot 3) = (6 \cdot 7)3$ (b) $7 + 2 = 2 + 7$

(c) $7(8 + 5) = 7 \cdot 8 + 7 \cdot 5$ (d) $9 \cdot 3 = 27$

(ē) $4 + 5 = 9$ (f) $(9 + 2)6 = 9 \cdot 6 + 2 \cdot 6$

*2. Name the property used in going from each step to the next one.

(a) $8(3 + 4) = 8 \cdot 3 + 8 \cdot 4$

(b) $\qquad\qquad = 3 \cdot 8 + 8 \cdot 4$

(c) $\qquad\qquad = 24 + 32$

(d) $\qquad\qquad = 32 + 24$

(e) $\qquad\qquad = 56$

Arithmetic of Fractions

6.1 WHAT IS A FRACTION?

If we divide the rectangle below into *two* equal parts, then *one* part represents $\frac{1}{2}$ of the rectangle.

If we divide the rectangle into *three* equal parts, then *one* part represents $\frac{1}{3}$ of the whole rectangle.

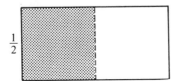

If we divide the rectangle into *five* equal parts, then *two* parts represents $\frac{2}{5}$ of the whole rectangle. Thus the shaded area is $\frac{2}{5}$, the unshaded area is $\frac{3}{5}$, and the whole area (shaded plus unshaded) is $\frac{2}{5} + \frac{3}{5} = \frac{5}{5} = 1$ rectangle.

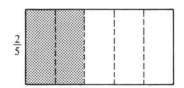

numerator
denominator

The top portion of a fraction is called the *numerator*. The bottom portion of a fraction is called the *denominator*. For example, in $\frac{2}{5}$, 2 is the numerator and 5 is the denominator.

$$\frac{2}{5} \quad \begin{array}{l} \longleftarrow \text{ numerator} \\ \longleftarrow \text{ denominator} \end{array}$$

proper fraction

improper fraction

A fraction is called a *proper fraction* when the numerator is less than the denominator. When the numerator is greater than the denominator or equal to the denominator, the fraction is called an *improper fraction*. Here are a few examples.

$$\text{Proper fractions:} \quad \frac{1}{2}, \frac{5}{6}, \frac{3}{11}$$

$$\text{Improper fractions:} \quad \frac{2}{2}, \frac{7}{6}, \frac{17}{11}$$

Improper fractions will arise throughout the chapter and they will be given special attention in Section 6.7.

6.2 MULTIPLICATION OF FRACTIONS

> To multiply fractions, multiply numerator by numerator and denominator by denominator.

To help you justify this procedure, consider $\frac{1}{2} \times \frac{1}{3}$. You can visualize that half of one-third is one-sixth, since two-sixths equal one-third, and half of two-sixths is one-sixth.

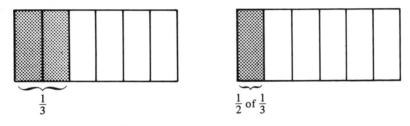

$$\underbrace{}_{}\\ \frac{1}{3}$$

$$\frac{1}{2} \text{ of } \frac{1}{3}$$

By the rule for multiplication of fractions,

$$\frac{1}{2} \times \frac{1}{3} = \frac{1 \times 1}{2 \times 3} = \frac{1}{6}.$$

Example 1. *Multiply* $\frac{2}{3} \times \frac{1}{7}$.

$$\frac{2}{3} \times \frac{1}{7} = \frac{2 \times 1}{3 \times 7} = \frac{2}{21} \quad \checkmark$$

Example 2. *Multiply* $\frac{5}{6} \times \frac{7}{8}$.

$$\frac{5}{6} \times \frac{7}{8} = \frac{5 \times 7}{6 \times 8} = \frac{35}{48} \quad \checkmark$$

Example 3. *Multiply* $3 \times \frac{2}{7}$.

$$3 \times \frac{2}{7} = \frac{3}{1} \times \frac{2}{7} = \frac{3 \times 2}{1 \times 7} = \frac{6}{7} \quad \checkmark$$

Writing 3 as $\frac{3}{1}$ is merely a convenient notation that may make it easier to see what is happening. You will probably want to use this convention.

Example 4. *Multiply* $0 \times \frac{3}{8}$.

0 times any number is 0, so

$$0 \times \frac{3}{8} = 0 \quad \checkmark$$

EXERCISES 6.2

Answers to starred exercises are given in the back of the book.

1. Multiply.

    *(a) $\frac{3}{5} \times \frac{1}{2}$ *(b) $\frac{4}{9} \times \frac{2}{3}$ *(c) $\frac{6}{7} \times \frac{1}{5}$ (d) $\frac{1}{6} \times \frac{5}{8}$

*(e) $\dfrac{3}{4} \times \dfrac{5}{7}$ (f) $\dfrac{3}{11} \times \dfrac{5}{7}$ *(g) $\dfrac{1}{9} \times \dfrac{1}{9}$ *(h) $\dfrac{4}{5} \times \dfrac{3}{5}$

(i) $\dfrac{9}{14} \times \dfrac{3}{4}$ *(j) $\dfrac{10}{13} \times \dfrac{6}{7}$ *(k) $\dfrac{15}{23} \times \dfrac{12}{13}$ (l) $\dfrac{11}{15} \times \dfrac{11}{17}$

*(m) $7 \times \dfrac{1}{8}$ (n) $6 \times \dfrac{1}{7}$ *(o) $4 \times \dfrac{2}{9}$ *(p) $\dfrac{3}{8} \times 1$

(q) $\dfrac{9}{100} \times 7$ *(r) $\dfrac{10}{99} \times 5$ *(s) $0 \times \dfrac{3}{5}$ (t) $\dfrac{7}{8} \times 0$

2. Use the formula $A = l \cdot w$ to determine the area of a rectangle whose length (l) and width (w) are given below.

*(a) $l = \dfrac{2}{3},\ w = \dfrac{5}{7}$ (b) $l = \dfrac{1}{5},\ w = \dfrac{1}{5}$

*(c) $l = \dfrac{4}{7},\ w = \dfrac{2}{5}$ (d) $l = \dfrac{5}{9},\ w = \dfrac{2}{3}$

6.3 REDUCTION OF FRACTIONS

The multiplication process can be reversed in order to reduce fractions. Also used in the reducing process is the fact that any nonzero number divided by itself is equal to 1. For instance,

$$\frac{5}{5} = 1, \qquad \frac{4}{4} = 1, \qquad \frac{17}{17} = 1$$

To see how fractions are reduced, observe the following example in which the fraction $\frac{20}{36}$ is reduced.

$$\frac{20}{36} = \frac{5 \times 4}{9 \times 4} \qquad \text{writing 20 as } 5 \times 4 \text{ and 36 as } 9 \times 4$$

$$= \frac{5}{9} \times \frac{4}{4} \qquad \left\{ \begin{array}{l} \text{separating } \dfrac{4}{4} \text{ by reversing the steps of} \\ \text{multiplication of fractions} \end{array} \right.$$

$$= \frac{5}{9} \times 1 \qquad \text{since } \dfrac{4}{4} \text{ is equal to 1}$$

$$= \frac{5}{9} \qquad \text{since } \dfrac{5}{9} \times 1 = \dfrac{5}{9}$$

The reason that $\frac{20}{36}$ can be reduced to $\frac{5}{9}$ is that both numerator 20 and denominator 36 have a common factor—4. And so we chose to write 20 as 5×4 and 36 as 9×4. If a *factor* appears in both the numerator and denominator, then both can be removed because they form a fraction equal to one and multiplication by one has no effect. This point can be seen once again in the next examples.

Example 5. *Reduce* $\frac{12}{15}$.

$$\frac{12}{15} = \frac{3 \times 4}{3 \times 5}$$

$$= \frac{3}{3} \times \frac{4}{5}$$

$$= 1 \times \frac{4}{5}$$

$$= \frac{4}{5} \checkmark$$

Example 6. *Reduce* $\frac{18}{72}$.

$$\frac{18}{72} = \frac{9 \times 2}{9 \times 8} = \frac{\cancel{9}}{\cancel{9}} \times \frac{2}{8} = \frac{2}{8}$$

But $\frac{2}{8}$ can be reduced!

$$\frac{2}{8} = \frac{2 \times 1}{2 \times 4} = \frac{2}{2} \times \frac{1}{4} = \frac{1}{4} \checkmark$$

Actually, this could have been done in one step, as

$$\frac{18}{72} = \frac{18 \times 1}{18 \times 4} = \frac{18}{18} \times \frac{1}{4} = \frac{1}{4} \checkmark$$

Either approach is correct; it pays to look for shortcuts.

Example 7. *Multiply* $\frac{4}{5} \times \frac{3}{4}$.

$$\frac{4}{5} \times \frac{3}{4} = \frac{4 \times 3}{5 \times 4} = \frac{12}{20}$$

And $\frac{12}{20}$ can be reduced.

$$\frac{12}{20} = \frac{4 \times 3}{4 \times 5} = \frac{4}{4} \times \frac{3}{5} = \frac{3}{5} \checkmark$$

Actually, reduction earlier in the process was possible.

$$\frac{4}{5} \times \frac{3}{4} = \frac{4 \times 3}{5 \times 4} = \frac{3 \times 4}{5 \times 4} = \frac{3}{5} \times \frac{4}{4} = \frac{3}{5} \checkmark$$

For convenience, we often write

$$\frac{4}{5} \times \frac{3}{4} = \frac{\overset{1}{\cancel{4}}}{5} \times \frac{3}{\underset{1}{\cancel{4}}} = \frac{3}{5}$$

Note the common factor 4 in both numerator and denominator.

where it is *observed* that we could manipulate the numbers to get $\frac{4}{4}$, which is 1, but we don't actually write all the steps.

Example 8. *Multiply* $\frac{15}{14} \times \frac{7}{11}$.

$$\frac{15}{14} \times \frac{7}{11} = \frac{15}{\overset{\cancel{7}}{1} \times 2} \times \frac{\overset{1}{\cancel{7}}}{11} = \frac{15 \times 1}{1 \times 2 \times 11} = \frac{15}{22} \checkmark$$

EXERCISES 6.3

1. Reduce each fraction.

   *(a) $\frac{4}{6}$ *(b) $\frac{15}{15}$ *(c) $\frac{6}{8}$ *(d) $\frac{9}{12}$ (e) $\frac{10}{15}$

   (f) $\frac{8}{10}$ *(g) $\frac{20}{30}$ *(h) $\frac{14}{28}$ (i) $\frac{14}{21}$ (j) $\frac{24}{30}$

   *(k) $\frac{24}{32}$ *(l) $\frac{20}{48}$ *(m) $\frac{18}{54}$ (n) $\frac{14}{105}$

2. Multiply; reduce if possible.

   *(a) $\frac{4}{9} \times \frac{1}{4}$ *(b) $\frac{7}{16} \times \frac{4}{5}$ *(c) $\frac{1}{9} \times \frac{3}{4}$ *(d) $\frac{3}{5} \times \frac{6}{7}$

   *(e) $\frac{12}{15} \times \frac{5}{6}$ *(f) $2 \times \frac{1}{2}$ (g) $\frac{2}{3} \times \frac{3}{8}$ *(h) $\frac{2}{5} \times \frac{10}{11}$

   (i) $\frac{2}{5} \times \frac{10}{44}$ *(j) $\frac{3}{4} \times 4$ *(k) $\frac{8}{5} \times \frac{5}{16}$ (l) $18 \times \frac{2}{9}$

   *(m) $\frac{18}{24} \times \frac{16}{81}$ (n) $\frac{24}{32} \times 40$ *(o) $\frac{13}{35} \times \frac{21}{26}$ (p) $\frac{33}{48} \times \frac{32}{68}$

6.4 DIVISIBILITY TESTS

In this section we present special tests that will make it easier for you to reduce fractions. The tests will also be useful in later applications as well.

We begin with the concept of *divisibility*. Using your present terminology, you would say that 3 is a *factor* of 15, since $15 = 3 \cdot 5$; that is, 15 can be written as 3 times some whole number. That 3 is a **divisor** factor of 15 can also be viewed as 3 is a *divisor* of 15; that is, when 15 is divided by 3, the result is a whole number and no remainder. We say **divisible** that 15 *divisible* by 3. Observe a few examples.

15 is divisible by 3,
since 15 ÷ 3 = 5   (no remainder in the division)

14 is divisible by 7,
since 14 ÷ 7 = 2   (no remainder in the division)

14 is divisible by 2,
since 14 ÷ 2 = 7   (no remainder in the division)

But

15 is *not* divisible by 6, since 15 ÷ 6 ≠ whole number.
In fact, 15 ÷ 6 = 2 with remainder 3.

14 is not divisible by 5, since 14 ÷ 5 ≠ whole number.
In fact, 14 ÷ 5 = 2 with remainder 4.

It should be clear that factors and divisors are the same thing. The factors of 12 are 1, 2, 3, 4, 6, and 12. The divisors of 12 are also 1, 2, 3, 4, 6, and 12. We have merely changed our terminology from factors to divisors because the tests that we will be using are traditionally called *divisibility tests*. And here they are.

divisibility tests

2 *A number is divisible by* 2 *if it is an even number—that is, if its rightmost digit is* 0, 2, 4, 6, *or* 8.
   Example.   84, 32, 5376, 100, and 88 are each divisible by 2; their rightmost digits are 4, 2, 6, 0, and 8, respectively.

3 *A number is divisible by* 3 *if the sum of its digits is divisible by* 3.
   Example.   825 is divisible by 3 because 8 + 2 + 5 = 15, and 15 is divisible by 3. Similarly, 414 is divisible by 3, since 4 + 1 + 4 = 9, and 9 is divisible by 3.

4 *A number is divisible by* 4 *if the number formed by its rightmost two digits is divisible by* 4.
   Example.   740 is divisible by 4 because 40 is divisible by 4. 512 is divisible by 4 because 12 is divisible by 4.

5 *A number is divisible by* 5 *if its rightmost digit is* 0 *or* 5.
   Example.   20, 685, 700, 810, and 415 are each divisible by 5.

6 *A number is divisible by* 6 *if it is divisible by both* 2 *and* 3.
   Example.   714 is divisible by 6 because it is divisible by 2 (it ends in 4, which is even) and is also divisible by 3 (7 + 1 + 4 = 12, which is divisible by 3).

9   *A number is divisible by 9 if the sum of its digits is divisible by 9.*
Example.   864 is divisible by 9 because $8 + 6 + 4 = 18$, and 18 is divisible by 9.

10   *A number is divisible by 10 if its rightmost digit is 0.*
Example.   50, 90, 300, and 1040 are each divisible by 10.

The test for divisibility suggests the following. *If a number is divisible by two numbers having no common factors, then it is divisible by the product of those two numbers.* For instance, if a number is divisible by 2 and divisible by 3, then it is divisible by $2 \cdot 3$, or 6, because 2 and 3 have no common factors. On the other hand, if a number is divisible by 2 and by 4, it is not necessarily divisible by $2 \cdot 4$, or 8, because 2 and 4 have a common factor—2. Consider that 12 is divisible by 2 and by 4, but 12 is not divisible by 8.

Let's apply the divisibility tests to the reduction of fractions.

Example 9. *Reduce* $\dfrac{27}{126}$.

To reduce a fraction, you must find a factor (or divisor) common to both numerator and denominator. In this instance, both the numerator and denominator are divisible by 9 (check this). So divide both 27 and 126 by 9 to see what the factored form will be.

$$9\,\overline{)\,27}^{\,3}, \qquad 9\,\overline{)\,126}^{\,14}$$

This means that $27 = 3 \cdot 9$ and $126 = 14 \cdot 9$. Thus

$$\frac{27}{126} = \frac{3 \times 9}{14 \times 9}$$

$$= \frac{3}{14} \times \frac{9}{9}$$

$$= \frac{3}{14} \times 1$$

$$= \frac{3}{14} \;\checkmark$$

Example 10. *Reduce* $\dfrac{14}{21}$.

Unfortunately, our divisibility tests don't help us directly on this one, although they do lead us to $14 = 2 \cdot 7$ and $21 = 3 \cdot 7$ as we try various tests. And this, of course, demonstrates that both numerator and denominator are divisible by 7.

$$\frac{14}{21} = \frac{2 \times 7}{3 \times 7}$$

$$= \frac{2}{3} \times \frac{7}{7}$$

$$= \frac{2}{3} \times 1$$

$$= \frac{2}{3} \checkmark$$

Example 11. *Reduce* $\frac{84}{132}$.

Both 84 and 132 are divisible by 4. (Check this.) Also, both 84 and 132 are divisible by 3. (Check this.) The numbers 4 and 3 have no common factors; so 84 and 132 are each divisible by 4 · 3, or 12. Moreover,

$$12 \overline{)\,84}^{\,7} \quad \text{and} \quad 12 \overline{)\,132}^{\,11}$$

So $84 = 7 \cdot 12$ and $132 = 11 \cdot 12$. Thus

$$\frac{84}{132} = \frac{7 \cdot 12}{11 \cdot 12}$$

$$= \frac{7}{11} \times \frac{12}{12}$$

$$= \frac{7}{11} \times 1$$

$$= \frac{7}{11} \checkmark$$

EXERCISES 6.4

1. Apply all the divisibility tests to determine by which of the numbers 2, 3, 4, 5, 6, 9, and 10 the given number is divisible.

   *(a) 20   *(b) 30   *(c) 45   (d) 48   (e) 28
   *(f) 51   *(g) 19   *(h) 22   (i) 65   *(j) 42
   (k) 72   *(l) 75   *(m) 23   (n) 1000   *(o) 111
   *(p) 96   *(q) 252   (r) 414   *(s) 188   (t) 315

2. Reduce each fraction if possible.

   *(a) $\frac{4}{10}$ *(b) $\frac{21}{33}$ *(c) $\frac{2}{6}$ *(d) $\frac{28}{30}$ *(e) $\frac{12}{15}$

   (f) $\frac{3}{21}$ *(g) $\frac{35}{40}$ (h) $\frac{20}{65}$ *(i) $\frac{18}{72}$ *(j) $\frac{10}{240}$

   (k) $\frac{14}{33}$ (l) $\frac{78}{90}$ *(m) $\frac{15}{81}$ *(n) $\frac{58}{135}$ (o) $\frac{50}{80}$

   *(p) $\frac{42}{48}$ (q) $\frac{85}{90}$ *(r) $\frac{117}{153}$

3. Reduce each fraction.

*(a) $\dfrac{60}{105}$ *(b) $\dfrac{12}{300}$ *(c) $\dfrac{30}{135}$ *(d) $\dfrac{105}{120}$ (e) $\dfrac{90}{126}$

*(f) $\dfrac{84}{132}$ (g) $\dfrac{42}{210}$ *(h) $\dfrac{78}{144}$ *(i) $\dfrac{135}{720}$ (j) $\dfrac{210}{600}$

*(k) $\dfrac{160}{620}$ (l) $\dfrac{82}{205}$

4. Multiply. Reduce when possible.

*(a) $\dfrac{30}{31} \times \dfrac{7}{30}$ *(b) $\dfrac{30}{35} \times \dfrac{1}{7}$ *(c) $\dfrac{30}{35} \times \dfrac{5}{7}$ (d) $\dfrac{30}{35} \times \dfrac{1}{2}$

*(e) $\dfrac{21}{5} \times \dfrac{10}{42}$ *(f) $\dfrac{1}{3} \times \dfrac{114}{19}$ (g) $\dfrac{4}{105} \times \dfrac{15}{22}$ *(h) $\dfrac{17}{100} \times \dfrac{5}{34}$

(i) $\dfrac{3}{7} \times \dfrac{14}{48}$ *(j) $\dfrac{30}{72} \times \dfrac{9}{144}$ *(k) $15 \times \dfrac{32}{1200}$ (l) $\dfrac{27}{35} \times \dfrac{21}{81}$

6.5 DIVISION OF FRACTIONS

To divide two fractions, invert (that is, flip over) the second fraction and then multiply.

Example 12. *Divide* $\dfrac{3}{5} \div \dfrac{4}{7}$.

$$\frac{3}{5} \div \frac{4}{7} = \frac{3}{5} \times \frac{7}{4} \qquad \begin{cases} \text{invert: } \dfrac{4}{7} \longrightarrow \dfrac{7}{4} \\ \text{and change } \div \text{ to } \times \end{cases}$$

$$= \frac{3 \times 7}{5 \times 4}$$

$$= \frac{21}{20} \quad \checkmark$$

Note that you must invert the *second* fraction, not the first one. To see why, consider another approach to the example.

$$\frac{3}{5} \div \frac{4}{7} = \frac{\frac{3}{5}}{\frac{4}{7}} \qquad \text{another way of writing it}$$

$$= \frac{\frac{3}{5}}{\frac{4}{7}} \cdot \frac{\frac{7}{4}}{\frac{7}{4}} \qquad \begin{cases} \text{mutliplying both numerator } \left(\dfrac{3}{5}\right) \text{ and denominator} \\ \left(\dfrac{4}{7}\right) \text{ by the same number } \left(\dfrac{7}{4}\right) \end{cases}$$

$$= \frac{\frac{3}{5} \cdot \frac{7}{4}}{\frac{4}{7} \cdot \frac{7}{4}}$$

$$= \frac{\frac{3}{5} \cdot \frac{7}{4}}{1} \qquad \text{since } \frac{4}{7} \cdot \frac{7}{4} = 1$$

$$= \frac{3}{5} \cdot \frac{7}{4} \qquad \left\{ \begin{array}{l} \text{This completes the steps showing why we invert} \\ \text{and change to multiplication.} \end{array} \right.$$

$$= \frac{21}{20} \quad \checkmark$$

Example 13. *Divide* $\frac{4}{7} \div \frac{4}{3}$.

$$\frac{4}{7} \div \frac{4}{3} = \frac{4}{7} \times \frac{3}{4}$$

$$= \frac{\overset{1}{\cancel{4}}}{7} \times \frac{3}{\underset{1}{\cancel{4}}}$$

$$= \frac{1 \times 3}{7 \times 1}$$

$$= \frac{3}{7} \quad \checkmark$$

Example 14. *Divide* $\frac{7}{13} \div \frac{6}{8}$.

$$\frac{7}{13} \div \frac{6}{8} = \frac{7}{13} \div \frac{3}{4} \qquad \text{The } \frac{6}{8} \text{ can be reduced to } \frac{3}{4}.$$

$$= \frac{7}{13} \times \frac{4}{3} \qquad \left\{ \begin{array}{l} \text{invert and change to multi-} \\ \text{plication} \end{array} \right.$$

$$= \frac{7 \times 4}{13 \times 3}$$

$$= \frac{28}{39} \quad \checkmark$$

Example 15. *Divide* $\frac{5}{28} \div \frac{1}{5}$.

$$\frac{5}{28} \div \frac{1}{5} = \frac{5}{28} \times \frac{5}{1} \qquad \left(\begin{array}{l} \text{Note that these fives} \\ \text{cannot be removed.} \\ \text{You cannot reduce} \\ \text{across the } \div \text{ symbol.} \end{array} \right)$$

$$= \frac{5 \times 5}{28 \times 1}$$

$$= \frac{25}{28} \quad \checkmark$$

EXERCISES 6.5

1. Divide each and reduce if possible.

*(a) $\dfrac{5}{7} \div \dfrac{3}{4}$ *(b) $\dfrac{1}{3} \div \dfrac{1}{2}$ (c) $\dfrac{1}{5} \div \dfrac{1}{4}$ *(d) $\dfrac{5}{9} \div \dfrac{2}{3}$

(e) $\dfrac{11}{16} \div \dfrac{4}{5}$ *(f) $\dfrac{7}{8} \div \dfrac{3}{5}$ *(g) $\dfrac{5}{6} \div \dfrac{7}{12}$ *(h) $\dfrac{2}{7} \div \dfrac{7}{12}$

(i) $\dfrac{2}{7} \div \dfrac{12}{7}$ *(j) $\dfrac{5}{9} \div \dfrac{5}{9}$ (k) $\dfrac{4}{9} \div \dfrac{1}{2}$ *(l) $\dfrac{24}{30} \div \dfrac{5}{4}$

(m) $\dfrac{8}{25} \div \dfrac{2}{5}$ *(n) $\dfrac{2}{7} \div 3$ (o) $\dfrac{6}{11} \div 6$ *(p) $\dfrac{17}{12} \div 2$

*(q) $12 \div \dfrac{4}{5}$ (r) $4 \div \dfrac{2}{3}$

6.6 ADDITION AND SUBTRACTION OF FRACTIONS

You probably know that $\frac{2}{8} + \frac{3}{8} = \frac{5}{8}$. But do you know why we can merely add the numerators 2 and 3 to get 5 and why the denominator remains an 8? The following illustration may provide some insight.

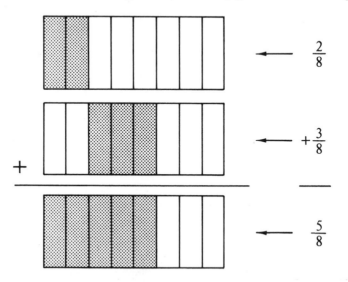

Mathematically, we can apply the distributive property (recall from Chapter 5) to see how the addition can be done.

$$\frac{2}{8} + \frac{3}{8} = 2 \times \frac{1}{8} + 3 \times \frac{1}{8} \qquad \begin{cases} \frac{2}{8} \text{ means } 2 \times \frac{1}{8} \text{(two-eights)} \\[2mm] \frac{3}{8} \text{ means } 3 \times \frac{1}{8} \text{(three-eighths)} \end{cases}$$

$$= (2 + 3) \times \frac{1}{8}$$

$$= (5) \times \frac{1}{8}$$

$$= \frac{5}{8} \ \checkmark$$

using the distributive property

As another example, consider

$$\frac{2}{13} + \frac{5}{13} + \frac{4}{13} = 2 \times \frac{1}{13} + 5 \times \frac{1}{13} + 4 \times \frac{1}{13}$$

$$= (2 + 5 + 4) \times \frac{1}{13}$$

$$= (11) \times \frac{1}{13}$$

$$= \frac{11}{13} \ \checkmark$$

As you can see, to add fractions: add the numerators to obtain the new numerator, and keep the same denominator. This addition can be expressed symbolically. If a, b, and c are used to represent whole numbers, then

$$\boxed{\frac{a}{c} + \frac{b}{c} = \frac{a + b}{c}}$$

Since addition of fractions uses the distributive property, the denominators must be the same in order to complete the process. We cannot add $\frac{1}{3} + \frac{1}{5}$ unless we can change their denominators so that both are the same; then we can apply the distributive property. The "same denominator" we seek for fractions being added is called the *common denominator*.

common denominator

One sure way to always make the denominators the same is to change each to a denominator equal to the product of the two denominators. With $\frac{1}{3}$ and $\frac{1}{5}$, the desired denominator is 3×5 or 15.

As you know, if a number is multiplied by 1 it is unchanged. Let us take the fractions $\frac{1}{3}$ and $\frac{1}{5}$ that we wish to add and multiply each by an appropriate form of 1.

$$\frac{1}{3} \times 1 = \frac{1}{3} \times \frac{5}{5} = \frac{1 \times 5}{3 \times 5} = \frac{5}{15}$$

$$\frac{1}{5} \times 1 = \frac{1}{5} \times \frac{3}{3} = \frac{1 \times 3}{5 \times 3} = \frac{3}{15}$$

Note that 1 can be written as $\frac{5}{5}$ and as $\frac{3}{3}$, since any nonzero number divided by itself is equal to 1.

The forms of 1 that were chosen, $\frac{5}{5}$ and $\frac{3}{3}$, were selected so that the end result in each case would be a fraction with a denominator of 15, 5 times 3. The fractions can now be added.

$$\frac{1}{3} + \frac{1}{5} = \frac{1}{3} \cdot \frac{5}{5} + \frac{1}{5} \cdot \frac{3}{3}$$

$$= \frac{5}{15} + \frac{3}{15}$$

$$= 5 \times \frac{1}{15} + 3 \times \frac{1}{15}$$

$$= (5 + 3) \times \frac{1}{15}$$

$$= \frac{8}{15} \quad \checkmark$$

Example 16. *Add* $\frac{1}{7} + \frac{1}{4}$.

Make each denominator 28—that is, 7×4.

$$\frac{1}{7} + \frac{1}{4} = \frac{1}{7} \cdot \frac{4}{4} + \frac{1}{4} \cdot \frac{7}{7}$$

$$= \frac{4}{28} + \frac{7}{28}$$

$$= \frac{11}{28} \quad \checkmark$$

Example 17. *Add* $\frac{1}{4} + \frac{3}{8}$.

Note that the denominator 4 can be changed to 8 very easily. And so 8 need not be changed. Although 4×8, or 32, will work as a denominator, there is no need to use the larger number, since 8 is sufficient.

$$\frac{1}{4} + \frac{3}{8} = \frac{1}{4} \cdot \frac{2}{2} + \frac{3}{8}$$

$$= \frac{2}{8} + \frac{3}{8}$$

$$= \frac{5}{8} \quad \checkmark$$

Example 18. *Add* $\frac{1}{4} + \frac{1}{3} + \frac{2}{5}$.

To add three fractions, all denominators must be the same; that is, the fractions must have a common denominator. The common denominator here is $4 \times 3 \times 5 = 60$, the product of the three denominators.

$$\frac{1}{4} + \frac{1}{3} + \frac{2}{5} = \frac{1}{4} \cdot \frac{15}{15} + \frac{1}{3} \cdot \frac{20}{20} + \frac{2}{5} \cdot \frac{12}{12}$$

$$= \frac{15}{60} + \frac{20}{60} + \frac{24}{60}$$

$$= \frac{59}{60} \quad \checkmark$$

Subtraction of fractions resembles addition, as shown in the examples below.

Example 19. *Combine* $\frac{5}{8} - \frac{4}{8}$.

$$\frac{5}{8} - \frac{4}{8} = 5 \times \frac{1}{8} - 4 \times \frac{1}{8}$$

$$= (5 - 4) \times \frac{1}{8}$$

$$= 1 \times \frac{1}{8}$$

$$= \frac{1}{8} \quad \checkmark$$

Example 20. *Combine* $\dfrac{5}{9} - \dfrac{1}{4}$.

$$\frac{5}{9} - \frac{1}{4} = \frac{5}{9} \cdot \frac{4}{4} - \frac{1}{4} \cdot \frac{9}{9}$$

$$= \frac{20}{36} - \frac{9}{36}$$

$$= \frac{11}{36} \quad \checkmark$$

In the next section we develop techniques that will help us to add and subtract fractions in cases where the common denominator is more difficult to determine.

EXERCISES 6.6

*1. Add the fractions.

(a) $\dfrac{4}{13} + \dfrac{2}{13}$ (b) $\dfrac{3}{7} + \dfrac{1}{7}$ (c) $\dfrac{1}{9} + \dfrac{4}{9}$ (d) $\dfrac{2}{5} + \dfrac{2}{5}$

*2. Subtract.

(a) $\dfrac{5}{7} - \dfrac{2}{7}$ (b) $\dfrac{10}{11} - \dfrac{8}{11}$ (c) $\dfrac{2}{3} - \dfrac{1}{3}$ (d) $\dfrac{8}{9} - \dfrac{7}{9}$

3. Add the fractions. Reduce the answer if possible.

*(a) $\dfrac{2}{5} + \dfrac{3}{5}$ (b) $\dfrac{4}{7} + \dfrac{3}{7}$ *(c) $\dfrac{1}{4} + \dfrac{1}{4}$

(d) $\dfrac{1}{6} + \dfrac{1}{6}$ *(e) $\dfrac{1}{3} + \dfrac{1}{6}$ (f) $\dfrac{5}{12} + \dfrac{1}{3}$

*(g) $\dfrac{2}{5} + \dfrac{3}{4}$ (h) $\dfrac{3}{10} + \dfrac{1}{2}$ *(i) $\dfrac{4}{13} + \dfrac{3}{7}$

(j) $\dfrac{1}{9} + \dfrac{3}{4}$ *(k) $\dfrac{3}{11} + \dfrac{5}{12}$ (l) $\dfrac{2}{9} + \dfrac{2}{5}$

*(m) $\dfrac{3}{8} + \dfrac{5}{16} + \dfrac{3}{4}$ (n) $\dfrac{1}{3} + \dfrac{5}{12} + \dfrac{1}{4}$ *(o) $\dfrac{4}{7} + \dfrac{2}{3} + \dfrac{1}{2}$

(p) $\dfrac{2}{9} + \dfrac{1}{3} + \dfrac{2}{5}$

4. Subtract. Reduce the answer if possible.

*(a) $\dfrac{4}{9} - \dfrac{1}{9}$ (b) $\dfrac{7}{10} - \dfrac{3}{10}$ *(c) $\dfrac{6}{7} - \dfrac{2}{5}$ (d) $\dfrac{4}{5} - \dfrac{2}{3}$

*(e) $\dfrac{3}{4} - \dfrac{2}{3}$ (f) $\dfrac{7}{9} - \dfrac{3}{7}$ *(g) $\dfrac{11}{12} - \dfrac{2}{3}$ (h) $\dfrac{11}{18} - \dfrac{4}{9}$

*(i) $\dfrac{11}{12} - \dfrac{3}{4}$ (j) $\dfrac{13}{16} - \dfrac{3}{8}$

6.7 IMPROPER FRACTIONS AND MIXED NUMBERS

The number $1\frac{7}{30}$ is an example of a *mixed number*. It is a whole number plus a fraction combined. $1\frac{7}{30}$ is an abbreviated form of $1 + \frac{7}{30}$.   **mixed number**

Mixed Number	Its Meaning
$1\frac{7}{30}$	$1 + \frac{7}{30}$
$2\frac{1}{2}$	$2 + \frac{1}{2}$
$3\frac{7}{8}$	$3 + \frac{7}{8}$

As mentioned in Section 6.1, a fraction is called an *improper fraction* if the numerator is greater than the denominator or equal to the denominator. If the numerator is less than the denominator, the fraction is a *proper fraction*.

Some proper fractions: $\frac{1}{2}, \frac{5}{6}, \frac{3}{11}$

Some improper fractions: $\frac{2}{2}, \frac{7}{6}, \frac{13}{11}$

Often it is desirable to change an improper fraction to the form of a mixed number. If you divide the denominator into the numerator, the quotient will be the whole-number part of the mixed number and the remainder will be the numerator of the fractional part.

Example 21. *Change $\frac{14}{3}$ to a mixed number.*

$$\begin{array}{r} 4 \text{ quotient} \\ 3\overline{)14} \\ 12 \\ \hline 2 \text{ remainder} \end{array} \longrightarrow 4\frac{2}{3} \checkmark$$

Since the quotient is 4, the whole-number part is 4. The remainder 2 is the numerator of the fraction. The denominator, of course, is 3, since the original fraction $\frac{14}{3}$ has a denominator of 3.

To change a mixed number to an improper fraction, multiply the denominator of the fraction by the whole number and then add the numerator of the fraction. Place the total over the denominator.

Example 22. *Change $7\frac{2}{5}$ to an improper fraction.*

$$7\frac{2}{5} \longrightarrow 5 \times 7 + 2 = 37$$

So

$$7\frac{2}{5} = \frac{37}{5} \checkmark$$

Example 23. *Multiply $2\frac{7}{8} \times 1\frac{2}{3}$.*

To multiply (or divide) mixed numbers, begin by changing the mixed numbers to improper fractions.

$$2\frac{7}{8} \times 1\frac{2}{3} = \frac{23}{8} \times \frac{5}{3}$$

$$= \frac{115}{24} \checkmark$$

If you prefer a mixed-number answer, change the improper fraction to a mixed number.

$$\frac{115}{24} = 4\frac{19}{24} \checkmark$$

Example 24. *Add $5\frac{3}{4} + 3\frac{1}{2}$.*

When addition (or subtraction) of mixed numbers is required, change them to improper fractions and then add (or subtract) the improper fractions.

$$5\frac{3}{4} + 3\frac{1}{2} = \frac{23}{4} + \frac{7}{2}$$

$$= \frac{23}{4} + \frac{14}{4}$$

$$= \frac{37}{4} \checkmark \quad \text{or} \quad 9\frac{1}{4} \checkmark$$

EXERCISES 6.7

1. Change each improper fraction to a mixed number.

*(a) $\frac{7}{6}$ (b) $\frac{5}{3}$ *(c) $\frac{9}{2}$ (d) $\frac{15}{7}$ *(e) $\frac{29}{4}$

(f) $\frac{22}{4}$ *(g) $\frac{19}{5}$ (h) $\frac{14}{3}$ *(i) $\frac{51}{2}$ (j) $\frac{25}{7}$

2. Change each mixed number to an improper fraction.

*(a) $1\frac{2}{3}$ (b) $1\frac{4}{5}$ *(c) $2\frac{6}{7}$ *(d) $3\frac{1}{9}$ *(e) $5\frac{2}{5}$

*(f) $8\frac{1}{2}$ *(g) $4\frac{5}{12}$ *(h) $7\frac{2}{9}$ *(i) 12   *(j) $9\frac{3}{7}$

3. Multiply or divide as indicated.

   *(a) $3\frac{1}{2} \times 2\frac{5}{8}$ (b) $3\frac{2}{3} \times 2\frac{5}{8}$ *(c) $1\frac{2}{3} \times 1\frac{2}{5}$

   (d) $4\frac{1}{8} \times 5\frac{1}{3}$ *(e) $3\frac{3}{7} \div 1\frac{1}{3}$ (f) $7\frac{1}{4} \div 2\frac{3}{8}$

   *(g) $5\frac{3}{4} \div 5\frac{1}{9}$ (h) $3\frac{6}{7} \div 2\frac{2}{3}$

4. Add or subtract as indicated. Reduce if possible.

   *(a) $2\frac{1}{2} + 1\frac{3}{4}$ (b) $3\frac{5}{8} + 2\frac{1}{4}$ *(c) $4\frac{5}{12} + 7\frac{1}{6}$

   (d) $5\frac{2}{3} + 1\frac{1}{5}$ *(e) $2\frac{2}{5} - 2\frac{1}{7}$ (f) $6\frac{1}{2} - 4\frac{2}{3}$

   *(g) $5\frac{2}{3} - 3\frac{3}{4}$ (h) $3\frac{7}{10} - 1\frac{1}{5}$

6.8 PRIME FACTORIZATION AND LCM

The terminology and techniques developed in this section will be applied to the addition of fractions.

Such numbers as 2, 3, 5, 7, 11, 13, 17, 19, and 23 are *prime numbers*; none can be written as the product of two whole numbers other than 1 and the number itself. On the other hand, 4, 6, 8, 9, 10, 12, 14, 15, 16, 18, 20, 21, and 22 are *not prime*. For instance, $4 = 2 \cdot 2$, $6 = 2 \cdot 3$, $15 = 3 \cdot 5$, and so on. Numbers that can be written in this factored form are called *composite numbers*. The number 1 is not considered a prime.

prime number

composite number

Example 25. *Write 30 as the product of prime numbers.*

One approach to this problem is to examine 30 in search of a divisor (factor), preferably one that is a prime number. The divisibility tests should prove helpful in such problems. Since 30 is an even number, 2 is a divisor. So divide 30 by 2 and write the factored form.

$$30 = 2 \cdot 15$$

But 15 is not a prime number; so we must next write 15 as a product of primes. Since 15 ends in 5, it is divisible by 5. In fact, $15 = 3 \cdot 5$. Thus the factored form of 30 becomes

$$30 = 2 \cdot 3 \cdot 5 \checkmark$$

This, then, is the desired *prime factorization*, the number (30) is now written as the product of prime numbers.

Example 26. *Write* 56 *as the product of prime numbers.*

Here are the steps in one approach to the prime factorization. Although you may prefer to go in a different direction, we will both end up with the same prime factorization. Such factorization is unique for any given number.

$$56 = 2 \cdot 28$$
$$= 2 \cdot 2 \cdot 14$$
$$= 2 \cdot 2 \cdot 2 \cdot 7$$

We can use exponent notation to compact the $2 \cdot 2 \cdot 2$ as 2^3. Thus the desired prime factorization is

$$56 = 2^3 \cdot 7 \ \checkmark$$

Now let's look at two other prime factorizations.

$$12 = 2^2 \cdot 3$$
$$15 = 3 \cdot 5$$

Suppose that we select and multiply all the different prime factors present in the above factorizations $(2, 3, 5)$ and include the largest exponents that appear on any factors with exponents (2^2). The result is

$$2^2 \cdot 3 \cdot 5, \text{ or } 60$$

This number 60 is a multiple of 12 (it is $5 \cdot 12$), and it is a multiple of 15 (it is $4 \cdot 15$). It is also the *smallest* number that is a multiple of both 12 and 15. As a result, it is called the *least common multiple*, or LCM for short. That the least common multiple of 12 and 15 is 60 can be written in the following notation:

LCM

$$\text{LCM } (12, 15) = 60$$

In general, the LCM of two or more numbers is the product of all the different prime factors of those numbers, including the highest power of each that appears.

To determine the LCM:

1. Obtain the prime factorization of each number.
2. Select all the different prime factors that appear, including the highest power of each, and multiply them together.

Example 27. *Determine the LCM of* 35 *and* 10.

The prime factorizations:

$$35 = 5 \cdot 7$$
$$10 = 2 \cdot 5$$

So

$$\text{LCM} (35, 10) = 2 \cdot 5 \cdot 7 = 70 \quad \checkmark$$

Example 28. *Determine the LCM of* 30 *and* 56.

From prime factorizations determined earlier in the section we know that

$$30 = 2 \cdot 3 \cdot 5$$
$$56 = 2^3 \cdot 7$$

So

$$\text{LCM} (30, 56) = 2^3 \cdot 3 \cdot 5 \cdot 7 = 840 \quad \checkmark$$

Example 29. *Determine LCM* (24, 30, 45).

The prime factorizations:

$$24 = 2^3 \cdot 3$$
$$30 = 2 \cdot 3 \cdot 5$$
$$45 = 3^2 \cdot 5$$

So

$$\text{LCM} (24, 30, 45) = 2^3 \cdot 3^2 \cdot 5 = 360 \quad \checkmark$$

EXERCISES 6.8

1. Write each number as a product of prime numbers.

*(a) 21	*(b) 10	(c) 15	*(d) 33	*(e) 42
*(f) 105	(g) 20	*(h) 40	(i) 18	*(j) 50
(k) 100	*(l) 36	*(m) 64	(n) 81	(o) 98
*(p) 90	(q) 120	(r) 150	*(s) 144	*(t) 216
(u) 210				

2. Determine the least common multiple (LCM) of each set of numbers.

*(a) 6, 10	*(b) 6, 35	*(c) 20, 25
*(d) 30, 32	*(e) 21, 35	(f) 96, 108
*(g) 4, 20, 30	*(h) 2, 3, 18	*(i) 10, 24, 30
(j) 15, 20, 25	(k) 18, 40, 50	*(l) 20, 36, 60
*(m) 50, 75, 120	(n) 4, 18, 20, 35	*(o) 12, 15, 20, 36
(p) 2, 4, 6, 8, 10		

6.9 THE LCM AND MORE ADVANCED ADDITION OF FRACTIONS

Examine the following *addition* problems to determine what should be used as the common denominator in each case.

(a) $\dfrac{1}{6} + \dfrac{3}{6}$

(b) $\dfrac{1}{2} + \dfrac{3}{8}$

(c) $\dfrac{2}{3} + \dfrac{1}{4}$

(d) $\dfrac{1}{2} + \dfrac{2}{3} + \dfrac{5}{7}$

(e) $\dfrac{5}{12} + \dfrac{2}{15} + \dfrac{3}{7}$

(f) $\dfrac{5}{8} + \dfrac{3}{10} + \dfrac{5}{14}$

The common denominators are (in order): 6, 8, 12, 42, 420, and 280. And you will probably agree that the last two were the most difficult to determine. Were you able to determine them? Your knowledge of LCM should have helped you. After all, what you seek as a common denominator is the smallest number that is a multiple of each of the denominators. In other words, the least common multiple (LCM) of the denominators is the smallest common denominator.

Example 30. *Determine the common denominator to be used in adding:*
$$\frac{5}{8} + \frac{3}{10} + \frac{5}{14}.$$

The prime factorizations of the denominators:
$$8 = 2^3$$
$$10 = 2 \cdot 5$$
$$14 = 2 \cdot 7$$

So LCM (8, 10, 14) $= 2^3 \cdot 5 \cdot 7 = 280$. The common denominator is 280. ✓

Example 31. *Add* $\dfrac{4}{15} + \dfrac{5}{6} + \dfrac{1}{4}$.

The prime factorization of the denominators:
$$15 = 3 \cdot 5$$
$$6 = 2 \cdot 3$$
$$4 = 2^2$$

The LCM (15, 6, 4) is $2^2 \cdot 3 \cdot 5$, or 60. So the common denominator is 60. We now proceed to change all denominators to 60 and then add the fractions.

$$\frac{4}{15} + \frac{5}{6} + \frac{1}{4} = \frac{4}{15} \cdot \frac{4}{4} + \frac{5}{6} \cdot \frac{10}{10} + \frac{1}{4} \cdot \frac{15}{15}$$

$$= \frac{16}{60} + \frac{50}{60} + \frac{15}{60}$$

$$= \frac{81}{60} \qquad \text{which can be reduced}$$

$$= \frac{3 \cdot 27}{3 \cdot 20}$$

$$= \frac{27}{20}, \text{ or } 1\frac{7}{20} \quad \checkmark$$

Example 32. *Add* $\frac{7}{12} + 5 + \frac{11}{14}$.

The whole number 5 is what makes this example different. One approach is to rewrite the 5 as the fraction $\frac{5}{1}$ and then determine the least common multiple of the denominators 12, 1, and 14. And LCM (12, 1, 14) = $2^2 \cdot 3 \cdot 7 = 84$. So we proceed as

$$\frac{7}{12} + \frac{5}{1} + \frac{11}{14} = \frac{7}{12} \cdot \frac{7}{7} + \frac{5}{1} \cdot \frac{84}{84} + \frac{11}{14} \cdot \frac{6}{6}$$

$$= \frac{49}{84} + \frac{420}{84} + \frac{66}{84}$$

$$= \frac{535}{84}, \text{ or } 6\frac{31}{84} \quad \checkmark$$

Another approach is to combine the whole number 5 with one of the two fractions to make a mixed number, change the mixed number to an improper fraction, and proceed to add. This process is shown next.

$$\frac{7}{12} + 5 + \frac{11}{14} = 5\frac{7}{12} + \frac{11}{14}$$

$$= \frac{67}{12} + \frac{11}{14}$$

$$= \frac{67}{12} \cdot \frac{7}{7} + \frac{11}{14} \cdot \frac{6}{6}$$

$$= \frac{469}{84} + \frac{66}{84}$$

$$= \frac{535}{84}, \text{ or } 6\frac{31}{84} \quad \checkmark$$

EXERCISES 6.9

1. Add the fractions below. Reduce all results to simplest form.

*(a) $\dfrac{3}{18} + \dfrac{1}{9}$ *(b) $\dfrac{17}{30} + \dfrac{3}{40}$ *(c) $\dfrac{5}{18} + \dfrac{3}{24}$

(d) $\dfrac{4}{9} + \dfrac{11}{30}$ (e) $\dfrac{5}{12} + \dfrac{4}{15}$ *(f) $\dfrac{7}{30} + \dfrac{11}{24}$

(g) $\dfrac{10}{21} + \dfrac{11}{25}$ *(h) $\dfrac{2}{15} + \dfrac{5}{28}$ (i) $\dfrac{13}{48} + \dfrac{13}{35}$

*(j) $\dfrac{25}{72} + \dfrac{19}{42}$ (k) $\dfrac{18}{49} + \dfrac{5}{42}$ *(l) $\dfrac{11}{81} + \dfrac{25}{54}$

*(m) $\dfrac{3}{10} + \dfrac{3}{20} + \dfrac{7}{25}$ (n) $\dfrac{1}{5} + \dfrac{7}{20} + \dfrac{13}{30}$ *(o) $\dfrac{5}{16} + \dfrac{17}{40} + \dfrac{9}{10}$

(p) $\dfrac{3}{4} + \dfrac{7}{18} + \dfrac{5}{36}$ *(q) $\dfrac{1}{4} + \dfrac{5}{9} + \dfrac{19}{30}$ (r) $\dfrac{11}{24} + \dfrac{1}{32} + \dfrac{13}{40}$

*(s) $\dfrac{1}{12} + \dfrac{7}{50} + \dfrac{4}{15}$ (t) $\dfrac{5}{48} + \dfrac{13}{72} + \dfrac{17}{60}$ *(u) $\dfrac{13}{45} + 6 + \dfrac{7}{18}$

(v) $\dfrac{5}{16} + \dfrac{17}{44} + 3$

2. Determine the perimeter of a triangle $(P = a + b + c)$ whose sides a, b, and c are $\frac{2}{9}$, $\frac{5}{6}$, and $\frac{7}{15}$.

3. Determine the perimeter of a rectangle $(P = 2l + 2w)$ whose length is $\frac{4}{11}$ and whose width is $\frac{1}{7}$.

6.10 A NOTE ON DECIMALS IN MULTIPLICATION AND DIVISION

In Chapter 1 a rule was given for multiplication of numbers containing decimal points. The product will have after its decimal point the total number of digits found after the decimal points of the numbers being multiplied. Here is an example.

$$
\begin{array}{r}
14.37 \quad \longleftarrow \quad \text{two digits after the decimal} \\
21.2 \quad \longleftarrow \quad \text{one digit after the decimal} \\
\hline
304.644 \quad \longleftarrow \quad \text{three (two + one) digits after decimal}
\end{array}
$$

To see *why* this rule works, begin by writing 14.37 and 21.2 as fractions and multiplying them that way.

$$14.37 = 14\frac{37}{100} = \frac{1437}{100}$$

$$21.2 = 21\frac{2}{10} = \frac{212}{10}$$

and $\dfrac{1437}{100} \times \dfrac{212}{10} = \dfrac{1437 \times 212}{100 \times 10} = \dfrac{304644}{1000} = 304.644$

Notice why the number of digits after the decimal point is determined to be the total number of digits after the decimal points of the two numbers being multiplied. When 14.37 is written as a fraction with denominator 100, you can see by the 100 that there are two decimal places. Similarly, when 21.2 is written as a fraction with denominator 10, you can see by the 10 that there is one decimal place. So when the two fractions are multiplied, you can see the three decimal places evolving as two from the first fraction and one from the second.

In Chapter 1 a rule was given for shifting the decimal points in order to make the divisor a whole number. Here is an example of such a shift. Consider the division

$$7.42 \overline{)\, 269.3516}$$

The mechanical process is to shift the decimal two places to the right in both numbers. This step will make the divisor a whole number, 742.

$$7.42 \overline{)\, 269.3516}$$

or

$$742 \overline{)\, 26935.16}$$

To see why this new form is equivalent to the original, observe the following use of fractions. First of all,

$$7.42 \overline{)\, 269.3516}$$

is the same as

$$\dfrac{269.3516}{7.42}$$

And the value of this fraction will not change if it is multiplied by 1. Choose 1 in the form

$$1 = \dfrac{100}{100}$$

because this will eliminate the decimal point in the denominator. So let's proceed to multiply the fraction by $\frac{100}{100}$.

$$\dfrac{269.3516}{7.42} \times \dfrac{100}{100} = \dfrac{26935.16}{742}$$

And this resulting fraction is the same as the division

$$742 \overline{)\, 26935.16}$$

Chapter 6. REVIEW EXERCISES

1. Reduce each fraction if possible.

*(a) $\dfrac{15}{25}$ *(b) $\dfrac{22}{80}$ *(c) $\dfrac{24}{40}$ (d) $\dfrac{48}{64}$ *(e) $\dfrac{48}{120}$

(f) $\dfrac{34}{57}$ *(g) $\dfrac{21}{98}$ (h) $\dfrac{64}{81}$ *(i) $\dfrac{112}{150}$

2. Multiply; reduce if possible.

*(a) $\dfrac{2}{3} \times \dfrac{3}{2}$ (b) $\dfrac{3}{5} \times \dfrac{7}{9}$ *(c) $\dfrac{12}{15} \times \dfrac{5}{8}$

(d) $\dfrac{15}{24} \times \dfrac{6}{5}$ *(e) $\dfrac{32}{50} \times \dfrac{25}{14}$ (f) $\dfrac{22}{9} \times \dfrac{27}{33}$

*(g) $\dfrac{17}{13} \times \dfrac{23}{29}$ (h) $\dfrac{26}{39} \times \dfrac{15}{11}$ *(i) $\dfrac{14}{98} \times \dfrac{14}{35}$

(j) $\dfrac{64}{81} \times \dfrac{27}{12}$ *(k) $5 \times \dfrac{13}{44}$ (l) $\dfrac{7}{20} \times 5$

*(m) $2\dfrac{3}{4} \times 3\dfrac{2}{5}$ *(n) $1\dfrac{2}{7} \times 2\dfrac{4}{9}$

3. Divide; reduce if possible.

*(a) $\dfrac{2}{3} \div \dfrac{2}{3}$ (b) $\dfrac{3}{5} \div \dfrac{5}{3}$ *(c) $\dfrac{15}{17} \div \dfrac{25}{34}$

(d) $\dfrac{100}{120} \div \dfrac{20}{15}$ *(e) $\dfrac{65}{85} \div \dfrac{39}{34}$ (f) $\dfrac{31}{37} \div \dfrac{11}{19}$

*(g) $\dfrac{38}{39} \div \dfrac{11}{26}$ (h) $\dfrac{45}{78} \div \dfrac{20}{19}$ *(i) $\dfrac{46}{54} \div \dfrac{69}{48}$

(j) $\dfrac{124}{105} \div \dfrac{88}{21}$ *(k) $\dfrac{15}{31} \div 4$ (l) $7 \div \dfrac{14}{19}$

*(m) $5\dfrac{3}{4} \div 2\dfrac{1}{2}$ *(n) $3\dfrac{2}{3} \div 2\dfrac{3}{7}$

4. Write each number as product of prime numbers.

*(a) 135   (b) 72   *(c) 88   (d) 84
*(e) 96   (f) 54   *(g) 110   (h) 125
*(i) 243

5. Determine the least common multiple (LCM) of each set of numbers.

*(a) 2, 3, 5   (b) 15, 24   *(c) 30, 40
(d) 26, 65   *(e) 8, 12, 20   (f) 18, 30, 36
*(g) 35, 45, 63   (h) 36, 72, 81

6. Add or subtract as indicated. Reduce all results to simplest form.

*(a) $\dfrac{11}{20} + \dfrac{1}{10}$ (b) $\dfrac{13}{22} + \dfrac{10}{33}$ *(c) $\dfrac{13}{24} - \dfrac{9}{32}$

(d) $\dfrac{12}{42} + \dfrac{11}{12}$ *(e) $\dfrac{15}{56} + \dfrac{16}{35}$ (f) $\dfrac{17}{50} - \dfrac{13}{55}$

*(g) $\dfrac{15}{84} + \dfrac{17}{63}$

(h) $\dfrac{15}{34} + \dfrac{1}{6} + \dfrac{6}{17}$

*(i) $\dfrac{5}{18} + \dfrac{4}{21} + \dfrac{7}{15}$

(j) $\dfrac{17}{28} + \dfrac{3}{14} + \dfrac{9}{16}$

*(k) $7 + \dfrac{3}{20} + \dfrac{5}{36}$

(l) $\dfrac{11}{49} + 3 + \dfrac{4}{21}$

*(m) $5\dfrac{3}{4} + 4\dfrac{1}{8}$

*(n) $3\dfrac{2}{5} - 1\dfrac{2}{3}$

Fractions, Decimals, and Percent

In this chapter we explore the relationships between fractions and decimals, decimals and percent, and fractions and percent. We begin with a look at fractions and decimals.

7.1 FORMS OF FRACTIONS AND DECIMALS

The first place to the right of the decimal point is the *tenths* place. The second place is *hundredths*. The third is *thousandths*. The fourth is *ten thousandths*. The fifth is *hundred thousandths*. The sixth is *millionths*. Thus

.7	is read	7 tenths
.23	is read	23 hundredths
.785	is read	785 thousandths
.6409	is read	6409 ten thousandths

All this is consistent with the way you would read or write the decimal in fraction form.

$$.7 = \frac{7}{10} \qquad \text{one zero in the denominator}$$

$$.23 = \frac{23}{100} \qquad \text{two zeros in the denominator}$$

$$.785 = \frac{785}{1000} \qquad \text{three zeros in the denominator}$$

$$.6409 = \frac{6409}{10,000} \qquad \text{four zeros in the denominator}$$

In should be clear how to write a decimal form as a fraction. The numerator of the fraction will be the original number without the decimal point, and there are as many zeros after the one in the denominator as there are digits after the decimal of the original.

Example 1. *Write .49 and .9521 as fractions.*

$$.49 = \frac{49}{100} \quad \checkmark$$

$$.9521 = \frac{9521}{10,000} \quad \checkmark$$

Often the fractional form can be simplified; we can *reduce* it. For instance,

$$.5 = \frac{5}{10} = \frac{1}{2}$$

$$.75 = \frac{75}{100} = \frac{3}{4}$$

$$.24 = \frac{24}{100} = \frac{6}{25}$$

$$.720 = \frac{720}{1000} = \frac{18}{25}$$

Consider the reverse process, that of changing a fraction to its decimal form.

Example 2. *Change $\frac{3}{8}$ to its decimal equivalent.*

The fraction $\frac{3}{8}$ can be written as 3/8 or $3 \div 8$. And division of 3 by 8 gives a decimal result.

$$\frac{3}{8} = 3 \div 8 \longrightarrow \begin{array}{r} .375 \\ 8 \overline{)\, 3.000} \\ \underline{24} \\ 60 \\ \underline{56} \\ 40 \\ \underline{40} \end{array}$$

So $\frac{3}{8} = .375$, exactly.

Example 3. *Change $\frac{5}{19}$ to a decimal.*

$$\frac{5}{19} = 5 \div 19 \longrightarrow 19 \overline{)\begin{array}{l}.26315 \\ 5.00000\end{array}}$$

$$\begin{array}{r}
38 \\ \hline
120 \\
114 \\ \hline
60 \\
57 \\ \hline
30 \\
19 \\ \hline
110 \\
95 \\ \hline
\end{array}$$

We have decided to quit with five digits after the decimal point. When the result does not come out exact after a few decimal places, you must decide when to quit. This decision usually depends on the application. Here the decision was arbitrary.

EXERCISES 7.1

Answers to starred exercises are given in the back of the book.

1. Change each decimal to a fraction and reduce if possible.

*(a) .17	*(b) .31	*(c) .50	*(d) .34
*(e) .75	(f) .44	*(g) .2	*(h) .315
*(i) .704	(j) .9	*(k) .95	(l) .85
*(m) .48	*(n) .3000	*(o) .5655	(p) .454
*(q) .825	(r) .285	*(s) .950	*(t) .999999

2. Change each fraction to a decimal form. Use four decimal places.

*(a) $\frac{1}{8}$	*(b) $\frac{3}{5}$	*(c) $\frac{7}{9}$	(d) $\frac{5}{12}$	*(e) $\frac{4}{7}$
*(f) $\frac{2}{3}$	*(g) $\frac{11}{13}$	(h) $\frac{5}{6}$	*(i) $\frac{5}{8}$	(j) $\frac{3}{14}$
*(k) $\frac{2}{11}$	(l) $\frac{3}{7}$	*(m) $\frac{1}{9}$	*(n) $\frac{4}{15}$	

7.2 PERCENT

percent   Percents are fractions in disguise. As you probably know, 50% means the same as $\frac{1}{2}$; 75% and $\frac{3}{4}$ are the same. Actually, *percent means hundredths.*

$$50\% = \frac{50}{100} \quad \text{which reduces to } \frac{1}{2}$$

$$75\% = \frac{75}{100} \quad \text{which reduces to } \frac{3}{4}$$

$$40\% = \frac{40}{100} \quad \text{which reduces to } \frac{2}{5}$$

$$35\% = \frac{35}{100} \quad \text{which reduces to } \frac{7}{20}$$

$$29\% = \frac{29}{100} \quad \text{which cannot be reduced}$$

Example 4. *Change 64% to a fraction.*

$$64\% = \frac{64}{100} \quad \text{which can be reduced to } \frac{16}{25} \checkmark$$

Percents are easily represented as decimals, since the fraction form of the percent has 100 in the denominator.

$$30\% = \frac{30}{100} = .30$$

$$89\% = \frac{89}{100} = .89$$

$$50\% = \frac{50}{100} = .50$$

$$25\% = \frac{25}{100} = .25$$

Dividing by 100 merely moves the decimal point two places to the left. This is shown in more detail below.

$$89\% = \frac{89}{100} = 89 \div 100 \longrightarrow 100 \overline{)\begin{array}{r} .89 \\ 89.00 \\ \underline{80\ 0} \\ 9\ 00 \\ \underline{9\ 00} \end{array}}$$

Thus

> To change a percent to a decimal, move the decimal point two places to the left and remove the % sign.

Again, a percent is a fraction with denominator 100.

Example 5. *Change 15% to a decimal.*

$$15\% = \frac{15}{100} = .15 \quad \checkmark$$

Example 6. *Change 2% to a decimal.*

$$2\% = \frac{2}{100} = .02 \quad \checkmark$$

Example 7. *Change 17% to a decimal.*

$$17\% = \frac{17}{100} = .17 \quad \checkmark$$

You can change a decimal to a percent by reversing this process.

> To change a decimal to a percent, multiply by 100 (or move the decimal point two places to the right) and attach a % sign.

Here are some examples.

Example 8. *Change .29 to a percent.*

$$.29 = .29 = 29\% \quad \checkmark \qquad \text{As a check, note that } 29\% = \frac{29}{100} = .29$$

Example 9. *Change .84 to a percent.*

$$.84 = .84 = 84\% \quad \checkmark$$

Example 10. *Change .816 to a percent.*

$$.816 = .816 = 81.6\% \quad \checkmark$$

How can we change a fraction to a percent? If the fraction has 100 as its denominator (or can be made to have 100 as its denominator), then the corresponding numerator is the percent.

Example 11. *Change $\frac{53}{100}$ to a percent.*

$$\frac{53}{100} = 53\% \quad \checkmark \qquad \text{Recall, percent means hundredths.}$$

Example 12. *Change* $\frac{9}{20}$ *to a percent.*

The denominator 20 is easily changed to 100.

$$\frac{9}{20} = \frac{9}{20} \times \frac{5}{5} = \frac{45}{100} = 45\% \quad \checkmark$$

Example 13. *Change* $\frac{3}{10}$ *to a percent.*

The denominator 10 is easily changed to 100.

$$\frac{3}{10} = \frac{3}{10} \times \frac{10}{10} = \frac{30}{100} = 30\% \quad \checkmark$$

But what if the denominator is 17 or 83 or 42? In such cases, change the fraction to a decimal and then change the decimal to a percent. Often the result will be approximate, because the division does not come out exact, yet we choose to quit after three or four decimal places. We will be using the symbol $\doteq$ to mean "approximately equal to."

Example 14. *Change* $\frac{5}{12}$ *to a percent.*

$$\frac{5}{12} = 5 \div 12 \longrightarrow 12 \overline{)5.000} \doteq .4166 \doteq 41.66\% \quad \checkmark$$

$$\begin{array}{r} .4166 \\ 12 \overline{)5.000} \\ \underline{4\,8} \\ 20 \\ \underline{12} \\ 80 \\ \underline{72} \\ 80 \\ \underline{72} \end{array}$$

Example 15. *Change* $\frac{1}{3}$ *to a percent.*

$$\frac{1}{3} = 1 \div 3 \longrightarrow 3 \overline{)1.000} \doteq .333 \doteq 33.3\% \quad \checkmark$$

$$\begin{array}{r} .333 \\ 3 \overline{)1.000} \end{array}$$

Example 16. *Change* $\dfrac{7}{80}$ *to a percent.*

$$\frac{7}{80} = 7 \div 80 \longrightarrow 80 \overline{\smash{)}\ 7.0000} = .0875 = 8.75\% \quad \checkmark$$

$$\begin{array}{r} .0875 \\ 80\ \overline{\smash{)}\ 7.0000} \\ \underline{6\ 40} \\ 600 \\ \underline{560} \\ 400 \\ \underline{400} \end{array}$$

EXERCISES 7.2

1. Change each percent to a fraction and reduce it if possible.
   *(a) 35%   (b) 40%   *(c) 28%   (d) 47%   *(e) 20%
   (f) 65%   *(g) 36%   *(h) 5%   *(i) 3%   *(j) 64%
   *(k) 42%   (l) 75%   *(m) 85%   (n) 12%   *(o) 8%
   (p) 4%

2. Change each percent to a decimal.
   *(a) 40%   (b) 60%   *(c) 25%   (d) 18%   *(e) 15%
   *(f) 6%   *(g) 19%   (h) 24%   *(i) 54%   (j) 88%

3. Change each decimal to a percent.
   *(a) .73   (b) .49   *(c) .65   (d) .58   *(e) .07
   (f) .03   *(g) .08   *(h) .80   *(i) .8   (j) .3
   *(k) .723   (l) .518   *(m) .062   (n) .091   *(o) .901
   (p) .003   *(q) .007   (r) .807

4. Change each fraction to a percent. If you use division and decimals arise, continue the division to three decimal places only.

   *(a) $\dfrac{4}{5}$ (b) $\dfrac{3}{10}$ *(c) $\dfrac{1}{3}$ (d) $\dfrac{2}{9}$ *(e) $\dfrac{4}{15}$

   (f) $\dfrac{5}{12}$ *(g) $\dfrac{6}{7}$ *(h) $\dfrac{4}{11}$ *(i) $\dfrac{2}{5}$ (j) $\dfrac{5}{6}$

   *(k) $\dfrac{7}{12}$ (l) $\dfrac{3}{4}$ *(m) $\dfrac{9}{13}$ (n) $\dfrac{9}{10}$ *(o) $\dfrac{3}{14}$

   *(p) $\dfrac{3}{8}$

7.3 *MORE ADVANCED PERCENTS*

   A few important cases were omitted in the preceding section. Let's consider them now by using examples.

   The next two examples show that even if a percent involves a number having a decimal point, you can move the decimal point two places to the left in order to change it from a percent to a decimal.

Example 17. *Change 58.4% to a decimal.*

$$58.4\% = 58.4 = .584 \quad \checkmark$$

Example 18. *Change 2.3% to a decimal.*

$$2.3\% = 02.3 = .023 \quad \checkmark$$

Example 19. *Change $2\frac{1}{4}\%$ to a decimal.*

$$2\tfrac{1}{4}\% = 2.25\% = 02.25 = .0225 \quad \checkmark$$

To change such percents to fractions, proceed as in Examples 17 and 18. Then, once the decimal form is obtained, write the decimal as a fraction. This is demonstrated in the next examples.

Example 20. *Change 58.4% to a fraction.*

$$58.4\% = 58.4 = .584 = \frac{584}{1000} \quad \checkmark$$

Example 21. *Change 2.3% to a fraction.*

$$2.3\% = 02.3 = .023 = \frac{23}{1000} \quad \checkmark$$

Sometimes you will encounter percents that are greater than 100%. What do such percents mean? For instance, what is meant by 200%, 180%, or 354%? The following example will show you.

Example 22. *Change 250% to a fraction and to a decimal.*

$$250\% = \frac{250}{100} = \frac{5}{2} \quad \checkmark$$

The fraction 5/2 is, of course, the same as the mixed number $2\frac{1}{2}$.

$$250\% = \frac{250}{100} = 2.5 \quad \checkmark$$

This result may seem strange. If so, note that percents between 0% and 100% represent numbers between 0 and 1; specifically, if we change the number 1 to a percent, the result is 100%.

$$1 = \frac{1}{1} = \frac{100}{100} = 100\%$$

So you should expect *any percent larger than* 100% *to be equivalent to a fraction larger than 1.*

Also, *any fraction greater than 1 is equivalent to a percent greater than* 100%.

Example 23. *Change* $3\frac{3}{4}$ *to a percent.*

$$3\frac{3}{4} = 3.75 = 3.75 = 375\% \quad \checkmark$$

Example 24. *Change* $\frac{9}{5}$ *to a percent.*

$$\frac{9}{5} = 1.80 = 1.80 = 180\% \quad \checkmark$$

EXERCISES 7.3

1. Change each percent to a fraction. Then reduce the fraction if possible.
   - *(a) 5.9%
   - *(b) 200%
   - *(c) 15.3%
   - (d) 32.5%
   - (e) 175%
   - *(f) 0%
   - (g) 50.8%
   - *(h) 1.6%
   - *(i) 400%
   - (j) 365%
   - *(k) .45%
   - *(l) .03%
   - *(m) 240%
   - (n) 10.9%
   - *(o) 134.5%
   - (p) 152.7%
   - *(q) 313%
   - (r) 100%
   - *(s) 1000%
   - (t) 1300%

2. Change each percent to a decimal.
   - *(a) $3\frac{1}{2}\%$
   - (b) $5\frac{1}{2}\%$
   - *(c) $7\frac{1}{4}\%$
   - (d) $8\frac{1}{4}\%$
   - *(e) $2\frac{3}{4}\%$
   - (f) $7\frac{3}{4}\%$
   - *(g) $5\frac{3}{8}\%$
   - (h) $4\frac{1}{8}\%$

3. Change each percent to a decimal.
   - *(a) 300%
   - (b) 700%
   - *(c) 150%
   - (d) 160%
   - *(e) 173%
   - (f) 942%
   - *(g) 1350%
   - (h) 1150%
   - *(i) 1425%
   - (j) 1392%

4. Change each fraction or decimal to a percent.
   - *(a) $2\frac{3}{10}$
   - (b) $4\frac{1}{2}$
   - *(c) $\frac{3}{2}$
   - (d) 1.6
   - *(e) 3.5
   - (f) $6\frac{1}{4}$
   - *(g) $\frac{1}{100}$
   - (h) $\frac{1}{1000}$
   - *(i) .2
   - (j) .02
   - *(k) .002
   - (l) 2
   - *(m) .071
   - (n) .086
   - *(o) 9.99
   - (p) 1.01
   - *(q) $7\frac{2}{5}$
   - (r) $5\frac{3}{5}$
   - *(s) $2\frac{1}{8}$
   - (t) $3\frac{5}{8}$

7.4 PERCENT PROBLEMS

In this section we see how to solve three basic types of percentage problems. An example of each is given.

Example 25. *What number is* 16% *of* 62*?*

Here we are asked to compute 16% of 62—that is, 16% times 62.

$$16\% \times 62 = .16 \times 62$$
$$= 9.92 \checkmark$$

Thus 16% of 62 is 9.92. Notice that 16% was changed to the decimal form .16 before being used in the calculation. It would also be correct to use the fractional form $\frac{16}{100}$ in the calculation, but the form 16% cannot be used in calculations.

Example 26. 16% *of what number is* 62*?*

When compared with the preceding example, we see that here we must determine which number should be multiplied by .16 in order to produce 62.

$$.16 \times (?) = 62$$

The number that we seek can be determined by dividing 62 by .16. The result is

$$\frac{62}{.16} = 387.5 \checkmark$$

To convince you that this division does, in fact, produce the correct number, let's check it. Is 16% of 387.5 equal to 62? Indeed it is.

$$.16 \times 387.5 = 62$$

Once you have studied the algebra of Chapters 12 and 13, the reason for dividing will be more apparent.

Example 27. *What percent of* 8 *is* 3*?*

Percentages are related to fractions. So begin by changing the question temporarily to "What fraction of 8 is 3?" Of course, the answer is

$$\frac{3}{8}$$

But we want a percentage, not a fraction. So change $\frac{3}{8}$ to a percent.

$$\frac{3}{8} \longrightarrow 8 \overline{) 3.000} \quad \overset{.375}{}$$

So

$$\frac{3}{8} = .375 = 37.5\% \checkmark$$

Thus 3 is 37.5% of 8.

The following word problem involves one of the three basic types of percent problems explained in this section. The exercises include word problems involving all three types.

Example 28. *Bill bought* 40 *light bulbs and* 3 *were defective. What percent of the* 40 *bulbs were defective?*

Since 3 of the 40 were defective, it follows that the fraction of the 40 that were defective is

$$\frac{3}{40}$$

So to find the percent of the 40 bulbs that were defective, merely change this fraction to a percent. The result is

$$7.5\% \quad \checkmark$$

EXERCISES 7.4

1. Solve each problem after first deciding which of the three types it is. When decimals are involved, three decimal places is sufficient for all answers.
   *(a) What number is 43% of 75?
   *(b) 58% of 80 is ____.
   *(c) 25% of what number is 41?
   *(d) 18% of ____ is 90.
   *(e) What percent of 72 is 9?
   *(f) 15 is ____% of 105.
   *(g) 19% of what number is 50?
    (h) What number is 24% of 91?
    (i) What percent of 42 is 5?
    (j) 29 is ____% of 74.
    (k) 52% of ____ is 100.
    (l) 79% of 18 is ____.
    (m) 15 is ____% of 5.

2. Solve each word problem.
   *(a) A team has won 14 of the 20 games it has played. What percent of its games has it won?
    (b) A team has lost 5 of the 8 games it has played. What percent of its games has it lost?
   *(c) Carol's house has 15 windows and she has cleaned 40% of them. How many windows has she cleaned?
    (d) Joan invested 24% of her $15,000 in U.S. Treasury notes. How much money does she have invested in Treasury notes?

*(e) Bill has invested 15% of his life savings in a municipal bond fund. The dollar amount of his investment is $6000. How much is his life savings?

(f) A father decides to give his son 18% of the land he owns. If this means that his son will get 7.2 acres, how much land does the father own?

Chapter 7. REVIEW EXERCISES

1. Change each number to a decimal (four decimal places if needed) and then to a percent.

*(a) $\dfrac{4}{5}$ *(b) $\dfrac{1}{6}$ (c) $\dfrac{5}{7}$

(d) $\dfrac{3}{11}$ *(e) 6   (f) $7\dfrac{1}{5}$

*(g) $\dfrac{5}{3}$ (h) .68   *(i) 9.514

2. Change each percent to a decimal and to a fraction. Reduce the fraction if possible.

*(a) 55%   (b) 32%   *(c) 72%
(d) 88%   *(e) 725%   *(f) 16.3%
(g) .51%   *(h) .09%   (i) 214.8%

3. Solve each problem. When decimals are involved, three decimal places is sufficient for all answers.
   *(a) What percent of 81 is 9?
   *(b) What number is 35% of 116?
   (c) 16 is ____ % of 120.
   (d) 30% of what number is 17?

*4. Solve each word problem.
   (a) Susan has sold 12 of the 40 paintings she has made. What percent of the paintings has she sold?
   (b) A town has 40,000 men and 70% of them are married. How many men are married?
   (c) Donna has invested 35% of her life savings in a certificate of deposit. The amount of this investment is $3850. How much is her life savings?

CHAPTER **8** _____

Units
of Measure

8.1 BASIC CONVERSIONS

If you were given a measurement in inches and asked to change it to feet, you would simply divide by 12 to get the result. For example, 39 inches = 39 ÷ 12 = 3.25 feet. To change from feet to inches, multiply by 12. For example, 7 feet = 7 · 12 = 84 inches. To change from hours to minutes, multiply by 60. And there are other situations in which you can easily change from one unit to another. However, there are also situations in which you would have considerable difficulty. For instance, how would you change miles per hour to feet per second? When using inches in computations, when is the result inches, when is it square inches, and when is it cubic inches? Our purpose here is to answer these questions and others like them. Metric units will not be used until Chapter 9, where the metric system will be explained. Our approach is demonstrated in the two examples that follow.

Example 1. *Change* 39 *inches to feet.*

We know that 1 ft = 12 in. or that there is 1 foot in every 12 inches.

That is,

$$\frac{1 \text{ ft}}{12 \text{ in.}} = 1$$

So if we write 39 inches as

$$\frac{39 \text{ in.}}{1}$$

and then multiply it by 1 in the form $\frac{1 \text{ ft}}{12 \text{ in.}}$, we get

$$39 \text{ in.} = \frac{39 \text{ in.}}{1} \times \frac{1 \text{ ft}}{12 \text{ in.}}$$

$$= \frac{39 \text{ i̶n̶.}}{1} \times \frac{1 \text{ ft}}{12 \text{ i̶n̶.}} \qquad \begin{cases} \text{The inches} \\ \text{can be removed.} \end{cases}$$

$$= \frac{39 \times 1 \text{ ft}}{1 \times 12}$$

$$= \frac{39 \text{ ft}}{12}$$

$$= 3.25 \text{ ft} \quad \checkmark$$

Example 2. *Change 7 feet to inches.*

$$7 \text{ ft} = \frac{7 \text{ ft}}{1}$$

$$= \frac{7 \text{ ft}}{1} \times \frac{12 \text{ in.}}{1 \text{ ft}} \qquad 12 \text{ inches per foot}$$

$$= \frac{7 \text{ f̶t̶}}{1} \times \frac{12 \text{ in.}}{1 \text{ f̶t̶}} \qquad \text{The feet divide out.}$$

$$= 84 \text{ in.} \quad \checkmark$$

In summary, the method to be used to change units is presented in the box below.

> When changing units, get the units to line up so that only the desired ones are left; that is, all others divide out.

Here are a few more examples.

Example 3. *Change 5 hours to seconds.*

$$1 \text{ hr} = 60 \text{ min} \qquad 1 \text{ min} = 60 \text{ sec}$$

$$5 \text{ hr} = \frac{5 \text{ hr}}{1}$$

$$= \frac{5 \text{ hr}}{1} \times \frac{60 \text{ min}}{1 \text{ hr}} \times \frac{60 \text{ sec}}{1 \text{ min}} \qquad \begin{cases} 60 \text{ minutes per hour} \\ 60 \text{ seconds per minute} \end{cases}$$

$$= \frac{5 \cancel{\text{hr}}}{1} \times \frac{60 \cancel{\text{min}}}{1 \cancel{\text{hr}}} \times \frac{60 \text{ sec}}{1 \cancel{\text{min}}} \qquad \begin{cases} \text{The hours and minutes} \\ \text{divide out.} \end{cases}$$

$$= \frac{5 \times 60 \times 60 \text{ sec}}{1}$$

$$= 18{,}000 \text{ sec} \; \checkmark$$

Example 4. *Change 17 miles to feet.*

$$1 \text{ mi} = 5280 \text{ ft}$$

$$17 \text{ mi} = \frac{17 \text{ mi}}{1}$$

$$= \frac{17 \text{ mi}}{1} \times \frac{5280 \text{ ft}}{1 \text{ mi}} \qquad 5280 \text{ ft per mile}$$

$$= \frac{17 \cancel{\text{mi}}}{1} \times \frac{5280 \text{ ft}}{1 \cancel{\text{mi}}} \qquad \text{The miles divide out.}$$

$$= \frac{17 \times 5280 \text{ ft}}{1}$$

$$= 89{,}760 \text{ ft} \; \checkmark$$

Example 5. *Change 30 miles per hour (mph) to feet per second.*

Here we combine the concepts developed in Example 3 and 4. In Example 3 we changed hours to seconds. In Example 4 we changed miles to feet.

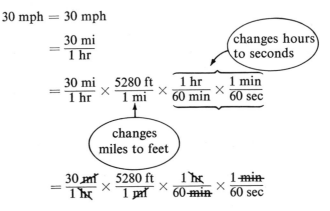

$$30 \text{ mph} = 30 \text{ mph}$$

$$= \frac{30 \text{ mi}}{1 \text{ hr}}$$

$$= \frac{30 \text{ mi}}{1 \text{ hr}} \times \frac{5280 \text{ ft}}{1 \text{ mi}} \times \frac{1 \text{ hr}}{60 \text{ min}} \times \frac{1 \text{ min}}{60 \text{ sec}}$$

changes hours to seconds

changes miles to feet

$$= \frac{30 \cancel{\text{mi}}}{1 \cancel{\text{hr}}} \times \frac{5280 \text{ ft}}{1 \cancel{\text{mi}}} \times \frac{1 \cancel{\text{hr}}}{60 \cancel{\text{min}}} \times \frac{1 \cancel{\text{min}}}{60 \text{ sec}}$$

The miles, hours, and minutes divide out.

$$= \frac{30 \times 5280 \text{ ft}}{60 \times 60 \text{ sec}}$$

$$= 44 \text{ ft/sec} \checkmark$$

The fractions were chosen to eliminate unwanted units and introduce the desired units.

When you *add* inches and inches, you get a sum of *inches*. Yet when you *multiply* inches by inches, you get *square inches*. Let's see why this happens.

Example 6. (a) *Add 5 inches + 6 inches.*
        (b) *Multiply 5 inches × 6 inches.*

  (a) 5 in. + 6 in. = (5 + 6) in.   using the "distributive property"
                   = 11 in. $\checkmark$
  (b) 5 in. × 6 in. = 5 × 6 × in. × in.
                   = 30 in.^{2}, or 30 sq in. $\checkmark$

Linear (distance) measure is expressed in such units as inches and feet, whereas area is expressed in square inches (in.^{2}) and square feet (ft^2). Volumes are expressed in such units as cubic inches (in.^{3}) and cubic feet (ft^3). These ideas are presented in the next three examples.

Example 7. *The area A of a triangle is computed as $A = \frac{1}{2} bh$, where b is the base and h is the height. Determine the area of the triangle that has a base of 24 feet and a height of 10 feet.*

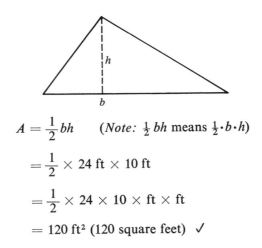

$$A = \frac{1}{2} bh \quad (\textit{Note: } \tfrac{1}{2} bh \textit{ means } \tfrac{1}{2} \cdot b \cdot h)$$

$$= \frac{1}{2} \times 24 \text{ ft} \times 10 \text{ ft}$$

$$= \frac{1}{2} \times 24 \times 10 \times \text{ft} \times \text{ft}$$

$$= 120 \text{ ft}^2 \ (120 \text{ square feet}) \ \checkmark$$

Example 8. *The volume of a box is computed by multiplying the length times the width times the height ($V = lwh$). Find the volume of a box that has length 10 inches, width 5 inches, and height 4 inches.*

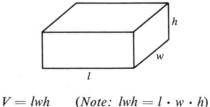

$V = lwh$ (*Note: lwh = l · w · h*)

$\quad = 10 \text{ in.} \times 5 \text{ in.} \times 4 \text{ in.}$

$\quad = 10 \times 5 \times 4 \times \text{in.} \times \text{in.} \times \text{in.}$

$\quad = 200 \text{ in.}^3$ (200 cubic inches) ✓

Suppose the length and width of the box above were again given as 10 inches and 5 inches, but the height was given as 2 *feet*. To use the formula $V = l \cdot w \cdot h$, we must have the same units for all dimensions l, w, and h. The easiest way to accomplish this would be to change the height of 2 feet to 24 inches. Then

$V = l \cdot w \cdot h$

$\quad = 10 \text{ in.} \times 5 \text{ in.} \times 24 \text{ in.}$

$\quad = 10 \times 5 \times 24 \times \text{in.} \times \text{in.} \times \text{in.}$

$\quad = 1200 \text{ in.}^3$ ✓

Example 9. *Change 7 cubic feet to cubic inches.*

$7 \text{ cubic feet} = 7 \text{ ft}^3$

$\quad = 7 \times \text{ft} \times \text{ft} \times \text{ft}$

$\quad = 7 \times \text{ft} \times \text{ft} \times \text{ft} \times \dfrac{12 \text{ in.}}{1 \text{ ft}} \times \dfrac{12 \text{ in.}}{1 \text{ ft}} \times \dfrac{12 \text{ in.}}{1 \text{ ft}}$

$\quad = 7 \times 12 \times 12 \times 12 \times \text{in.} \times \text{in.} \times \text{in.}$

$\quad = 12{,}096 \text{ in.}^3$ (12,096 cubic inches) ✓

Here is a brief table of units of measure that will be useful in doing the exercises that follow.

12 inches = 1 foot
3 feet = 1 yard
5280 feet = 1 mile

---

60 seconds = 1 minute
60 minutes = 1 hour
24 hours = 1 day
7 days = 1 week

---

16 ounces = 1 pound
2000 pounds = 1 ton

---

16 fluid ounces = 1 pint
2 pints = 1 quart
4 quarts = 1 gallon

EXERCISES 8.1

Answers to starred exercises are given in the back of the book.

1. Change the units as requested.
   *(a) Change 1476 inches to feet.
   *(b) Change 23 feet to inches.
   *(c) Change 12 hours to minutes.
   *(d) Change 5640 minutes to hours.
    (e) Change 73,920 feet to miles.
    (f) Change 54,000 pounds to tons.
   *(g) Change 20 gallons to quarts.
   *(h) Change 18 yards to feet.
    (i) Change 496 ounces to pounds.

2. Change the units as requested.
   *(a) Change 3 hours to seconds.
   *(b) Change 5 miles to inches.
   *(c) Change 10 weeks to hours.
   *(d) Change 8 tons to ounces.
   *(e) Change 25,200 seconds to hours.
    (f) Change 150,000 ounces to tons.
    (g) Change 300 minutes to seconds.
   *(h) Change 65,536 fluid ounces to gallons.
    (i) Change 1,108,800 inches to miles.

*3. Change the units as requested.
    (a) Change 180 miles per hour to miles per minute.
    (b) Change 180 miles per hour to feet per minute.

(c) Change 180 miles per hour to feet per second.
(d) Change 66 feet per second to miles per hour.
(e) Change 34 inches per second to feet per minute.
(f) Change 85 feet per minute to inches per second.

4. Change the units as requested.
   *(a) Change 5 square feet to square inches.
   *(b) Change 2304 square inches to square feet.
   (c) Change 135 square yards to square feet.
   (d) Change 135 square feet to square yards.
   *(e) Change 17 cubic feet to cubic inches.
   (f) Change 90 cubic feet to cubic yards.

5. Indicate the units as well as the numerical answer in each exercise below.
   *(a) Given $A = lw$, find A if $l = 14$ inches and $w = 13$ inches.
   *(b) Given $V = lwh$, find V if $l = 10$ feet, $w = 6.8$ feet, and $h = 4$ feet.
   *(c) Given $A = \pi r^2$, find A if $\pi = 3.14$ and $r = 5.4$ miles.
   (d) Given $A = \frac{1}{2}bh$, find A if $b = 16$ feet and $h = 7$ feet.
   *(e) Given $A = \frac{1}{2}(b_1 + b_2)h$, find A if $b_1 = 13$ inches, $b_2 = 7$ inches, and $h = 10$ inches.
   (f) Given $C = 2\pi r$, find C if $r = 19$ yards and $\pi = 3.14$.
   *(g) Given $V = \frac{4}{3}\pi r^3$, find V if $r = 10$ inches and $\pi = 3.14$.
   (h) Given $P = 2l + 2w$, find P if $l = 6.3$ inches and $w = 4.1$ inches.
   *(i) Given $V = lwh$, find V if $l = 5$ feet, $w = 3$ feet, and $h = 2$ *yards*.
   (j) Given $A = \frac{1}{2}bh$, find A if b is 7 feet and $h = 1$ *yard*.

Chapter 8.   REVIEW EXERCISES

*1. Change the units as requested.
   (a) Change 522 feet to yards.
   (b) Change 1140 seconds to minutes.
   (c) Change 9 weeks to seconds.
   (d) Change 846,720 seconds to weeks.
   (e) Change 90 cubic yards to cubic feet.

*2. Indicate the units as well as the numerical answer in each exercise below.
   (a) Given $V = \frac{1}{3}\pi r^2 h$, find V if $r = 9$ feet, $h = 4$ feet, and $\pi = 3.14$.
   (b) Given $P = 2l + 2w$, find P if $l = 15$ inches and $w = 4$ feet.

Metric System

9.1 BASIC CONCEPTS

The United States is now in the process of converting from the English system of units to the metric system. In the English system length is measured in inches, feet, yards, and miles. Which unit is used depends on the length or distance measured. In the metric system one basic unit is used, and all lengths are measured in terms of that unit. The basic unit of length in the metric system is the *meter* (abbreviated m). A meter is about 39.37 inches, a little more than a yard. The symbol $\doteq$ is used to mean *approximately equal to*. Thus,

> 1 meter $\doteq$ 39.37 inches

Example 1. *Change 27 inches to meters.*

We'll use the techniques of Chapter 8 and the fact that 1 meter is approximately 39.37 inches.

$$27 \text{ in.} \doteq \frac{27 \cancel{\text{ in.}}}{1} \times \frac{1 \text{ m}}{39.37 \cancel{\text{ in.}}}$$

$$\doteq \frac{27}{39.37} \text{ m}$$

$$\doteq .69 \text{ m} \quad \checkmark$$

Example 2. *Change* 3.4 *meters to inches.*

$$3.4 \text{ m} \doteq \frac{3.4 \text{ m}}{1} \times \frac{39.37 \text{ in.}}{1 \text{ m}}$$

$$\doteq 3.4 \times 39.37 \text{ in.}$$

$$\doteq 133.86 \text{ in.} \checkmark$$

Example 3. *Change* 35 *feet to meters.*

$$35 \text{ ft} \doteq \frac{35 \text{ ft}}{1} \times \frac{12 \text{ in.}}{1 \text{ ft}} \times \frac{1 \text{ m}}{39.37 \text{ in.}}$$

$$\doteq \frac{35 \times 12}{39.37} \text{ m}$$

$$\doteq 10.67 \text{ m} \checkmark$$

Prefixes are used in the metric system to specify lengths that are much larger or much smaller than a meter. The most common prefixes used in the metric system are given here, together with their meanings. Three others (hecto, deca, and deci) are introduced in Exercise 8.

kilo	1000	thousand
centi	$\frac{1}{100}$	hundredth
milli	$\frac{1}{1000}$	thousandth

A kilometer (km) is a thousand meters, which is about six-tenths of a mile. The kilometer is the metric unit used on road signs and maps.

$$1 \text{ kilometer} \doteq .6 \text{ mile}$$

Example 4. *Change* 50 *kilometers to miles.*

$$50 \text{ km} \doteq \frac{50 \text{ km}}{1} \times \frac{.6 \text{ mi}}{1 \text{ km}}$$

$$\doteq 50 \times .6 \text{ mi}$$

$$\doteq 30 \text{ mi} \checkmark$$

Example 5. *Change* 48 *miles per hour to kilometers per hour.*

$$48\frac{\text{mi}}{\text{hr}} \doteq \frac{48 \ \text{mi}}{1 \ \text{hr}} \times \frac{1 \ \text{km}}{.6 \ \text{mi}}$$

$$\doteq \frac{48}{.6} \times \frac{\text{km}}{\text{hr}}$$

$$\doteq 80 \ \text{km/hr} \quad \checkmark$$

A centimeter (cm) is one hundredth of a meter or about the diameter of an aspirin tablet. A millimeter (mm) is a tenth of a centimeter or a thousandth of a meter.

$$\boxed{1 \ \text{inch} \doteq 2.54 \ \text{centimeters}}$$

Example 6. *Change a measure of* 19 *centimeters to inches.*

$$19 \ \text{cm} \doteq \frac{19 \ \text{cm}}{1} \times \frac{1 \ \text{in.}}{2.54 \ \text{cm}}$$

$$\doteq \frac{19}{2.54} \ \text{in.}$$

$$\doteq 7.5 \ \text{in.} \quad \checkmark$$

The next two examples point out the simplicity of conversions done within the metric system. This is a real advantage of using the metric system.

Example 7. *Change* 3 *meters to millimeters.*

$$3 \ \text{m} = \frac{3 \ \text{m}}{1} \times \frac{1000 \ \text{mm}}{1 \ \text{m}}$$

$$= 3000 \ \text{mm} \quad \checkmark \quad \text{exactly}$$

Notice the simplicity of this conversion within the metric system. If you become familiar enough with the metric system so that you need not rely on the English system for reference, conversions between units will be simple.

Example 8. *Change* 5 *centimeters to meters.*

$$5 \ \text{cm} = \frac{5 \ \text{cm}}{1} \times \frac{1 \ \text{m}}{100 \ \text{cm}}$$

$$= \frac{5}{100} \ \text{m, or .05 m} \quad \checkmark \quad \text{exactly}$$

So far we have been concerned with units of length. In the metric system the *liter* is the basic unit of volume. A liter (1) is approximately 1.06 quarts (qt) in volume. In the English system fluid ounces (oz), quarts, and gallons (gal) are used. In metric it's liters, milliliters (ml), and kiloliters (kl). *One fluid ounce is about 30 milliliters.*

$$1 \text{ liter} \doteq 1.06 \text{ quarts}$$

Example 9. *Change 15 quarts to liters.*

$$15 \text{ qt} \doteq \frac{15 \text{ qt}}{1} \times \frac{1 \text{ liter}}{1.06 \text{ qt}}$$

$$\doteq \frac{15}{1.06} \text{ liters}$$

$$\doteq 14.15 \text{ liters } \checkmark$$

Example 10. *Change 85 milliliters to fluid ounces.*

$$85 \text{ ml} \doteq \frac{85 \text{ ml}}{1} \times \frac{1 \text{ oz}}{30 \text{ ml}}$$

$$\doteq \frac{85}{30} \text{ oz}$$

$$\doteq 2.83 \text{ oz } \checkmark$$

The next two examples show conversions within metric. Again, note the simplicity of such conversions.

Example 11. *Change 450 milliliters to liters.*

A milliliter is one-thousandth of a liter. Thus

$$450 \text{ ml} = \frac{450 \text{ ml}}{1} \times \frac{1 \text{ liter}}{1000 \text{ ml}}$$

$$= \frac{450}{1000} \text{ liter}$$

$$= .450 \text{ liter } \checkmark \text{ exactly}$$

Example 12. *Change 12 liters to milliliters.*

$$12 \text{ liters} = \frac{12 \text{ liters}}{1} \cdot \frac{1000 \text{ ml}}{1 \text{ liter}}$$

$$= 12{,}000 \text{ ml } \checkmark \text{ exactly}$$

In the English system weights are measured in ounces (oz), pounds (lb), and tons (ton). In metric it's *grams* (g) [and milligrams (mg) and kilograms (kg)].† An ounce is equal to about 28.35 grams; a pound is about 454 grams; a kilogram is about 2.2 pounds.

1 ounce $\doteq$ 28.35 grams
1 pound $\doteq$ 454 grams

1 kilogram $\doteq$ 2.2 pounds

Example 13. *Change* 15 *kilograms to pounds.*

Use the fact that 1 kg $\doteq$ 2.2 lb.

$$15 \text{ kg} \doteq \frac{15 \text{ kg}}{1} \times \frac{2.2 \text{ lb}}{1 \text{ kg}}$$

$$\doteq 15 \times 2.2 \text{ lb}$$

$$\doteq 33 \text{ lb} \quad \checkmark$$

The next three examples show conversions within metric. Such conversions are simple—involving only a shift of the decimal point.

Example 14. *Change* 5 *kilograms to grams.*

$$5 \text{ kg} = \frac{5 \text{ kg}}{1} \times \frac{1000 \text{ g}}{1 \text{ kg}}$$

$$= 5000 \text{ g} \quad \checkmark \text{ exactly}$$

Example 15. *Change* 7 *kilograms to milligrams.*

$$7 \text{ kg} = \frac{7 \text{ kg}}{1} \times \frac{1000 \text{ g}}{1 \text{ kg}} \times \frac{1000 \text{ mg}}{1 \text{ g}}$$

$$= 7,000,000 \text{ mg} \quad \checkmark \text{ exactly}$$

Example 16. *Change* 38 *milligrams to grams.*

$$38 \text{ mg} = \frac{38 \text{ mg}}{1} \times \frac{1 \text{ g}}{1000 \text{ mg}}$$

$$= \frac{38}{1000} \text{ g, or .038 g} \quad \checkmark \text{ exactly}$$

---

†Physicists make the distinction that pounds are a measure of force, whereas grams are a measure of mass.

Users of the English system usually measure temperature in degrees *Fahrenheit* (F). Water freezes at 32°F and boils at 212°F. Users of the metric system usually measure temperature in degrees *Celsius* (C). Water freezes at 0°C and boils at 100°C. The relationship between Fahrenheit and Celsius is given next.

$$C = \frac{5}{9}(F - 32)$$

$$F = \frac{9}{5}C + 32$$

Example 17. *Change 68°F to Celsius.*

$$C = \frac{5}{9}(F - 32°)$$

$$= \frac{5}{9}(68° - 32°)$$

$$= \frac{5}{9}(36°)$$

$$= \frac{5 \cdot 36°}{9}$$

$$= 20° \checkmark$$

Example 18. *Change 15°C to Fahrenheit.*

$$F = \frac{9}{5}C + 32°$$

$$= \frac{9}{5} \cdot 15° + 32°$$

$$= 27° + 32°$$

$$= 59° \checkmark$$

EXERCISES 9.1

Answers to starred exercises are given in the back of the book.

In Exercises 1 through 6, two decimal places is sufficient for answers that require digits after the decimal point. You will probably need to refer to the conversions given throughout this chapter and in the table on page 95.

1. Change each length from English to metric.
   *(a) 53 inches to meters
   (b) 93 inches to meters
   *(c) 29 feet to meters
   (d) 153 feet to meters
   *(e) 90 miles to kilometers
   (f) 123 miles to kilometers
   *(g) 19 inches to centimeters
   (h) 34 inches to centimeters
   *(i) 3 inches to millimeters
   (j) 7 inches to millimeters

2. Change each length from metric to English.
   *(a) 7 meters to inches
   (b) 9.7 meters to inches
   *(c) 43 meters to feet
   (d) 80 meters to feet
   *(e) 24 centimeters to inches
   (f) 76.3 centimeters to inches
   *(g) 140 kilometers to miles
   (h) 103 kilometers to miles
   *(i) 235 millimeters to inches
   (j) 254 millimeters to inches

3. Change each volume from English to metric.
   *(a) 23 quarts to liters
   (b) 50 quarts to liters
   *(c) 12 fluid ounces to milliliters
   (d) 34 fluid ounces to milliliters
   *(e) 428 fluid ounces to liters
   (f) 168 fluid ounces to liters
   *(g) 18 gallons to liters
   (h) 26 gallons to liters
   *(i) 300 gallons to kiloliters
   (j) 245 gallons to kiloliters

4. Change each volume from metric to English.
   *(a) 37 liters to quarts
   (b) 56 liters to quarts
   *(c) 90 milliliters to fluid ounces
   (d) 214 milliliters to fluid ounces
   *(e) 15 liters to fluid ounces
   (f) 23 liters to fluid ounces
   *(g) 23 liters to gallons
   (h) 19 liters to gallons
   *(i) 9 kiloliters to gallons
   (j) 17 kiloliters to gallons

5. Change each weight from English to metric.
   *(a) 23 pounds to kilograms
   (b) 65 pounds to kilograms
   *(c) 7 ounces to grams
   (d) 12 ounces to grams
   *(e) 3 ounces to milligrams
   (f) 10 ounces to milligrams

6. Change each weight from metric to English.
   *(a) 28 kilograms to pounds
   (b) 34 kilograms to pounds
   *(c) 97 grams to ounces
   (d) 115 grams to ounces
   *(e) 753 milligrams to ounces
   (f) 1987 milligrams to ounces

7. Change each measure within metric.
   *(a) 7 meters to millimeters
   (b) 16 meters to centimeters
   *(c) 158 millimeters to meters
   (d) 387 centimeters to meters
   *(e) 45 liters to milliliters
   (f) 852 liters to kiloliters
   *(g) 700 milliliters to liters
   *(h) 19 kilograms to grams
   *(i) 70 grams to milligrams
   (j) 43 kilograms to milligrams
   (k) 60 milligrams to grams
   *(l) 3 kiloliters to milliliters
   *(m) 7 kilometers to centimeters
   *(n) 7 kilometers to millimeters

*8. Here are three other prefixes used in the metric system.

hecto	100	hundred
deca	10	ten
deci	$\frac{1}{10}$	tenth

This means that a hectogram is 100 grams, a decagram is 10 grams, and a decigram is one-tenth of a gram.

Change each measure within metric.

(a) 3 hectograms to grams   (b) 19 hectograms to grams
(c) 5 decagrams to grams   (d) 17 decagrams to grams
(e) 250 decigrams to grams   (f) 33 decigrams to grams
(g) 74 decaliters to liters   (h) 150 liters to decaliters
(i) 200 decimeters to meters   (j) 300 meters to decimeters
(k) 400 hectoliters to decaliters   (l) 3 decaliters to deciliters

9. Change each Fahrenheit (F) temperature to Celsius (C) and each Celsius temperature to Fahrenheit. If decimals arise, round answers to the nearest degree.

*(a) 95°F   (b) 59°F   *(c) 41°F
  (d) 122°F   *(e) 20°C   (f) 80°C
*(g) 45°C   (h) 30°C   *(i) 75°F
  (j) 35°F   *(k) 59°C   (l) 99°C

Chapter 9. REVIEW EXERCISES

*1. Change from English to metric.
  (a) 5 inches to centimeters   (b) 19 ounces to grams
  (c) 13 quarts to liters   (d) 300 miles to kilometers

*2. Change from metric to English.
  (a) 40 kilograms to pounds   (b) 5 liters to fluid ounces
  (c) 320 millimeters to inches   (d) 17 meters to feet

*3. Change within metric.
  (a) 852 centimeters to millimeters   (b) 852 millimeters to centimeters
  (c) 52,000 millimeters to kilometers (d) 5000 milligrams to kilograms

*4. Change from Fahrenheit to Celsius or from Celsius to Fahrenheit, as appropriate. If decimals arise, round answers to the nearest degree.
  (a) 52°F   (b) 71°C

REVIEW PROBLEMS FOR PART I

1. Perform each operation.

    *(a) Add 294
            865
            473

    (b) Add 152.9
            761.1
            605.4

    *(c) Subtract 8937
                7959

    (d) Subtract 42.85
                38.99

    *(e) Multiply 951
                827

    (f) Multiply 19.25
                6.3

    *(g) 62) 74196

    (h) 3.8) 29.46

    *(i) 7^3

    (j) 4^5

2. Evaluate each expression.

    *(a) $9 + 4 \cdot 6 + 2 + 3^2$

    *(b) $15 - 16 \div 2 + 2$

    (c) $(9 + 2)^2$

3. Evaluate each formula, using the values given.

    *(a) $A = \frac{1}{2}bh$; $b = 10$, $h = 4$

    *(b) $S = 4\pi r^2$; $r = 6$, $\pi = 3.14$

    (c) $A = \frac{1}{2}(b_1 + b_2)h$; $b_1 = 4$, $b_2 = 22$, $h = 11$

4. Use the Pythagorean theorem to determine the length of the hypotenuse
   c when the lengths of the two legs a and b are those given.

    *(a) $a = 5$, $b = 12$

    (b) $a = 9$, $b = 12$

    *(c) $a = 1$, $b = 1$

5. Name the property used in going from each step to the next one.

    *(a) $(5 + 3)6 = (3 + 5)6$

    (b) $= 3 \cdot 6 + 5 \cdot 6$

    *(c) $= 18 + 5 \cdot 6$

    *(d) $= 18 + 6 \cdot 5$

    (e) $= 6 \cdot 5 + 18$

6. Carry out each operation. Reduce when possible.

    *(a) $\frac{8}{9} \times \frac{18}{17}$

    (b) $\frac{4}{15} \times \frac{25}{10}$

    *(c) $\frac{18}{32} \div \frac{5}{16}$

    (d) $\frac{35}{40} \div \frac{11}{12}$

    *(e) $\frac{77}{10} + \frac{4}{25}$

    (f) $\frac{1}{6} + \frac{3}{8} + \frac{5}{12}$

    *(g) $\frac{9}{14} - \frac{11}{42}$

    (h) $\frac{15}{24} - \frac{7}{32}$

7. Determine the least common multiple (LCM) of each set of numbers.

    *(a) 3, 5, 7

    (b) 6, 8

    *(c) 24, 90

    (d) 32, 40

8. Change each fraction to a percent.

   *(a) $\dfrac{2}{5}$ *(b) $\dfrac{7}{8}$ (c) $\dfrac{17}{25}$

9. Change each percent to a fraction.
   *(a) 45%   *(b) 83%   (c) 1.3%

10. Change units as requested.
    *(a) Change 5 hours to seconds.
    (b) Change 2304 inches to feet.
    *(c) Change 240 miles per hour to miles per minute.
    (d) Change 270 cubic feet to cubic yards.

11. Change units as requested.
    *(a) Change 15 gallons to liters.
    *(b) Change 500 grams to pounds.
    *(c) Change 5 meters to millimeters.
    (d) Change 120 kilometers to miles.
    (e) Change 135 centimeters to meters.

PART 2

Signed Numbers

10.1 INTRODUCTION

On the thermometer shown there are temperatures both above zero degrees and below zero degrees. The temperatures below zero are preceded by minus (or negative) signs to differentiate them from temperatures above zero. Thus 5 degrees below zero is −5°, whereas 5 degrees above zero is just 5° or perhaps +5° for emphasis.

Signed numbers are used in other settings, too. Positive numbers are used to indicate gain, increase, profit, and similar items. Negative numbers are used to indicate loss, decrease, deficit, and so forth.

The use of signed numbers in algebra is so common that you must understand them if you are to progress in the study of arithmetic and its generalized form—algebra.

Before beginning the study of signed numbers, it is worthwhile to distinguish between different kinds of numbers—natural numbers (counting numbers), whole numbers, and integers.

Natural numbers: 1, 2, 3, 4, 5, 6, etc.

Counting numbers: 1, 2, 3, 4, 5, 6, etc.

Whole numbers: 0, 1, 2, 3, 4, 5, 6, etc.

natural number

whole number

109

integer

Integers: 0, 1, −1, 2, −2, 3, −3, 4, −4, etc.

Positive integers: 1, 2, 3 ,4, 5, 6, etc.

Negative integers: −1, −2, −3, −4, −5, −6, etc.

number line

    The thermometer represents one form of the *number line*. We shall use the number line to introduce basic signed-number concepts. The number line can be pictured as

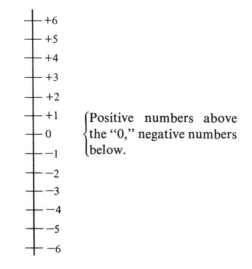

{Positive numbers above the "0," negative numbers below.

or as

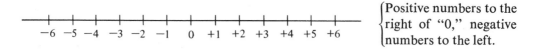

{Positive numbers to the right of "0," negative numbers to the left.

We shall use the second line, the horizontal one.

10.2 ADDITION OF SIGNED NUMBERS

directed number

    Signed numbers are also called *directed numbers*. The sign (+ or −) indicates direction on the number line; + is right and − is left. Plus and minus are opposite signs; they indicate opposite directions.

magnitude The unsigned portion of a number is called the *magnitude* of the number.

For example, the magnitude of $+17$ is 17. The magnitude of -17 is also 17.

We can add two signed numbers by placing one of them on a number line and moving (from that point) the number of units in the direction indicated by the second number. Here are some examples.

Example 1. *Add $+3$ and $+4$.*

First, place $+3$ on the line.

Next, move $+4$ units from the point $+3$. This means move 4 units to the right, since $+$ indicates the right direction.

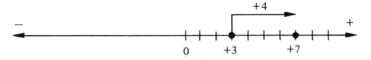

Moving $+4$ takes us to $+7$. So $+7$ is the sum of $+3$ and $+4$.

$$\begin{array}{r} +3 \\ +4 \\ \hline +7 \end{array} \checkmark$$

Example 2. *Add -3 and -4.*

First, place -3 on the line.

Next, move -4 units from the point -3. That is, move 4 units to the left, since $-$ indicates the left direction.

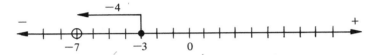

Moving -4 takes us to -7. So -7 is the sum of -3 and -4.

$$\begin{array}{r} -3 \\ -4 \\ \hline -7 \end{array} \checkmark$$

Example 3. *Add −3 and +4.*

First, place −3 on the line.

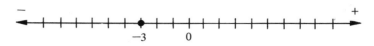

Next, move +4 units from the point −3. In other words, move 4 units to the right.

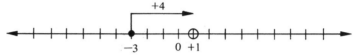

Moving +4 takes us to +1. So the sum of −3 and +4 is +1.

$$\begin{array}{r} -3 \\ +4 \\ \hline +1 \end{array} \checkmark$$

Example 4. *Add +3 and −4.*

First, place +3 on the line.

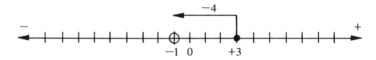

Next, move −4 units from the point +3. That is, move 4 units to the left.

Moving −4 takes us to −1. Thus the sum of +3 and −4 is −1.

$$\begin{array}{r} +3 \\ -4 \\ \hline -1 \end{array} \checkmark$$

The number line is useful in gaining an understanding of addition of signed numbers, but it is slow and not always practical. For example, to add +1945 and −2813 on a number line would be difficult. And how about adding −243.79 and +175.85? So we need a mechanical method that does not use a number line.

After studying the preceding four examples, we can make some important observations about the addition of signed numbers.

Addition of Signed Numbers

1. When both numbers have the same sign (both + or both —),
the result has that same sign. The magnitude of the result is
simply the sum of the magnitudes.

Two examples:

$$\begin{array}{r} +3 \\ +4 \\ \hline +7 \end{array} \qquad \begin{array}{r} -3 \\ -4 \\ \hline -7 \end{array}$$

Addition of Signed Numbers

2. When two numbers have opposite signs (one +, one —), the
sign of the sum is the sign of the number that is larger in
magnitude. The magnitude of the sum is the magnitude of the
difference of the two numbers.

Two examples:

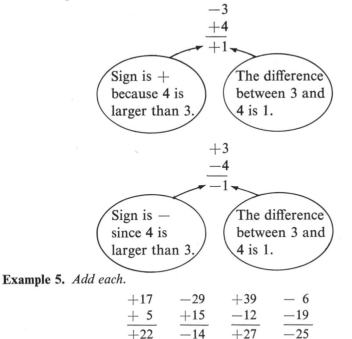

Example 5. *Add each.*

$$\begin{array}{r} +17 \\ +\ 5 \\ \hline +22 \end{array} \qquad \begin{array}{r} -29 \\ +15 \\ \hline -14 \end{array} \qquad \begin{array}{r} +39 \\ -12 \\ \hline +27 \end{array} \qquad \begin{array}{r} -\ 6 \\ -19 \\ \hline -25 \end{array}$$

Note that we added in each case. The negative (minus) signs indicate negative
numbers, not subtraction.

Often two numbers to be added are written on the same line. The addition of -6 and $+17$ would be so written as

$$(-6) + (+17)$$

Similarly, the addition of $+12$ and -10 is specified as

$$(+12) + (-10)$$

Verify the following additions.

$$(-6) + (+17) = +11$$
$$(+12) + (-10) = +2$$
$$(-5) + (+1) = -4$$
$$(-8) + (-3) = -11$$

If there are more than two numbers to add, add them two at a time.

Example 6. *Combine* $(+6) + (-3) + (+5) + (-6) + (+3)$.

Here we'll add the numbers two at a time, proceeding from left to right.

$$(+6) + (-3) + (+5) + (-6) + (+3) = (+3) + (+5) + (-6) + (+3)$$
$$= (+8) + (-6) + (+3)$$
$$= (+2) + (+3)$$
$$= +5 \checkmark$$

opposite   You may prefer to look for shortcuts if they exist. One shortcut is based on using opposites. The *opposite* of a number is a number that has the same magnitude but an opposite sign. The word "opposite" is short for "opposite-signed number." An important property is given in the box below.

$$\boxed{\text{number} + \text{opposite} = 0}$$

Consider these examples.

Number	Opposite
$+5$	-5
-5	$+5$
-7	$+7$
$+3.2$	-3.2
$+8\frac{1}{4}$	$-8\frac{1}{4}$
-19	$+19$
x	$-x$
$-x$	x

Observe that -5 is the opposite of $+5$; they add to zero.

$$\begin{array}{r} +5 \\ -5 \\ \hline 0 \end{array}$$

Example 7. *Combine* $(+6) + (-3) + (+5) + (-6) + (+3)$. *Look for shortcuts.*

Let's look for opposites and use them to simplify the computations. If we examine the expression $(+6) + (-3) + (+5) + (-6) + (+3)$, we find $+6$ and -6. They are opposites; so they add up to zero. Similarly, -3 and $+3$ are present, and they add up to zero. All that is left, then, is $+5$. Therefore the sum is $+5$. (Recall: $0 + x = x$ for all numbers x.)

Another approach to adding signed numbers is to add all the positive numbers together and add all the negative numbers together. Then add the two sums. This method is demonstrated in the next example. The advantage of this method is that you only add numbers of unlike sign once.

Example 8. *Combine* $(+7) + (-4) + (-5) + (+12) + (-1)$.

Separate $(+7) + (-4) + (-5) + (+12) + (-1)$ into positive and negative numbers and compute each sum separately. Then add the two sums.

$$\begin{array}{lll} \text{Positive:} & (+7) + (+12) & = +19 \\ \text{Negative:} & (-4) + (-5) + (-1) & = -10 \\ \hline & & +9 \text{ is the sum } \checkmark \end{array}$$

EXERCISES 10.2

Answers to starred exercises are given in the back of the book.

*1. Add.

(a) $+19$ $+14$	(b) $+15$ $+24$	(c) $+13$ $-\ 8$	(d) $+16$ -25
(e) -19 $+23$	(f) -35 $+28$	(g) -15 -12	(h) -12 -15
(i) $+14$ -19	(j) -18 $+23$	(k) -49 -28	(l) -33 $+15$

2. Add.

(a) $(+8) + (+3)$ *(b)* $(-7) + (-4)$

(c) $(-6) + (+2)$ *(d)* $(-3) + (+9)$

*(e) $(+7) + (-7)$ (f) $(-12) + (-19)$
*(g) $(-6) + (+17)$ (h) $(+13) + (-5)$
*(i) $(-16) + (-7)$ (j) $(-19) + (-11)$
*(k) $(+16) + (-9)$ (l) $(-12) + (-8)$
*(m) $(+17) + (+28)$ (n) $(-19) + (+3)$
*(o) $(-23) + (+14)$ (p) $(+30) + (-14)$
*(q) $(-16) + (-15)$ (r) $(+13) + (+18)$
*(s) $(-6) + (-25)$ (t) $(-12) + (-12)$

*3. Add.

(a) $\left(-\frac{2}{3}\right) + \left(+\frac{3}{4}\right)$ (b) $\left(-\frac{1}{2}\right) + \left(+\frac{2}{5}\right)$

(c) $\left(-\frac{3}{5}\right) + \left(-\frac{2}{3}\right)$ (d) $\left(+\frac{6}{7}\right) + \left(-\frac{1}{4}\right)$

4. Determine the sum; that is, combine the numbers.
   *(a) $(+16) + (+2) + (-5) + (+3)$
   *(b) $(-7) + (+1) + (+3) + (-4) + (-3) + (+7)$
   *(c) $(+9) + (+2) + (+4) + (+6) + (-7) + (-3) + (-5)$
   *(d) $(+6) + (-6) + (+9) + (-9) + (-4) + (+4)$
   *(e) $(+3) + (+4) + (-6) + (-7) + (0)$
   (f) $(+3) + (-2) + (+1) + (-4) + (+5) + (-2) + (+1) + (-1)$
   *(g) $(+15) + (-12) + (+60) + (-5) + (-30) + (-17)$
   (h) $(+52) + (+13) + (-42) + (+12) + (-1) + (+49)$
   *(i) $(0) + (-5) + (+1) + (-2) + (+3) + (-1) + (-2) + (+1)$
   (j) $(+16) + (-19) + (-6) + (+9) + (-28) + (-34) + (-12)$
   *(k) $(-18) + (-3) + (+12) + (-6) + (-1) + (+9) + (-20)$
   (l) $(-5) + (0) + (+17) + (-10) + (-9) + (+5) + (-1)$

*5. Combine the numbers.
   (a) $(+3.6) + (+2.1) + (-2.8) + (+1.3) + (-7.3)$
   (b) $(-1.05) + (3.46) + (-2.91) + (8.2)$

10.3 SUBTRACTION OF SIGNED NUMBERS

If you add 3 to 8, the result is 11. But if you subtract 3 from 8, the result is 5. On the number line you would *add* $+3$ to $+8$ by placing the $+8$ on the line and moving 3 units to the right. The result is, of course, $+11$.

In order to obtain a result of $+5$ when *subtracting* $+3$ from $+8$ on the number line, place $+8$ on the line and move in the opposite direction as you would if adding. In other words, move 3 units to the left rather than to the right.

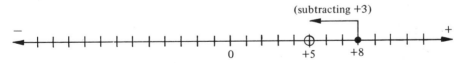

(subtracting $+3$)

In this sense, subtraction can be considered the opposite of addition. When subtracting a positive number, move in the negative (left) direction. When subtracting a negative number, move in the positive (right) direction.

Example 9. *Use a number line to show the result of subtracting* -5 *from* -6.

Place -6 on the line and subtract -5 from it.

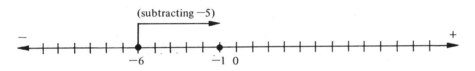

(subtracting -5)

The result is -1. Notice that in the subtraction of -5 we moved in the opposite direction we would have if we were adding -5. We moved right for -5 instead of left, as we would have done in addition.

Moving in the opposite direction on the number line is the same as adding the opposite signed number. In other words,

> To subtract a signed number (from another), add its opposite.
>
> or
>
> To subtract a signed number (from another), change its sign and add it.

subtract	*add*	
$+8$	$+8$	
$+3$ $\longrightarrow$	-3	change $+3$ to -3 and add
	$+5$ ✓	

$$
\begin{array}{cc}
subtract & add \\
-6 & -6 \\
\underline{-5} & \underline{+5} \\
 & -1 \checkmark
\end{array}
\qquad \text{change } -5 \text{ to } +5 \text{ and add}
$$

Example 10. *Subtract* -35 *from* $+180$.

$$
\begin{array}{cc}
subtract & add \\
+180 & +180 \\
\underline{-35} & \underline{+\ 35} \\
 & +215 \checkmark
\end{array}
$$

The difference between a and b, $a - b$, can be *defined* as the sum of a and the opposite of b. That is, $a - b = a +$ opposite of b, or

$$
\boxed{(a) - (b) = (a) + (-b)}
$$

where $-b$ means the opposite of b.

In subtraction of signed numbers, you change two things:

1. Change the operation to addition.
2. Change the sign of the number you were subtracting.

Example 11. *Subtract* $+5$ *from* -2.

Since we are subtracting $+5$ from -2, this should be written, for calculation purposes, as

$$
(-2) - (+5)
$$

Now, instead of subtracting $+5$, let us add -5.

$$
(-2) + (-5)
$$

The result is

$$
-7 \checkmark
$$

Example 12. *Compute* $(+5) - (-3)$.

This is, of course, a subtraction. So proceed accordingly.

$$
\begin{aligned}
&(+5) - (-3) \\
={}&(+5) + (+3) \\
={}&+8 \checkmark
\end{aligned}
$$

Example 13. *Compute* $(-8) - (-5)$.

$$(-8) - (-5)$$
$$= (-8) + (+5)$$
$$= -3 \ \checkmark$$

Example 14. *Compute* $(+5) - (+1)$.

$$(+5) - (+1)$$
$$= (+5) + (-1)$$
$$= +4 \ \checkmark$$

Addition can be used as a check for subtraction, and it can also be used indirectly to perform the subtraction. Consider the subtraction $(5) - (2)$. The result is 3. And 3 checks. That is, 3 is the number that when added to 2 produces 5. $(5) - (2) = 3$, since $2 + 3 = 5$. Similarly, $(+7) - (-6) = +13$, since $(-6) + (+13) = +7$. Consider another example. $(-8) - (+6) = ?$ The result is the number that when added to $+6$ produces -8. $(+6) + (\quad) = -8$. The number is -14. Thus $(-8) - (+6) = -14$.

In these last examples we have performed the subtraction indirectly by determining what number would check. If this method is applied to the earlier examples of the section, the results will be the same as those obtained by "adding the opposite" or "changing the sign of the number being subtracted." Yet this procedure is less mechanical and may perhaps be more convincing. It is shown here for understanding, but the mechanical approaches are generally more efficient.

EXERCISES 10.3

*1. Subtract.

    (a) $+30$ (b) -30 (c) -43
           $+\ 5$ $+\ 5$ $+17$

    (d) -15 (e) $+32$ (f) $+54$
           -17 -19 $+25$

    (g) -57 (h) -15 (i) -43
           $+41$ -94 $+72$

    (j) $+17$ (k) -34 (l) $+387$
           $+29$ $+42$ -524

    (m) -341 (n) -196 (o) $+385$
         -102 $+145$ $+290$

*2. Subtract.
    (a) $+68.52$ (b) $+359.16$ (c) -157.23
       -49.23 $+267.06$ -251.95

3. Perform the indicated subtractions.
    *(a) $(+7) - (+4)$ *(b) $(+9) - (+18)$
    *(c) $(+6) - (-3)$ *(d) $(-9) - (+6)$
    *(e) $(+5) - (-12)$ *(f) $(-7) - (-4)$
     (g) $(-6) - (-10)$ *(h) $(+7) - (-8)$
     (i) $(+3) - (-1)$ (j) $(-1) - (-3)$
    *(k) $(-16) - (+19)$ (l) $(+176) - (+761)$

*4. Perform the indicated subtractions.
    (a) $(-1.67) - (+3.45)$ (b) $(+78.45) - (-34.98)$
    (c) $\left(+\frac{3}{7}\right) - \left(-\frac{2}{7}\right)$ (d) $\left(-\frac{1}{2}\right) - \left(+\frac{1}{4}\right)$
    (e) $\left(+\frac{2}{5}\right) - \left(+\frac{2}{3}\right)$ (f) $\left(-\frac{2}{3}\right) - \left(-\frac{3}{4}\right)$

5. Compute.
    *(a) $(-5) - (-9)$ (b) $(-4) - (+7)$
    *(c) $(+7) - (-2)$ (d) $(+5) - (+9)$
    *(e) $(-9) - (+4)$ (f) $(+3) - (-11)$
    *(g) $(+7) - (+3)$ (h) $(-15) - (+3)$
    *(i) $(-25) - (-17)$ (j) $(+2) - (-56)$
    *(k) $(-16) - (+24)$ (l) $(+16) - (+24)$
    *(m) $(+2) - (-2)$ (n) $(+2) - (+2)$

10.4 MULTIPLICATION OF SIGNED NUMBERS

Four cases are possible in the multiplication of two signed numbers. Using the notation "$(+)$" for *positive-signed number* and "$(-)$" *for negative-signed number*, we list the four cases.

$$(+)(+) = ?$$
$$(+)(-) = ?$$
$$(-)(+) = ?$$
$$(-)(-) = ?$$

Let us consider each of these cases.

(+)(+)

Since positive numbers are the same as unsigned numbers, the product

$$(+6)(+7)$$

is the same as the product

$$(6)(7)$$

And $(6)(7) = 42$; so $(+6)(+7) = +42$.
In general,

$$\boxed{(+)(+) = (+)}$$

(+)(−)

The product

$$(+4)(-3)$$

is the same as

$$(4)(-3)$$

since $+4$ is the same as 4. This second form can be read as four -3's, which can be written and computed as

$$\left.\begin{array}{r} -3 \\ -3 \\ -3 \\ -3 \\ \hline \end{array}\right\} \text{ the sum of four } -3\text{'s}$$
$$-12$$

Thus

$$(+4)(-3) = -12$$

In general,

$$\boxed{(+)(-) = (-)}$$

(−)(+)

The product

$$(-3)(+4)$$

is the same as

$$(+4)(-3)$$

by the commutative property $(a \cdot b = b \cdot a)$. So

$$(-3)(+4) = -12$$

as determined in the previous case. Thus

$$\boxed{(-)(+) = (-)}$$

(−)(−)

This case must be shown less directly than the others. Consider some negative number, say -3, times 0. Since $x \cdot 0 = 0$ for any number x,

$$(-3)(0) = 0$$

and rewrite 0 as the sum of a number and its opposite, say $(+2) + (-2)$, since a number and its opposite add to zero.

$$(-3)[(+2) + (-2)] = 0$$

Applying the distributive property (see page 47), which says $a(b + c) = a \cdot b + a \cdot c$,

$$(-3)[(+2) + (-2)] = 0$$

becomes

$$(-3)(+2) + (-3)(-2) = 0$$

Since $(-)(+) = (-)$, we know that $(-3)(+2) = -6$. So we can replace $(-3)(+2)$ by -6 to obtain

$$(-6) + (-3)(-2) = 0$$

which tells us that $(-3)(-2)$ must multiply to give us the opposite of -6, since the sum of (-6) and $(-3)(-2)$ must add to 0. This means that $(-3)(-2) = +6$.

In general,

$$\boxed{(-)(-) = (+)}$$

In summary,

$$\boxed{\begin{array}{l} (+)(+) = (+) \\ (+)(-) = (-) \\ (-)(+) = (-) \\ (-)(-) = (+) \end{array}}$$

Example 15. *Observe the multiplication of signed numbers.*

$$(+7)(+4) = +28$$
$$(+6)(-2) = -12$$
$$(-5)(-4) = +20$$
$$(+5)(-3) = -15$$
$$(-7)(+2) = -14$$
$$(-3)(-4) = +12$$
$$(+2)(+5) = +10$$
$$(-5)(-3) = +15$$

In the event you have more than two numbers to multiply, multiply them two at a time, as shown in the next example.

Example 16. *Multiply* $(+6)(-2)(-4)(+3)$.

$$(+6)(-2)(-4)(+3)$$
$$(-12)(-4)(+3)$$
$$(+48)(+3)$$
$$+144 \checkmark$$

You can determine the *sign* of the product of several signed numbers by counting the number of negatives involved in the multiplication.

> *Multiplication*†
> even number of negatives → result is positive
> odd number of negatives → result is negative

To see that this is true, note that $(-)(-) = (+)$; that is, each pair of negatives (negative numbers) has no effect on the sign. So if there are four negatives, then we get $(-)(-)$ times $(-)(-)$, or $(+)$ times $(+)$.

$$(-)(-)(-)(-)$$
$$(+) \quad (+)$$
$$(+)$$

Similarly, six negatives, eight negatives—any even number of negatives —produce a positive result.

---

†0, 2, 4, 6, 8, 10, etc. are *even numbers*. 1, 3, 5, 7, 9, etc. are *odd numbers*.

Example 17. *Multiply* $(-5)(+2)(+3)(-1)(-2)(-4)$.

There are *four* negatives, an even number; so the sign of the result is $+$. The magnitude is $5 \cdot 2 \cdot 3 \cdot 1 \cdot 2 \cdot 4 = 240$, and so the product is $+240$.

Determining the sign in advance minimizes the chance for error; it's simpler.

Example 18. *Multiply* $(-1)(+6)(-2)(+1)(-5)(+3)(-2)(+1)(-3)$.

There are *five* negatives, an odd number; so the sign of the result is $-$. The magnitude is $1 \cdot 6 \cdot 2 \cdot 1 \cdot 5 \cdot 3 \cdot 2 \cdot 1 \cdot 3 = 1080$, and therefore the product is -1080.

What happens if we raise a negative number to a power? Let's use -2 as an example.†

$$(-2)^1 = (-2) = -2$$
$$(-2)^2 = (-2)(-2) = +4$$
$$(-2)^3 = (-2)(-2)(-2) = -8$$
$$(-2)^4 = (-2)(-2)(-2)(-2) = +16$$
$$(-2)^5 = (-2)(-2)(-2)(-2)(-2) = -32$$
$$(-2)^6 = (-2)(-2)(-2)(-2)(-2)(-2) = +64$$

The conclusion can be stated symbolically as

$$(-)^{\text{odd}} = (-)$$
$$(-)^{\text{even}} = (+)$$

In words, if a negative number is raised to an odd power, the result is negative; if it is raised to an even power, the result is positive. This is to be expected, since the exponent indicates the number of signs that we have in the product, and in this case they are all negatives. So here the exponent tells us how many negatives are in the product.

On the other hand, if a *positive* number is raised to *any* power, the result is positive, since there won't be any negative numbers in the product.

$$(+)^{\text{any}} = (+)$$

---

†The meaning of exponent and power was explained in Chapter 1. If these examples of powers of -2 are not clear, you might want to refer to page 25 for an in-depth explanation of exponents.

Example 19. *Find the value of* $(-3)^3 + (-4)^4 + (+2)^3$.

$$(-3)^3 = -27$$
$$(-4)^4 = +256$$
$$(+2)^3 = +8$$

so

$$(-3)^3 + (-4)^4 + (+2)^3 = (-27) + (+256) + (+8) = +237 \ \checkmark$$

EXERCISES 10.4

*1. Multiply.

(a) $(+3)(-2)$	(b) $(-5)(+2)$	(c) $(+7)(+5)$
(d) $(-3)(+6)$	(e) $(-8)(-3)$	(f) $(+9)(+7)$
(g) $(-1)(-9)$	(h) $(+8)(-5)$	(i) $(+2)(+10)$
(j) $(-4)(-9)$	(k) $(+4)(-1)$	(l) $(-6)(+8)$
(m) $(-3)(-3)$	(n) $(+7)(-7)$	(o) $(-9)(+10)$
(p) $(-5)(+16)$	(q) $(-12)(-7)$	(r) $(+19)(+4)$
(s) $(+15)(-8)$	(t) $(-16)(-4)$	(u) $(-19)(+3)$
(v) $(-11)(-9)$	(w) $(+16)(+15)$	(x) $(+15)(-7)$

*2. Multiply.

(a) $(-1.6)(+3.2)$ (b) $(+6.4)(+5.2)$ (c) $(+9.65)(-7.7)$

(d) $(-87.3)(-7.6)$ (e) $\left(+\frac{1}{3}\right)\left(-\frac{1}{7}\right)$ (f) $\left(+\frac{2}{5}\right)\left(+\frac{3}{11}\right)$

(g) $\left(-\frac{2}{3}\right)\left(-\frac{3}{4}\right)$ (h) $\left(-\frac{5}{6}\right)\left(+\frac{7}{8}\right)$

3. Multiply.

*(a) $(+4)(+3)(-2)$	*(b) $(+1)(-2)(+3)$
*(c) $(-8)(+9)(-3)$	*(d) $(-10)(-3)(-2)$
*(e) $(-4)(+8)(+3)$	(f) $(+5)(+7)(+6)$
*(g) $(-2)(-2)(+2)(+2)$	*(h) $(-1)(+1)(-1)(+1)(-1)$
*(i) $(+3)(-5)(+6)(-4)$	(j) $(-2)(+4)(-3)(+5)$
*(k) $(-4)(-3)(-2)(+1)$	(l) $(+3)(-5)(+2)(+1)$
*(m) $(+3)(+4)(-4)(-3)$	(n) $(-7)(-2)(+1)(-5)$
*(o) $(-2)(-1)(-3)(-4)$	(p) $(+9)(+1)(-2)(+1)$

*4. Determine the value of the expression.

(a) $(-1)^5$	(b) $(-2)^4$	(c) $(-3)^3$	(d) $(+2)^5$
(e) $(-3)^4$	(f) $(-3)^5$	(g) $(-2)^7$	(h) $(-1)^{50}$
(i) $(-1)^{101}$	(j) $(-4)^3$	(k) $(+1)^5$	(l) $(+3)^3$

*5. Determine the value of each expression; that is, simplify it.

(a) $(-1)^3 + (-2)^4$ (b) $(-2)^3 + (-1)^6$

(c) $(-4)^3 + (+2)^3 + (-2)^3$ (d) $(-5)^3 + (-4)^2 + (+3)^3$

(e) $(-1)^5 + (-1)^3 + (-1)^4$ (f) $(-9)^2 + (-4)^3 + (-10)^3$
(g) $(-2)^3 + (-1)^2 + (-2)^2$ (h) $(-1)^4 + (-5)^2 + (-2)^3$
(i) $(-4)^2 + (+3)^2 + (-1)^5$ (j) $(-2)^5 + (-1)^2 + (-1)^7$

read to ✳

10.5 NOTATION

Frequently it is too cumbersome to write

$$(+6) + (+2)$$

to specify the addition of $+6$ and $+2$. Instead it is written less formally as

$$+6 + 2$$

with the understanding that this is the *addition* of the two numbers $+6$ and $+2$. Similarly,

$$(+6) + (-2)$$

can be written

$$+6 - 2$$

with the understanding that this is the *addition* of the two numbers $+6$ and -2. As an extension of this notation,

$$-8 + 5 - 4 + 19$$

specifies the *sum* of the four numbers -8, $+5$, -4, and $+19$.

Example 20. *Combine* $-8 + 5 - 4 + 19$.

There are four numbers to be added. As done previously, we will separate the sum into positive and negative numbers and compute each sum separately. Then we'll add the two sums.

$$\begin{array}{ll} \text{Positive:} & +5 + 19 = +24 \\ \text{Negative:} & -8 - 4 = \underline{-12} \\ & +12 \checkmark \end{array}$$

The sum is $+12$.

As you will soon see, two signs can appear in front of a number. Such instances arise naturally, as demonstrated in Examples 22, 23, 24, and 26. There is a simple rule that explains how to replace the two signs by one sign, and that rule is given in a box at the end of this explanation. You may prefer to skip over the explanation and go directly to the rule. The rule is simple; the explanation is somewhat involved.

The four different cases possible of two signs in front of a number are illustrated next.

$$++ \quad \text{as in} \quad -8 + +6$$
$$+- \quad \text{as in} \quad -8 + -6$$
$$-+ \quad \text{as in} \quad -8 - +6$$
$$-- \quad \text{as in} \quad -8 - -6$$

The first two cases come from addition settings and the last two cases from subtraction settings. In the first case, the $+ +6$ indicates adding of a $+6$. So the $+ +6$ can be simplified to $+6$; that is, $-8 + +6 = -8 + 6$. In the second case, the $+ -6$ indicates the adding of a -6. So the $+ -6$ can be simplified to -6; that is, $-8 + -6 = -8 - 6$, the sum of the two numbers -8 and -6. In the third case, the $- +6$ indicates the subtraction of a $+6$. So the $- +6$ can be simplified to -6, since subtracting a $+6$ is the same as adding a -6. Thus $-8 - +6 = -8 - 6$, the sum of the two numbers -8 and -6. In the fourth case, the $- -6$ indicates the subtraction of a -6. So the $- -6$ can be simplified to $+6$, since subtracting a -6 is the same as adding a $+6$. Thus $-8 - -6 = -8 + 6$, the sum of the two numbers -8 and $+6$. In summary, the results are

$$-8 + +6 = -8 + 6 \qquad \begin{cases} \text{which simplifies to } -2 \\ \text{when the addition is done} \end{cases}$$

$$-8 + -6 = -8 - 6 \qquad \begin{cases} \text{which simplifies to } -14 \\ \text{when the addition is done} \end{cases}$$

$$-8 - +6 = -8 - 6 \qquad \begin{cases} \text{which simplifies to } -14 \\ \text{when the addition is done} \end{cases}$$

$$-8 - -6 = -8 + 6 \qquad \begin{cases} \text{which simplifies to } -2 \\ \text{when the addition is done} \end{cases}$$

The two connecting signs are replaced by one sign according to the following:

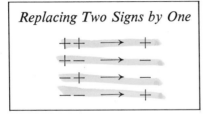

Replacing Two Signs by One

$$++ \longrightarrow +$$
$$+- \longrightarrow -$$
$$-+ \longrightarrow -$$
$$-- \longrightarrow +$$

This suggests that two signs in front of a number can be replaced by one sign according to the same pattern used for multiplication of signed numbers.

If a number has two signs in front of it, apply the rules for multiplication of signed numbers to replace the two signs by one sign.

Note the following examples.

$$+ -3 \quad \text{becomes} \quad -3, \text{ using } (+)(-) = (-)$$
$$- +7 \quad \text{becomes} \quad -7, \text{ using } (-)(+) = (-)$$
$$+ +2 \quad \text{becomes} \quad +2, \text{ using } (+)(+) = (+)$$
$$- -8 \quad \text{becomes} \quad +8, \text{ using } (-)(-) = (+)$$

Now let's consider some examples in which this rule will be applied.

Example 21. *Simplify $+8 + -3$.*

$$+8 + -3 = +8 - 3 \quad \text{since } + -3 = -3$$
$$= +5 \checkmark$$

Example 22. *Compute the value of $+7 - (-2)^3$.*

$$+7 - (-2)^3 = +7 - -8 \quad \text{since } (-2)^3 = -8$$
$$= +7 + 8 \quad \text{since } - -8 = +8$$
$$= +15 \checkmark$$

Example 23. *Simplify $+6 + (-1)(+2)$.*

$$+6 + (-1)(+2) = +6 + -2 \quad \text{since } (-1)(+2) = -2$$
$$= +6 - 2 \quad \text{since } + -2 = -2$$
$$= +4 \checkmark$$

Keep in mind the order in which different arithmetic operations are to be carried out. *Unless parentheses specify otherwise,* the order of operations is exponents first, then multiplication and division, then addition and subtraction.

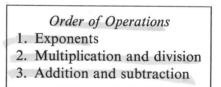

Order of Operations
1. Exponents
2. Multiplication and division
3. Addition and subtraction

Example 24. *Simplify* $(-8) + (-3)(+4)$.

$$(-8) + (-3)(+4) = -8 + -12 \qquad \begin{cases} \text{since multiplication is} \\ \text{done before addition} \end{cases}$$

$$= -8 - 12 \qquad \text{simplification of signs;} + - \text{ is } -$$

$$= -20 \ \checkmark \qquad \text{the addition of } -8 \text{ and } -12$$

Example 25. *Simplify* $(-3)(-2)^3$.

$$(-3)(-2)^3 = (-3)(-8) \qquad \begin{cases} \text{since exponentiation is} \\ \text{done before multiplication} \end{cases}$$

$$= 24 \ \checkmark$$

Example 26. *Compute the value of* $(-4)(-3) - (+6)(-2)$.

$$(-4)(-3) - (+6)(-2) = +12 + +12 \qquad \text{Multiply before subtracting.}$$

$$= 12 + 12 \qquad \qquad \ \ . \ -- = +$$

$$= 24 \ \checkmark$$

Example 27. *Determine the value of the expression* $-2^4 + (-2)^4$.

$-2^4 = -16$, since the 2 is raised to the fourth power before the sign is attached. On the other hand, $(-2)^4$ is 16 because the parentheses specify that it is -2 that is raised to the fourth power. Without parentheses, only the 2 is raised to the fourth power. Thus

$$-2^4 + (-2)^4 = -16 + 16$$

$$= 0 \ \checkmark$$

Still more examples of order of operation can be found in Chapter 2.

EXERCISES 10.5

*1. Simplify.

(a) $+8 - 7 + 3 - 9$
(b) $-8 - 5 - 3 - 2$
(c) $+4 + 3 + 8 + 7$
(d) $+9 - 7 - 5 - 2$
(e) $7 - 1 - 9 + 8 + 5$
(f) $16 + 3 - 5 + 2 + 0$
(g) $-2 + 4 - 5 + 19 - 2$
(h) $+15 - 12 + 45 - 16 + 2$

2. Simplify.

*(a) $-9 + 7 - 1 + 6 - 15 + 8 - 2$
(b) $+10 - 6 + 4 + 9 + 3 - 5 - 2 + 7$
*(c) $-11 + 6 + 3 - 5 + 11 - 3 + 7 - 6$
(d) $-5 + 17 - 3 + 10 + 5 + 3 - 17 + 6$

*3. Simplify.

   (a) $-7 + -6$ (b) $+8 - -7$
   (c) $-3 + +13$ (d) $-9 - +6$
   (e) $+8 + +9$ (f) $+5 + -2$
   (g) $-4 - -10$ (h) $+1 - +3$
   (i) $19 + -7$ (j) $12 - +17$
   (k) $6 + -9$ (l) $15 - +7$
   (m) $-8 - -10$ (n) $-15 - -12$
   (o) $+7 - +2$ (p) $+6 - +6$

4. Determine the value of each expression.

   *(a) $(-5) + (-2)(+6)$ *(b) $+4 - (-3)(-2)$
   *(c) $7 - (-3)(+5)$ (d) $18 + (-5)(-7)$
   *(e) $(-3)(-2) + (-9)(-8)$ (f) $(-3)(+2) - (+3)(-7)$
   *(g) $7(-3) + (+5)(-2)$ (h) $(-4)(-1)(-3) - (-9)$
   *(i) $-8 - +7 - (-8)(+1)$ *(j) $29 + 6(-5) + -3$

5. Compute the value of each expression.

   *(a) $(-4)(-2)^3$ *(b) $(-3)^2 - 3^2$
   *(c) $(+5)^2 - (-5)^2$ (d) $(-7)(+9) - 6^2 + -8$
   *(e) $(-3)(-1)^2(-2)$ (f) $(8)(-2)(+3) - (-4)(+16)$
   *(g) $(-2)^2(-3)^3 + (-1)^9(-5)$ *(h) $(-2)(3)(0) + (-4)(+5)(-1)$
   *(i) $-5^2 + (-5)^2$ *(j) $(-4)^2 + 7 - 4^2 - 9$
   *(k) $(-5)(-4)(-2) - (-3)^3 - (-3)^2$

10.6 DIVISION OF SIGNED NUMBERS

Since we already know the rules for multiplication of signed numbers

$$(+)(+) = (+)$$
$$(+)(-) = (-)$$
$$(-)(+) = (-)$$
$$(-)(-) = (+)$$

we can easily determine the rules for division of signed numbers.

Consider the case $\dfrac{(+)}{(+)}$, which can also be written $(+) \div (+)$ or

$$(+) \overline{)(+)}$$

Let's use $(+5) \overline{)(+10)}$ as an example. Since the result must check in order to be the correct quotient, we must have

$$(+5)\,|\overline{(+10)}$$
$$\underline{(+10)}$$

so that subtracting $+10$ from $+10$ produces a zero remainder. This means the quotient must be positive $(+)$ to produce it.

$$(+5)\,|\overline{\overset{(+\,2)}{(+10)}}\quad\text{or}\quad\frac{(+)}{(+)}=(+)$$
$$\underline{(+10)}$$

Similarly,

$$\frac{(+)}{(-)}=(-)\quad\text{since}\quad(-5)\,|\overline{\overset{(-2)}{(+10)}}$$
$$\underline{(+10)}$$

$$\frac{(-)}{(+)}=(-)\quad\text{since}\quad(+5)\,|\overline{\overset{(-2)}{(-10)}}$$
$$\underline{(-10)}$$

$$\frac{(-)}{(-)}=(+)\quad\text{since}\quad(-5)\,|\overline{\overset{(+2)}{(-10)}}$$
$$\underline{(-10)}$$

Thus the rules for division of signed numbers are the same as those for multiplication.

$$\frac{(+)}{(+)}=(+)\qquad\qquad\frac{(-)}{(-)}=(+)$$

$$\frac{(+)}{(-)}=(-)\qquad\qquad\frac{(-)}{(+)}=(-)$$

You should recall from arithmetic that division by zero is not defined; you cannot divide by zero. If you have any question about this, refer to Section 1.7.

Example 28. *Divide* (a) $(-8) \div (+1)$
(b) $(-5) \div (-10)$
(c) $(+9)/(-3)$

(a) $(-8) \div (+1) = \dfrac{-8}{+1} = -\dfrac{8}{1} = -8$ ✓

(b) $(-5) \div (-10) = \dfrac{-5}{-10} = +\dfrac{5}{10} = +\dfrac{1}{2}$ ✓

(c) $(+9)/(-3) = \dfrac{+9}{-3} = -\dfrac{9}{3} = -3$ ✓

Example 29. *Simplify* $\frac{+15}{-1} + \frac{-8}{-2} + \frac{+4}{+4}$.

$$\frac{+15}{-1} + \frac{-8}{-2} + \frac{+4}{+4} = -\frac{15}{1} + +\frac{8}{2} + +\frac{4}{4}$$

$$= -15 + +4 + +1$$

$$= -15 + 4 + 1$$

$$= -10 \ \checkmark$$

Be careful *not* to interpret a problem like this as addition of fractions. Although there may appear to be fractions here, when the division is done the results in all cases are integers. There will be no fractions. So *do not* seek a common denominator in order to add the "fractions." It's easy to make a mistake that way.

Example 30. *Simplify* $\frac{-6}{-3} + \frac{-14}{+2} - \frac{+10}{-5}$.

$$\frac{-6}{-3} + \frac{-14}{+2} - \frac{+10}{-5} = +\frac{6}{3} + -\frac{14}{2} - -\frac{10}{5}$$

$$= +2 + -7 - -2$$

$$= +2 - 7 + 2$$

$$= -3 \ \checkmark$$

Example 31. *Simplify* $\frac{(+6)(-2)}{(+4)} + (-2)(+3) - 5$.

$$\frac{(+6)(-2)}{(+4)} + (-2)(+3) - 5 = \frac{(-12)}{(+4)} + -6 - 5$$

$$= -3 - 6 - 5$$

$$= -14 \ \checkmark$$

Example 32. *Simplify* $\frac{(+10)(-2)}{(-4)(-5)} + \frac{(+9)(+4)}{(-1)(+1)}$.

$$\frac{(+10)(-2)}{(-4)(-5)} + \frac{(+9)(+4)}{(-1)(+1)} = \frac{-20}{+20} + \frac{+36}{-1}$$

$$= -\frac{20}{20} + -\frac{36}{1}$$

$$= -1 + -36$$

$$= -1 - 36$$

$$= -37 \ \checkmark$$

Example 33. *Simplify* $\dfrac{(+4)(-3) + (-2)(-5)}{-7 + 5 + 2 - 6 - 9}$.

Note that

$$(+4)(-3) = -12$$
$$(-2)(-5) = +10$$
$$-7 + 5 + 2 - 6 - 9 = -15$$

So the fraction simplifies to

$$\frac{-12 + +10}{-15} = \frac{-12 + 10}{-15} = \frac{-2}{-15} = +\frac{2}{15} \quad \checkmark$$

This last example points out that even though you do not actually carry out the division of 2 by 15, the sign of the fraction 2/15 should be determined. Similarly,

$$\frac{+4}{-7} = -\frac{4}{7}$$

$$\frac{-4}{+7} = -\frac{4}{7}$$

$$\frac{+4}{+7} = +\frac{4}{7}$$

The sign of the fraction should be determined and placed in front of it, as shown above.

EXERCISES 10.6

*1. Divide.

(a) $(+10) \div (+2)$ (b) $(-9) \div (+3)$ (c) $(+6) \div (-2)$

(d) $(-28) \div (-7)$ (e) $(-8)/(-2)$ (f) $(-18)/(+3)$

(g) $(+4)/(-1)$ (h) $(-100)/(+20)$ (i) $\dfrac{(-84)}{(-7)}$

(j) $\dfrac{(-3)}{(+3)}$ (k) $\dfrac{(+42)}{(-7)}$ (l) $\dfrac{-9}{-3}$

(m) $\dfrac{-81}{-3}$ (n) $\dfrac{54}{-9}$ (o) $\dfrac{0}{+3}$

(p) $\dfrac{0}{-5}$

*2. Simplify each expression. Your answer should take the form of a fraction with a sign in front of it, as explained in the note after Example 33.

(a) $\dfrac{-3}{+5}$ (b) $\dfrac{-4}{+7}$ (c) $\dfrac{-7}{-9}$

(d) $\dfrac{+6}{-7}$ (e) $\dfrac{+5}{+8}$ (f) $\dfrac{+1}{-3}$

(g) $\dfrac{-5}{+6}$ (h) $\dfrac{-3}{-4}$ (i) $\dfrac{+6}{+7}$

3. Simplify.

*(a) $\dfrac{+9}{-3} + \dfrac{+4}{+1}$ (b) $\dfrac{-15}{-3} + \dfrac{-20}{+4}$

*(c) $\dfrac{-10}{+2} - \dfrac{+6}{-1}$ (d) $\dfrac{+12}{-4} - \dfrac{-8}{-2}$

*(e) $\dfrac{-25}{+5} + \dfrac{+12}{-4} - \dfrac{-6}{-2}$ (f) $\dfrac{+8}{-4} - \dfrac{+30}{+10} + \dfrac{-2}{+1}$

*(g) $\dfrac{-16}{-4} + \dfrac{+18}{-2} + \dfrac{+10}{+2}$ (h) $\dfrac{+7}{+1} - \dfrac{+20}{-2} - \dfrac{-32}{-4}$

4. Determine the value of each expression.

*(a) $\dfrac{+10}{-2} + \dfrac{-8}{-4} + -9$ *(b) $\dfrac{(+2)(-5) - (+2)(-3)}{(+2)}$

*(c) $\dfrac{(-2)^2 - 8 + 6 - 5 - 3}{2 - 3 + 4 - 5}$ *(d) $\left(\dfrac{+7 - 2 + 5 - 3}{-5 - 12 + 8 + 2}\right)^{10}$

(e) $(-2)^3(-3) + \dfrac{+3 - 2 - 1}{(-3)^5(+2)^4}$ (f) $\dfrac{(+12)(-10)}{(+5)(-2)} - \dfrac{(+8)(-1)}{(-2)(+4)}$

*(g) $\dfrac{(+4)(-7) - (+8)(-11)}{+5 - 7 + 8 - 1 - 3}$ *(h) $\dfrac{(+3)(-2) - (-1)(-5)}{(-1)(-2)(-3)(-4)}$

(i) $\dfrac{(-1)(-7) - (-5)(+2)}{(-1)(+2) + (+4)(-2)}$

10.7 EVALUATING EXPRESSIONS (Optional)

When a specific value is used for x in an expression such as $x^2 - 5x + 2$, it becomes an arithmetic expression very similar to those that we have been evaluating in this chapter. Suppose, for example, that we let x be 6. Then

$$x^2 - 5x + 2 = (6)^2 - 5(6) + 2 \quad \textit{Note:} \quad 5x \text{ means 5 times } x.$$
$$= 36 - 30 + 2$$
$$= 8$$

If x is 2, then

$$x^2 - 5x + 2 = (2)^2 - 5(2) + 2$$
$$= 4 - 10 + 2$$
$$= -4$$

If x is 0, then

$$x^2 - 5x + 2 = (0)^2 - 5(0) + 2$$
$$= 0 - 0 + 2$$
$$= 2$$

If x is -3, then

$$x^2 - 5x + 2 = (-3)^2 - 5(-3) + 2$$
$$= +9 + 15 + 2$$
$$= 26$$

EXERCISES 10.7

*1. Determine the value of the expression $x^2 - 7x + 3$ for the following values of x.

(a) 9   (b) 5   (c) 0   (d) -4

2. Determine the value of the expression $x^2 + 3x - 8$ for the following values of x.

(a) 1   (b) -1 (c) -3 (d) 0

3. Determine the value of the expression $b^2 - 4ac$ for the values of a, b, and c given below. Note that $4ac$ means $4 \cdot a \cdot c$, the product of 4, a, and c.

  *(a) $a = 2$, $b = 7$, $c = 1$ (b) $a = 1$, $b = 5$, $c = 3$
  *(c) $a = 1$, $b = 2$, $c = -3$ (d) $a = 1$, $b = 4$, $c = -2$
  *(e) $a = 3$, $b = 2$, $c = 4$ (f) $a = 5$, $b = 1$, $c = 2$
  *(g) $a = 1$, $b = -5$, $c = -1$ (h) $a = 8$, $b = -1$, $c = 6$
  *(i) $a = 1$, $b = -3$, $c = 0$ (j) $a = 2$, $b = 0$, $c = -1$

4. Determine the value of m that is obtained when the values used for x_1, x_2, y_1, and y_2 are those given below.

$$m = \frac{y_2 - y_1}{x_2 - x_1}$$

  *(a) $x_1 = 1$, $x_2 = 2$, $y_1 = 1$, $y_2 = 5$
  *(b) $x_1 = 0$, $x_2 = 2$, $y_1 = 3$, $y_2 = 13$
  (c) $x_1 = -2$, $x_2 = 1$, $y_1 = 8$, $y_2 = 14$

*(d) $x_1 = -1,\ x_2 = 0,\ y_1 = 7,\ y_2 = 9$
(e) $x_1 = 3,\ x_2 = 1,\ y_1 = 8,\ y_2 = 2$
(f) $x_1 = -4,\ x_2 = -5,\ y_1 = -3,\ y_2 = -2$

10.8 INEQUALITY AND ABSOLUTE VALUE (Optional)

absolute value
    The expressions *magnitude* and *absolute value* mean the same thing —the value (or size) of the number—without sign. On the number line magnitude can be considered the undirected distance (number of units) from 0.

The magnitude of $+9$ is 9.
The magnitude of -9 is 9.
The absolute value of $+9$ is 9.
The absolute value of -9 is 9.

We use the notation | | for absolute value. Thus

$$|+9| = 9$$
$$|-9| = 9$$

Also,

$$|(+6) + (+4)| = |+10| = 10$$
$$|(+6) + (-4)| = |\ +2| = \ 2$$
$$|(-6) + (-4)| = |-10| = 10$$
$$|(-6) + (+4)| = |\ -2| = \ 2$$

Note that the absolute value indicator | | acts likes parentheses by grouping elements inside. The absolute value is taken *after* the sum inside is computed.

greater than
    The symbol $>$ means *greater than*. Read $5 > 2$ as "five is greater than two." On a number line the larger of two numbers is always farther right.

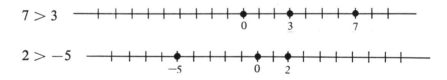

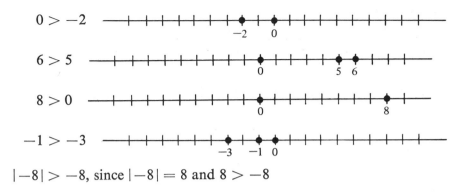

$|-8| > -8$, since $|-8| = 8$ and $8 > -8$

The symbol $<$ means *less than*. Read $2 < 5$ as "two is less than five." The relationships above can be written by using $<$ instead of $>$. less than

$$3 < 7$$

$$-5 < 2$$

$$-2 < 0$$

$$5 < 6$$

$$0 < 8$$

$$-3 < -1$$

$$-8 < |-8|, \text{ since } |-8| = 8 \text{ and } -8 < 8$$

The symbol $\geq$ means *greater than or equal to.* For example, $5 \geq 2$, $2 \geq 2$. $5 \geq 2$ is true, since 5 is greater than 2 (thus it is greater than or equal to 2). $2 \geq 2$ is true, since 2 is equal to 2 (thus it is greater than or equal to 2).

The symbol $\leq$ means *less than or equal to.* For example, $6 \leq 8$, $8 \leq 8$.

The symbol $\neq$ means *not equal to.* For example, $5 \neq 3$. **not equal to**

EXERCISES 10.8

*1. Determine the value of each expression.
   (a) $|-15| + |+15|$
   (b) $|(-15) + (+15)|$
   (c) $|(+3) + (-2) + (-3)|$
   (d) $|(-9) + (-5) + (-4) + (+1)|$

(e) $|(+6) + (-4) + (-3) + (-7) + (+1)|$

(f) $|(+5) + (-8) + (+8) + (-5)|$

(g) $|(-6) + (+2) + (-9) + (+3) + (-7) + (+5)|$

(h) $|(+1) + (-3) + (+7) + (-9) + (+1)|$

*2. Indicate whether each statement is true or false.

(a) $6 > 2$ (b) $5 > -3$

(c) $0 > -7$ (d) $-4 > -2$

(e) $-4 > -8$ (f) $-7 < 0$

(g) $0 < 0$ (h) $12 < -13$

(i) $-6 \geq 6$ (j) $0 \leq -2$

(k) $|-8| = 8$ (l) $|-3| \neq -3$

(m) $|(-5) + (+2)| = 7$ (n) $|(-5) + (-7)| \geq 12$

(o) $|(8) + (-12)| \neq |(0) + (+4)|$ (p) $7 + |-7| = 0$

(q) $-7 + |-7| = 0$ (r) $|(-7) + (7)| > 0$

Chapter 10. REVIEW EXERCISES

*1. Carry out each operation.

(a) $(+6) + (-3)$ (b) $(-8) + (-7)$

(c) $(-2) + (+1)$ (d) $(+6) + (+4)$

(e) $(-7) - (-4)$ (f) $(-9) - (+7)$

(g) $(+5) + (-1)$ (h) $(-6) - (-10)$

(i) $(-9)(+5)$ (j) $(-7)(-6)$

(k) $(-6) - (-7)$ (l) $(-9) \div (+9)$

(m) $(-54) \div (-9)$ (n) $(-7)^2$

(o) $(-17) + (+54)$ (p) $(-17) - (+54)$

(q) $(+100) \div (-5)$ (r) $(-1)^{50}$

(s) $(-2)^5$ (t) $(-5) + (+2)$

(u) $(-9)(+3)$ (v) $(-19) - (-14)$

*2. Simplify each expression.

(a) $(-9)(+2)(-1)$

(b) $+9 + -7 - -2$

(c) $(-3)(-4)(+1)(-1)$

(d) $+7 - 9 + 3 + 2 - 18$

(e) $(-2)(-9) + (-2)(+1)$

(f) $(+7)(-6) + (-4)(+5)$

(g) $(+5)(-6) - (-7)(+3)$

(h) $(-1)(-1)(-1) - (-1)(-2)(-4)$

(i) $-6^2 + (-1)$

(j) $(-1)(-2)^2(+3)^2(-1)^3$

(k) $(-3)^4 + (-4)^3 + 2^3(-1)$

(l) $(-3)^2 + (-2)^3 + 3^2(-2)$

(m) $(-6)(-5) - (+4)(-6)$

(n) $-9 + (-4)(-9) - (-7)^2$

(o) $(-8)(+4) - -7 + 1$

(p) $1 - (-3)(-2)(-1) - (-9)(-1)$

(q) $\dfrac{(-5)(-10)(-3)}{(+15)}$

(r) $\dfrac{(-3)(+8)}{(-2)} + \dfrac{(-1)}{(+1)}$

*3. Determine the value of each expression.

(a) $-9 + (-3)(+1) - (-1)^3 + (-1)^5 + (-4)$

(b) $-5 + (-2)^3 + (-5)^2 - (+8)(-1)$

(c) $(-2 + 3 - 6)^2 - (+4 - 2)^3$

(d) $(-5)(-3 + 2 - 1) + (2 - 5 + 6)(-1)^2 + -3$

(e) $-2 + 6 - 3 - 15 + 2 - 0 + 6 + 4 - 9 + 2 - 12$

(f) $+7 - 3 + 2 - (-3) + (-4) + -2 - (-2)^3 + (-2)(+1)$

CHAPTER **11** _____

Algebraic
Expressions

11.1 INTRODUCTION

constant   Throughout arithmetic you have worked with *constants*, numbers such as 2, −8, 13.569, 0, and $\frac{3}{4}$. In algebra we use letters to represent numbers whose values we do not know. For example, in the equation

unknown $x + 5 = 9$, the letter x is an *unknown*. The equation can be solved to determine that x is 4 in this example. Letters are also used to represent

variable   *variables*. For example, in $x + y = 10$, x and y are variables; their values can change or vary. Specifically, if x is 7 and y is 3, $x + y = 10$. Also, if x is 4 and y is 6, $x + y$ is 10. Many other values are also possible for the variables x and y in this example.

expression   An *algebraic expression* is a combination of constants and variables created by using arithmetic operations.

Algebraic Expression	Comments
$3x$	$3x$ is $3 \cdot x$, 3 times x
$9 + x^2$	uses exponentiation and addition
$\dfrac{8x}{c}$	uses multiplication and division

140

Technically speaking, even a constant alone or a variable alone is considered an expression. Thus, 7 is an expression and x is an expression.

An expression can be thought of as a sum of one or more quantities. A *term* of an expression is any constant, variable, product, or quotient that is added to form the expression. Here are some examples.

term

Expression	Terms of This Expression
$3x + 8$	$3x$ and 8
$9xy^2 - 17x + \dfrac{a}{b}$	$9xy^2$ and $-17x$ and $\dfrac{a}{b}$
$\dfrac{7abc}{5x}$	$\dfrac{7abc}{5x}$

An expression consisting of just one term is called a *monomial*. An expression having two terms is called a *binomial*. A three-term expression is known as a *trinomial*. Here are some examples.

monomial
binomial
trinomial

Monomials	Binomials	Trinomials
a	$a + b$	$a + b + c$
$5x^2$	$x^2 + 7$	$3x^2 - 9x + 2$
$\dfrac{2mn}{3}$	$2x^3 - y^2$	$m^2 + mn + n^2$

If we consider terms containing x, then in each term the *coefficient* of x is the number(s) multiplied by x.

coefficient

Term	Coefficient of x	
$5x$	5	
$-7x$	-7	
$7xy$	$7y$	
x	1	(x is $1 \cdot x$)
$-x$	-1	($-x$ is $-1 \cdot x$)

Coefficients play an important role when like terms of an expression are combined. Terms having the same variables and the same exponents are called *like terms*.

like terms

Like Terms	Unlike Terms
$6x$ and $7x$	$3x$ and $2x^2$
$3y^2$ and $-9y^2$	$5x$ and $4y$

Like terms can be combined by using the distributive property:

$$ba + ca = (b + c)a$$

You can see that if two terms contain the same factor, that factor can be removed from each term and the coefficients can be added.

Example 1. *Combine $3x + 8x$.*

By the distributive property,

$$3x + 8x = (3 + 8)x$$
$$= 11x \quad \checkmark$$

Example 2. *Combine $3x - 9x$.*

$$3x - 9x = (3 - 9)x$$
$$= -6x \quad \checkmark$$

Example 3. *Combine $3x + 5y$.*

$3x$ and $5y$ are not like terms; they cannot be combined. $3x + 5y$ is the simplest form of this expression.

Example 4. *Combine $5x + 8x - 12x + 2x$.*

$$5x + 8x - 12x + 2x = (5 + 8 - 12 + 2)x$$
$$= 3x \quad \checkmark$$

As you can see, like terms can be added simply by adding the coefficients.

Sometimes when we are asked to combine terms, not all the terms are like terms. Here are a few examples.

Example 5. *Combine $5x + 8x - 7x + 4y + x + 6y$.*

The expression contains x terms and y terms. All the x terms can be combined ($5x + 8x - 7x + x = 7x$) and all the y terms can be combined ($4y + 6y = 10y$). Thus

$$5x + 8x - 7x + 4y + x + 6y = 7x + 10y \quad \checkmark$$

Note, however, that $7x$ cannot be combined with $10y$; they are not like terms.

Example 6. *Combine like terms: $3x + 5y + 7 + 4x - 3$.*

$3x + 5y + 7 + 4x - 3$

$$= \underline{3x + 4x} + \underline{5y} + \underline{7 - 3} \qquad \begin{cases} 3x \text{ and } 4x \text{ are like terms.} \\ 7 \text{ and } -3 \text{ are like terms.} \end{cases}$$
$$= 7x + 5y + 4 \quad \checkmark$$

Example 7. *Combine like terms:* $5x + 2x - 7x + 5x^2 + x.$

All are like terms except $5x^2$, which is an x^2 term rather than an x term. Combine the x terms only.

$5x + 2x - 7x + 5x^2 + x$

$$= \underline{5x + 2x - 7x + x} + \underline{5x^2} \qquad \begin{cases} \text{after grouping like terms} \\ \text{together} \end{cases}$$

$$= (5 + 2 - 7 + 1)x + 5x^2$$

$$= 1x + 5x^2$$

$$= x + 5x^2 \ \checkmark$$

EXERCISES 11.1

Answers to starred exercises are given in the back of the book.

*1. Combine like terms.

(a) $5x + 3x$ (b) $8y + 9y$

(c) $3m + 4m + 7m$ (d) $6y + 3y - y$

(e) $8x - 4x + x$ (f) $4x - 7x + x$

(g) $6x - x + 2x$ (h) $-8t + t + 3t - 5t$

(i) $x + x + x + x + x$ (j) $y - y + y - y + y - y$

(k) $n - n + n - n + n$ (l) $7x - 8x + 2x - 6x - x + 3x$

2. Combine like terms, when possible.

*(a) $3x + 5x + 4x + 7$

(b) $7x + 3 + 2x + 9$

*(c) $9x + 3x - 6y - 12x + 4y + 2y$

(d) $6y - 15 + y$

*(e) $5mn + 2mn - mn + 5m$

(f) $5x - 8x + x - y + 7y + 3$

*(g) $5x^2 + 6x - x + x^2$

(h) $6x - 13x + 1 - 4x$

*(i) $2x + 3y + 9z$

(j) $9xy - xy + 2xy - 19xy$

*(k) $5 + x + 2 - x$

(l) $3 - y + x - 3 + x - y$

*3. Indicate the coefficient of x in each term.

(a) $2x$ (b) x (c) $-5x$

(d) $-x$ (e) xy (f) $3xy$

(g) $-5xy$ (h) $-xy$ (i) $25mxy^2$

(j) wx

4. Combine like terms.

*(a) $1.2x + 3.5x + 8.5x$ (b) $3.2y + 5.1y + 6.0y$

*(c) $\dfrac{2}{3}x + \dfrac{3}{4}x$ (d) $\dfrac{2}{5}x + \dfrac{3}{10}x$

*(e) $\dfrac{1}{2}x + \dfrac{1}{8}x$ (f) $\dfrac{3}{4}x + \dfrac{1}{8}x$

11.2 EXPONENTS, MONOMIALS, AND PARENTHESES

The concept of exponent was introduced in Chapter 1, has been used during the arithmetic presentation, and will continue to be used throughout the presentation of algebra. Here are a few arithmetic and algebra examples to serve as a brief review.

$$4^3 = 4 \cdot 4 \cdot 4 = 64$$
$$3^2 = 3 \cdot 3 = 9$$
$$2^5 = 2 \cdot 2 \cdot 2 \cdot 2 \cdot 2 = 32$$
$$x^5 = x \cdot x \cdot x \cdot x \cdot x$$
$$x^4 = x \cdot x \cdot x \cdot x$$
$$x^1 = x$$

To continue, let's consider $x^5 \cdot x^4$, a product of two powers of x.

$$x^5 \cdot x^4 = \underline{x \cdot x \cdot x \cdot x \cdot x} \cdot \underline{x \cdot x \cdot x \cdot x}$$
$$= x \cdot x \cdot x \cdot x \cdot x \cdot x \cdot x \cdot x \cdot x$$
$$= x^9$$

Thus we see that $x^5 \cdot x^4 = x^9$.

Similarly, $m^3 \cdot m^7 = \underline{m \cdot m \cdot m} \cdot \underline{m \cdot m \cdot m \cdot m \cdot m \cdot m \cdot m} = m^{10}$.

It should be apparent that if two numbers having the *same base* are *multiplied*, then the result has the same base and the exponent is the *sum* of the two exponents. In symbols,

$$\boxed{x^a \cdot x^b = x^{a+b}}$$

Example 8. *Simplify* $x^3 \cdot x^4$.

$$x^3 \cdot x^4 = x^{3+4} = x^7 \quad \checkmark$$

Example 9. *Simplify $y^2 \cdot y^9$.*

$$y^2 \cdot y^9 = y^{2+9} = y^{11} \quad \checkmark$$

Example 10. *Simplify $x^5 \cdot y^9$.*

This term cannot be simplified because the bases x and y are different.

Example 11. *Simplify $x^4 + x^6$.*

The property of exponents deals with *multiplication*, not addition, and so it cannot be used to simplify this expression.

Some knowledge of exponents is needed when multiplying monomials. Such multiplication is easy, although sometimes you must rearrange the factors involved. This is illustrated in the next examples.

Example 12. *Multiply $3x \cdot 4x$.*

$$3x \cdot 4x = 3 \cdot x \cdot 4 \cdot x$$
$$= 3 \cdot 4 \cdot x \cdot x$$
$$= 12 \cdot x^2$$
$$= 12x^2 \quad \checkmark$$

Example 13. *Multiply $(-5y^3)(-7y^7)$.*

$$(-5y^3)(-7y^7) = (-5)(-7) \cdot y^3 \cdot y^7$$
$$= 35y^{10} \quad \checkmark$$

Example 14. *Simplify $5x^2 \cdot x^{10} \cdot x^8$.*

$$5x^2 \cdot x^{10} \cdot x^8 = 5 \cdot x^2 \cdot x^{10} \cdot x^8$$
$$= 5 \cdot x^{2+10+8}$$
$$= 5 \cdot x^{20} \quad \text{or} \quad 5x^{20} \quad \checkmark$$

Example 15. *Simplify $(-2a^3b^2)(5ab^7)$.*

$$(-2a^3b^2)(5ab^7) = (-2)(5) \cdot a^3 \cdot a^1 \cdot b^2 \cdot b^7$$
$$= -10a^{3+1}b^{2+7}$$
$$= -10a^4b^9 \quad \checkmark$$

Multiplication of monomials can also occur when an expression to be simplified contains parentheses. If some of the terms you wish to combine are contained within parentheses, the parentheses must be removed before all the like terms can be combined. Note how the distributive property is used in the example that follows. Here there are two multiplications of monomial by monomial.

Example 16. *Simplify the expression* $7x + 3(4 + 2x) - 9$.

Here parentheses contain a binomial consisting of a constant term (4) and an x term ($2x$). Use the distributive property to multiply out $3(4 + 2x)$.

$$7x + 3(4 + 2x) - 9 = 7x + 12 + 6x - 9$$
$$= 7x + 6x + 12 - 9 \qquad \text{when rearranged}$$
$$= 13x + 3 \quad \checkmark$$

after like terms are combined.

Example 17. *Simplify the expression* $5x - 2(3 - 8x) + (4 - x)6$.

$$5x - 2(3 - 8x) + (4 - x)6 = 5x - 6 + 16x + 24 - 6x$$
$$= 15x + 18 \quad \checkmark$$

There is one special case that should be considered carefully in order to avoid misunderstanding. In the next example the expression $(7x + 5) - (4x - 3)$ will be simplified. To do this, you must know how to simplify the expression $-(4x - 3)$. Consider the minus in front of the expression in parentheses to mean "opposite of." Then since the opposite of any term is the same magnitude term with opposite sign, we have

$$-(4x - 3) = -4x + 3$$

If you prefer, $-(4x - 3)$ can be written as $-1(4x - 3)$ and multiplied out. The result is the same as obtained above.

$$-(4x - 3) = -1(4x - 3) = -4x + 3$$

In general,

> A minus sign in front of an expression in parentheses has the effect of changing the sign of each term within parentheses.

Examples:

$$-(4x - 3) = -4x + 3$$
$$-(1 - y) = -1 + y$$
$$-(a + b + c) = -a - b - c$$
$$-(-x^2 + 4x - 5) = x^2 - 4x + 5$$

Example 18. *Simplify the expression* $(7x + 5) - (4x - 3)$.

A minus in front of parentheses has the effect of changing the signs of all terms within the parentheses. That is, $-(4x - 3) = -4x + 3$. Thus

$$(7x + 5) - (4x - 3) = 7x + 5 - 4x + 3$$
$$= 3x + 8 \checkmark$$

Example 19. *Simplify* $(8x^2 + 7x - 1) - (3x^2 - 2x - 19)$.

$(8x^2 + 7x - 1) - (3x^2 - 2x - 19)$

$$= 8x^2 + 7x - 1 - 3x^2 + 2x + 19 \qquad \text{by boxed rule}$$
$$= 8x^2 - 3x^2 + 7x + 2x - 1 + 19 \qquad \text{grouping like terms}$$
$$= 5x^2 + 9x + 18 \checkmark \qquad \text{combining like terms}$$

Example 20. *Simplify* $(5x^2 - 8x + 1) + (2x^2 + x - 8)$.

$$(5x^2 - 8x + 1) + (2x^2 + x - 8) = 5x^2 - 8x + 1 + 2x^2 + x - 8$$
$$= 5x^2 + 2x^2 - 8x + x + 1 - 8$$
$$= 7x^2 - 7x - 7 \checkmark$$

The next example involves brackets and parentheses. To avoid errors, work *from the inside* out, as explained in the example.

Example 21. *Simplify the expression* $5 - 3[7 + (2x - 6)8 - 4x] + 19$.

Working from the inside out means simplifying the expression inside the brackets before multiplying by the -3. And you cannot combine the terms within the brackets until you multiply out $(2x - 6)8$. So the first step is to multiply out $(2x - 6)8$.

$$5 - 3[7 + (2x - 6)8 - 4x] + 19 = 5 - 3[7 + 16x - 48 - 4x] + 19$$

Now we can combine the like terms within the brackets. The result is

$$5 - 3[12x - 41] + 19$$

Next, multiply out by distributing the -3 to the terms within the brackets.

$$5 - 3[12x - 41] + 19 = 5 - 36x + 123 + 19$$

Finally, we can combine like terms to produce the simplest form of the expression, namely,

$$-36x + 147 \checkmark$$

Example 22. (optional). *Simplify the expression*

$$-7x - 2[3x - 4(y - 3) + 7] - y$$

$$-7x - 2[3x - 4(y - 3) + 7] - y$$
$$= -7x - 2[3x - 4y + 12 + 7] - y \qquad \text{after distributing } -4$$
$$= -7x - 2[3x - 4y + 19] - y \qquad \text{after combining } +12 \text{ and } +7$$
$$= -7x - 6x + 8y - 38 - y \qquad \text{after distributing } -2$$
$$= -13x + 7y - 38 \quad \checkmark$$

EXERCISES 11.2

*1. Simplify, if possible.

(a) $x^5 \cdot x^7$

(b) $y^9 \cdot y^4$

(c) $m^6 \cdot m^8$

(d) $t^2 \cdot t^3$

(e) $x^7 + x^4$

(f) $x^{17} \cdot x^4$

(g) $a^{20} \cdot a^{20}$

(h) $m^{10} + m^{15}$

(i) $y^{40} \cdot y^{80}$

(j) $x^{13} \cdot x^8$

2. Simplify.

*(a) $8y \cdot 3y$

(b) $2x \cdot 9x$

*(c) $5m^2 \cdot m^8$

(d) $3x^8 \, x^4$

*(e) $3x^2 \cdot x^5 \cdot x^4$

(f) $7x^4 \cdot x^3 \cdot x^6$

*(g) $8x^7 \cdot x^3 \cdot x$

(h) $2x \cdot x^9 \cdot x^{15}$

*(i) $(-3a^2b^3)(2ab^4)$

(j) $(-7m^7u^4)(4m^5u^3)$

*(k) $(-5x^2y^5)(-2x^8y^2)$

(l) $(-3t^4w)(-8tw^5)$

*(m) $(10u^2v)(-3u^8)$

(n) $(-a^3b^5)(-7ab^6)$

3. Simplify each expression.

*(a) $4x + 2(x + 5) - 3$

*(b) $7 + 14(7 - 3x) + 9$

*(c) $(2y - 9) - (3y + 1)$

(d) $(9x + 2) - (5x - 3)$

*(e) $(4x + 1) - (1 - 2x)$

(f) $-(2x - 7) - (x - 3)$

*(g) $3x - 4(x + 1) + x$

*(h) $9 - 3(1 - 5t) + t$

(i) $5x + 2(4x - 9) - 3$

(j) $x - (8 - x) - x - 8$

*(k) $5(3 - x) + 2x - 4(2x + 1)$

(l) $4(y - 5) - 3y + 3(2y + 6)$

4. Simplify.

*(a) $(4x^2 + 5x - 2) + (3x^2 + 7x + 1)$

(b) $(6x^2 - 8x + 1) + (x^2 + 3x - 9)$

*(c) $(x^2 - 19x + 5) + (5x^2 - 3x - 4)$

*(d) $(3x^2 - 8x + 14) - (5x^2 - 2x - 6)$
*(e) $(x^2 - 13x + 7) - (x^2 + 2x - 4)$
(f) $(4x^2 + 13x - 6) - (2x^2 - 3x - 4)$

5. Simplify each expression.
*(a) $-5[x - (8 - x) + 4] - 5x$
*(b) $7 - 2[6 + (3x - 2)9 - 5x] + 23$
*(c) $9 + 4[7 + (4 - 2x) - 9x] + x$
(d) $8 - 3[3x - (4 - 2x) + 13x] - x$
*(e) $9 - 7[4x - 2(3x - 5) + 2] - 9x$
*(f) $3 - 4[3 - (x + 2)] + 4(x + 2) - 3$
(g) $6(4 - 3x) + 5 - 2[3 + 2(x + 5)]$
*(h) $4x + 7 + 3[5 + (3x + 2) - 7]$

6. (optional) Simplify each expression.
*(a) $5x - 3[2x - 5(y - 4) + 6] - y$
(b) $7x + 4[5x - 3(4 - 3y) - 9] + 6x$
*(c) $5y - 7[2(3 - 2x) + 4y - 6] - 3y$
(d) $3y + 5[(5x - 3)7 - 3y + 2] + 2x$

11.3 MULTIPLICATION

In this section the concept of multiplication will be extended to involve binomials and trinomials. The distributive property will be used along with the property of exponents pressented in the preceding section.

Example 23. *Multiply $3x^2(7x^3 + 4)$.*

Apply the distributive property.

$$3x^2(7x^3 + 4) = 3x^2 \cdot 7x^3 + 3x^2 \cdot 4$$
$$= 3 \cdot 7 \cdot x^2 \cdot x^3 + 3 \cdot 4 \cdot x^2$$
$$= 21x^5 + 12x^2 \checkmark$$

We can handle $(x + 2)(y + 3)$ similarly if we consider $(x + 2)$ as one number and distribute it to each of y and 3. Thus

$$(x + 2)(y + 3) = (x + 2)y + (x + 2)3$$

Now distribute the y to both x and 2 and distribute the 3 to both x and 2.

$$(x + 2)y + (x + 2)3$$
$$x \cdot y + 2 \cdot y + x \cdot 3 + 2 \cdot 3$$

This can be rewritten as

$$xy + 2y + 3x + 6$$

Thus

$$(x + 2)(y + 3) = xy + 2y + 3x + 6 \;\checkmark$$

In effect, all that we have done is multiplied x by each of y and 3 and multiplied 2 by each of y and 3.

$$(x + 2)(y + 3) = xy + 3x + 2y + 6$$

Example 24. *Multiply $(a + b)(c + d)$.*

$$(a + b)(c + d) = a \cdot c + a \cdot d + b \cdot c + b \cdot d$$
$$= ac + ad + bc + bd \;\checkmark$$

Example 25. *Multiply $(x + 2)(x + 5)$.*

$$(x + 2)(x + 5) = x \cdot x + x \cdot 5 + 2 \cdot x + 2 \cdot 5$$
$$= x^2 + 5x + 2x + 10$$
$$= x^2 + 7x + 10 \;\checkmark$$

Example 26. *Multiply $(x + 3)(x - 3)$.*

$$(x + 3)(x - 3) = x \cdot x + x(-3) + 3 \cdot x + 3(-3)$$
$$= x^2 - 3x + 3x - 9$$
$$= x^2 + 0x - 9$$
$$= x^2 - 9 \;\checkmark$$

Example 27. *Multiply $(x + y)^2 = (x + y)(x + y)$ to show that $(x + y)^2 \neq x^2 + y^2$.*

$$
\begin{aligned}
(x + y)(x + y) &= x \cdot x + x \cdot y + y \cdot x + y \cdot y \\
&= x^2 + xy + yx + y^2 \\
&= x^2 + xy + xy + y^2 \qquad \text{since } yx = xy \\
&= x^2 + 2xy + y^2 \;\checkmark \qquad \text{since } xy + xy = 2xy
\end{aligned}
$$

So you can see that

$$(x + y)^2 = x^2 + 2xy + y^2$$

This means that

$$\boxed{(x + y)^2 \neq x^2 + y^2}$$

As an arithmetic example, we show that $(3 + 2)^2 \neq 3^2 + 2^2$.

$$(3 + 2)^2 \stackrel{?}{=} 3^2 + 2^2$$

$$(5)^2 \stackrel{?}{=} 9 + 4$$

$$25 \neq 13$$

Example 28. *Multiply* $(2x + 5)(7x - 4)$.

$$(2x + 5)(7x - 4) = 2x(7x) + 2x(-4) + 5(7x) + 5(-4)$$
$$= 14x^2 - 8x + 35x - 20$$
$$= 14x^2 + 27x - 20 \quad \checkmark$$

Example 29. *Multiply* $(3x^2 + 5x + 2)(4x + 1)$.

$(3x^2 + 5x + 2)(4x + 1)$

$$= 3x^2(4x + 1) + 5x(4x + 1) + 2(4x + 1)$$
$$= 3x^2 \cdot 4x + 3x^2 \cdot 1 + 5x \cdot 4x + 5x \cdot 1 + 2 \cdot 4x + 2 \cdot 1$$
$$= 12x^3 + 3x^2 + 20x^2 + 5x + 8x + 2$$
$$= 12x^3 + 23x^2 + 13x + 2 \quad \checkmark$$

Example 30. *Multiply* $(5x^2 + 3x + 1)(2x^2 + 7x - 4)$.

$(5x^2 + 3x + 1)(2x^2 + 7x - 4)$

$$= 5x^2(2x^2 + 7x - 4) + 3x(2x^2 + 7x - 4) + 1(2x^2 + 7x - 4)$$
$$= 10x^4 + 35x^3 - 20x^2 + 6x^3 + 21x^2 - 12x + 2x^2 + 7x - 4$$
$$= 10x^4 + 41x^3 + 3x^2 - 5x - 4 \quad \checkmark$$

EXERCISES 11.3

1. Multiply.
   *(a) $5x^2(3x^4 + 6)$
   *(c) $8y(7 + 3y^5)$
   *(e) $x^6(5x^3 + 8x)$
   *(g) $6m^7(8m^9 - 16)$
   (b) $7y^4(2y^6 + 1)$
   (d) $2m^6(m^3 + 4m)$
   (f) $n^3(n^6 + p^3)$
   (h) $6x^4(9x^7 - 15x)$

2. Multiply. Then simplify, if possible.
   *(a) $(x + 3)(x + 2)$
   *(c) $(x + 7)(x - 5)$
   *(e) $(y - 8)(y - 2)$
   *(g) $(c + 5)(e - 6)$
   *(i) $(a + 2)(b + 6)$
   *(k) $(m + 1)(m - 1)$
   (b) $(x + 1)(x + 4)$
   (d) $(x - 5)(x + 4)$
   (f) $(m - 4)(m - 3)$
   (h) $(u - 9)(v + 4)$
   (j) $(n + 5)(p + 3)$
   (l) $(y + 9)(y - 9)$

$*$(m) $(3 + x)(3 - x)$ $\cdot$

$*$(o) $(2 + x)(7 - x)$

(n) $(5 + m)(8 + m)$

(p) $(y - 3)(y + 5)$

3. Multiply. Then simplify, if possible.

$*$(a) $(5x + 3)(4x + 1)$

(b) $(5x + 3)(2x + 4)$

$*$(c) $(4x - 1)(2x + 1)$

(d) $(3x - 2)(x - 5)$

$*$(e) $(3x - 5)(x - 2)$

(f) $(2x + 7)(3x + 9)$

$*$(g) $(m + 5)(7m - 5)$

(h) $(5m + 2)(m - 6)$

$*$(i) $(2m + 7)(3n + 5)$

(j) $(7a - 2)(3b + 1)$

$*$(k) $(5 + 9x)(2 + x)$

(l) $(1 + y)(1 - 2y)$

$*$(m) $(6 + 3x)(2x - 4)$

(n) $(5t + 4)(t - 6)$

$*$(o) $(x + y)(2x + y)$

(p) $(x - y)(2x - y)$

4. Multiply. Then simplify.

$*$(a) $(m + n)^2$ (b) $(r + v)^2$ $*$(c) $(a - b)^2$ (d) $(t - w)^2$

$*$(e) $(x + 7)^2$ (f) $(y + 4)^2$ $*$(g) $(y - 5)^2$ (h) $(x - 6)^2$

5. Multiply. Then simplify, if possible.

$*$(a) $(x^2 + 2x + 4)(x + 3)$

(b) $(y^2 - 6y + 5)(y - 2)$

$*$(c) $(x - 6)(x^2 - 3x + 2)$

(d) $(m - 4)(m^2 - 7m - 8)$

$*$(e) $(x + 2)(x^2 + 4x - 5)$

(f) $(a + b + c)(d + e)$

$*$(g) $(2x^2 - 3x + 5)(x^2 - 4x - 1)$

(h) $(x^2 + 5x + 4)(x^2 + 7x + 3)$

$*$(i) $(3x - 4y + 7)(5x - 8y + 2)$

(j) $(2m + 3n + 4)(m - 5n - 7)$

6. (a) Verify that $x^3 - y^3 = (x - y)(x^2 + xy + y^2)$ by multiplying out and simplifying the product on the right-hand side of the equation.

   (b) Verify that $x^3 + y^3 = (x + y)(x^2 - xy + y^2)$ by multiplying out and simplifying the product on the right-hand side of the equation.

11.4 SHORT DIVISION, EXPONENTS, AND LONG DIVISION

Short division in algebra is division of an algebraic expression by a monomial. As a simple example, consider the division of $10x^3$ by $2x$, which can appear in either of two ways.

$$2x \overline{)\,10x^3} \quad \text{or} \quad \frac{10x^3}{2x}$$

The second form looks like a fraction to reduce, and in fact reduction of fractions like this and others is presented in Chapter 18. In this section we will use only the first form. A division such as

$$\frac{8x^3 + 9x^2 - 7x}{x}$$

will appear as

$$x \overline{)\, 8x^3 + 9x^2 - 7x}$$

Let us return now to consider the division $2x \overline{)\, 10x^3}$.

Example 31. *Divide* $2x \overline{)\, 10x^3}$.

The quotient is $5x^2$, since when $5x^2$ is multiplied by $2x$ the result is $10x^3$.

$$\begin{array}{r} 5x^2 \checkmark \\ 2x \overline{)\, 10x^3} \end{array}$$

since $5x^2 \cdot 2x = 10x^3$.

Example 32. *Divide* $2x \overline{)\, 10x^3 + 16x^2 - 6x}$.

This division can be done as three individual divisions. Divide each of the terms of the trinomial $10x^3 + 16x^2 - 6x$ by $2x$. From Example 31 we already know the first term of the quotient.

$$\begin{array}{r} 5x^2 \\ 2x \overline{)\, 10x^3 + 16x^2 - 6x} \end{array}$$

Next, divide $16x^2$ by $2x$. The result is $8x$, since $8x \cdot 2x = 16x^2$.

$$\begin{array}{r} 5x^2 + \ 8x \\ 2x \overline{)\, 10x^3 + 16x^2 - 6x} \end{array}$$

Finally, divide $-6x$ by $2x$, to get -3, since $-3 \cdot 2x = -6x$.

$$\begin{array}{r} 5x^2 + \ 8x \ - 3 \ \checkmark \\ 2x \overline{)\, 10x^3 + 16x^2 - 6x} \end{array}$$

As a check, multiply $5x^2 + 8x - 3$ by $2x$ to produce $10x^3 + 16x^2 - 6x$.

$$(2x)(5x^2 + 8x - 3) = 10x^3 + 16x^2 - 6x$$

This type of division can be used to determine another property of exponents. Consider, for example,

$$x^3 \overline{)\, x^{10}}$$

The quotient must be x^7, since $x^7 \cdot x^3 = x^{10}$. Thus

$$\frac{x^{10}}{x^3} = x^7$$

Similarly,

$$\frac{x^9}{x^5} = x^4$$

since $x^4 \cdot x^5 = x^9$.

In each of these examples the exponent in the result could have been obtained by subtracting the denominator exponent from the numerator exponent.

$$\boxed{\frac{x^a}{x^b} = x^{a-b}}$$

Example 33. *Simplify* $\dfrac{x^{15}}{x^4}$.

$$\frac{x^{15}}{x^4} = x^{15-4} = x^{11} \checkmark$$

When the divisor consists of two or more terms, then the division process becomes *long division*. The nature and sequence of the steps in algebraic long division are similar to those of the arithmetic long division presented in Chapter 1.

Example 34. *Divide.*

$$x + 3 \,\overline{)\, x^3 + 2x^2 - 5x - 6}$$

Begin by dividing the *first term* of $x + 3$ into the *first term* of $x^3 + 2x^2 - 5x - 6$. The result of dividing x into x^3 is x^2, of course.

$$
\begin{array}{r}
x^2 \\
x + 3 \,\overline{)\, x^3 + 2x^2 - 5x - 6}
\end{array}
$$

Next, *multiply back* the x^2 by the divisor $x + 3$.

$$
\begin{array}{r}
x^2 \\
x + 3 \,\overline{)\, x^3 + 2x^2 - 5x - 6} \\
x^3 + 3x^2
\end{array}
$$

Next, *subtract* $x^3 + 3x^2$ from $x^3 + 2x^2$. This is done by changing the signs of $x^3 + 3x^2$ (to $-x^3 - 3x^2$) and adding.

$$
\begin{array}{r}
x^2 \\
x + 3 \,\overline{)\, x^3 + 2x^2 - 5x - 6} \\
\underline{-x^3 - 3x^2 } \\
-x^2
\end{array}
$$

Now that the subtraction has been completed, *bring down* the next term $(-5x)$.

$$
\begin{array}{r}
x^2 \\
x + 3 \,\overline{)\, x^3 + 2x^2 - 5x - 6} \\
\underline{-x^3 - 3x^2 } \\
-x^2 - 5x
\end{array}
$$

Continue the process by dividing $x + 3$ into the new two-term expression $-x^2 - 5x$. This is done by dividing x (the first term of $x + 3$) into $-x^2$ (the first term of $-x^2 - 5x$). The result is $-x$.

$$\begin{array}{r} x^2 - x \\ x + 3\ \overline{)\ x^3 + 2x^2 - 5x - 6} \\ \underline{-x^3 - 3x^2} \\ -x^2 - 5x \end{array}$$

Now, multiply back $-x$ by $x + 3$. The result is $-x^2 - 3x$ and is shown below.

$$\begin{array}{r} x^2 \ \fbox{$-\ x$} \\ x + 3\ \overline{)\ x^3 + 2x^2 - 5x - 6} \\ \underline{-x^3 - 3x^2} \\ -x^2 - 5x \\ -x^2 - 3x \end{array}$$

Next, subtract $-x^2 - 3x$ by changing its signs ($-x^2 - 3x$ becomes $x^2 + 3x$) and adding it. Then bring down the next term (-6).

$$\begin{array}{r} x^2 - x \\ x + 3\ \overline{)\ x^3 + 2x^2 - 5x - 6} \\ \underline{-x^3 - 3x^2} \\ -x^2 - 5x \\ \underline{x^2 + 3x} \\ -2x - 6 \end{array}$$

Next, divide the x (of $x + 3$) into the $-2x$ (of $-2x - 6$) and get -2. Multiply it back and note that there is no remainder in the subtraction. The result of the division is $x^2 - x - 2$ exactly.

$$\begin{array}{r} x^2 - x \ - 2 \\ x + 3\ \overline{)\ x^3 + 2x^2 - 5x - 6} \\ \underline{-x^3 - 3x^2} \\ -x^2 - 5x \\ \underline{x^2 + 3x} \\ -2x - 6 \\ \underline{-2x - 6} \\ 0 \end{array}$$

As a check, you could multiply out $(x + 3)(x^2 - x - 2)$ to show that it is indeed equal to $x^3 + 2x^2 - 5x - 6$. Notice, too, that $(x + 3)$ is a factor of $x^3 + 2x^2 - 5x - 6$.

Example 35. *Divide.*

$$x - 2\ \overline{)\ 2x^4 + x^3 - 23x + 9}$$

Note that the x^2 term is missing. To avoid confusion later, insert the term $0x^2$ where the x^2 term would be. It will serve as a placeholder.

$$x - 2\ \overline{)\ 2x^4 + x^3 + 0x^2 - 23x + 9}$$

Now perform the division.

$$
\begin{array}{r}
2x^3 + 5x^2 + 10x - 3 \\
x - 2 \overline{)\ 2x^4 + x^3 + 0x^2 - 23x + 9} \\
\underline{2x^4 - 4x^3} \\
5x^3 + 0x^2 \\
\underline{5x^3 - 10x^2} \\
10x^2 - 23x \\
\underline{10x^2 - 20x} \\
-3x + 9 \\
\underline{-3x + 6} \\
3
\end{array}
$$

The remainder is 3. The result of the division can be written

$$2x^3 + 5x^2 + 10x - 3, \text{ R3} \quad \checkmark$$

or

$$2x^3 + 5x^2 + 10x - 3 + \frac{3}{x - 2} \quad \checkmark$$

As a check, you should show that

$$(x - 2)(2x^3 + 5x^2 + 10x - 3) + 3 = 2x^4 + x^3 - 23x + 9$$

In Example 35, $0x^2$ was inserted as a placeholder because no x^2 term was present in the dividend expression $2x^4 + x^3 - 23x + 9$. Here are two more examples of how placeholders should be inserted in dividends.

$$\overline{)\ x^3 + 3x^2 - 2} \quad \longrightarrow \quad \overline{)\ x^3 + 3x^2 + 0x - 2}$$
$$\overline{)\ x^3 + 5} \quad \longrightarrow \quad \overline{)\ x^3 + 0x^2 + 0x + 5}$$

EXERCISES 11.4

1. Divide. (short division)

   *(a) $x \overline{)\ x^3 + 6x^2 + x}$

   (b) $m \overline{)\ m^3 - 7m^2 + 3m}$

   *(c) $y \overline{)\ 3y^3 - 2y^2 - y}$

   (d) $t \overline{)\ 7t^4 + 5t^3 - 2t^2 - t}$

   *(e) $3x \overline{)\ 12x^4 + 3x^2 + 6x}$

   (f) $2y \overline{)\ 8y^3 + 4y^2 + 2y}$

   *(g) $5t \overline{)\ 20t^5 - 10t^4 - 5t^2 + 15t}$

   (h) $3x \overline{)\ 9x^5 - 3x^4 - 6x^2 + 15x}$

   *(i) $2x^2 \overline{)\ 6x^4 + 18x^3 - 20x^2}$

   (j) $3x^2 \overline{)\ 12x^4 - 9x^3 + 3x^2}$

*2. Simplify.

   (a) $\dfrac{x^{20}}{x^7}$

   (b) $\dfrac{x^{14}}{x^8}$

   (c) $\dfrac{y^6}{y^4}$

   (d) $\dfrac{m^{17}}{m^9}$

   (e) $\dfrac{a^{12}}{a^5}$

(f) $\dfrac{x^{11}}{x^8}$ (g) $\dfrac{b^8}{b^3}$ (h) $\dfrac{u^{16}}{u^7}$ (i) $\dfrac{n^{19}}{n^{11}}$ (j) $\dfrac{t^{18}}{t^{10}}$

3. Divide. (long division)

*(a) $x + 1 \,)\, \overline{x^2 + 9x + 8}$

(b) $y + 3 \,)\, \overline{y^2 - y - 12}$

*(c) $x - 3 \,)\, \overline{x^2 + 2x - 15}$

(d) $x - 4 \,)\, \overline{x^2 - 13x + 36}$

*(e) $x - 1 \,)\, \overline{x^3 + 4x^2 - 8x + 3}$

(f) $x - 2 \,)\, \overline{x^3 - 3x^2 + 4x - 4}$

*(g) $y + 5 \,)\, \overline{7y^3 + 35y^2 - 19y - 95}$

(h) $x + 2 \,)\, \overline{x^3 + 7x^2 + 7x - 6}$

*(i) $2x + 1 \,)\, \overline{8x^2 - 2x - 8}$

(j) $3n - 2 \,)\, \overline{6n^2 - 13n + 6}$

4. Divide. Be sure to use placeholders for missing terms. (long division)

*(a) $x - 2 \,)\, \overline{x^3 - x^2 - 4}$ (b) $x - 1 \,)\, \overline{x^3 + 5x + 17}$

*(c) $x + 1 \,)\, \overline{3x^4 + x^3 + 7x + 7}$ (d) $x - 2 \,)\, \overline{5x^4 - x^3 - 36x + 7}$

*(e) $x + 1 \,)\, \overline{x^3 + 1}$ (f) $x - 1 \,)\, \overline{x^3 - 1}$

*(g) $x - 1 \,)\, \overline{x^4 - 1}$ (h) $x - 1 \,)\, \overline{x^5 - 1}$

5. Perform each long division.

*(a) Divide $x^2 + 3x + 4$ by $x - 2$.

(b) Divide $x^2 + 17x + 60$ by $x + 5$.

*(c) Divide $x^3 + 5x^2 - 7x + 24$ by $x - 5$.

(d) Divide $x^3 + 8x^2 - 4x - 32$ by $x + 8$.

*(e) Divide $5x^3 - 19x^2 - 4x$ by $x - 4$.

(f) Divide $x^3 - 4x^2 - 31x + 70$ by $x - 7$.

*(g) Divide $x^3 + 3x^2 - 13x + 1$ by $x - 2$.

Chapter 11. REVIEW EXERCISES

1. Combine like terms, when possible.

*(a) $7x - x + 4x - 9x + x - 3x$ $-x$

*(b) $3x + 2y - 8x - 9y + x - y + 4y$

*(c) $9x - 1 + x - 15 + 3x + 4$

(d) $7r + 6 - 4w + r - w + 18 - 64$

2. Simplify.

   *(a) $x^{10} \cdot x^7$ (b) $m^{16} \cdot m^7$ *(c) $\dfrac{x^{15}}{x^8}$ (d) $\dfrac{n^9}{n^4}$

   *(e) $7x^8 \cdot x^4 \cdot x$ *(f) $3u^5 \cdot u^6 \cdot u^3$

*3. Simplify each expression.
- (a) $3x - 4(x + 2) - 9$
- (b) $8 - (1 - 4x) + 5x$
- (c) $(x^2 + 6x + 6) - (x^2 - 5)$
- (d) $6 - 2[2 + (3x - 1)7 + x] - 3x$

4. Multiply. Then simplify when possible.
- *(a) $3x^2(2x^3 + 7)$ *(b) $4x(x^5 + 9x)$
- *(c) $(m + 3)(m + 7)$ (d) $(m - 2)(m - 5)$
- *(e) $(x + 10)^2$ (f) $(p + q)^2$
- *(g) $(3y - 5)(3y - 1)$ (h) $(5m - 3)(m + 2)$
- *(i) $(x - 3)(x^2 + 8x - 9)$ *(j) $(5x^2 - 3x - 1)(2x + 7)$

5. Perform each short division.
- *(a) $4x \, \overline{)\, 8x^3 - 12x^2 - 4x}$ (b) $3m \, \overline{)\, 15m^3 + 9m^2 - 6m}$

6. Perform each long division.
- *(a) $x - 1 \, \overline{)\, x^2 - 11x + 10}$
- *(b) $x - 4 \, \overline{)\, 3x^3 - 10x^2 - 3x - 23}$
- *(c) $x - 2 \, \overline{)\, x^3 - 8}$
- (d) $x + 2 \, \overline{)\, x^3 - 8}$
- *(e) Divide $x + 3$ into $x^3 + 5x^2 + 4x + 6$.
- (f) Divide $x - 2$ into $x^3 - x^2 + 3x - 4$.

CHAPTER 12

Solving Linear Equations

12.1 INTRODUCTION

An *equation* is a statement of equality between two quantities. **equation**
The statement $4 = 3 + 1$ is a very simple equation. In algebra, equations usually contain variables or unknowns as well as constants. In this chapter we shall be concerned with *linear equations*, that is, equations **linear equation** in which the highest power of the variable or unknown is the first power. Here are some examples of linear equations in x:

$$x + 5 = 17$$
$$9x = 14$$
$$15x - 7 = 3x - 2$$
$$4(x - 3) + 15 = 31$$

A *solution* of an equation is a number which when substituted for the **solution** unknown will make the equation a true statement. For example, in the equation $x + 5 = 12$, the number 7 is a solution, since when 7 is substituted for x the equation becomes the true statement $12 = 12$. Sometimes a solution is referred to as a *root* of the equation. **root**

In order to solve linear equations, you must make use of your knowledge of signed numbers (Chapter 10) and simplifying expressions

(Chapter 11). Our approach to solving linear equations begins with the use of opposites.

12.2 ADDING OPPOSITES

The equation $x + 2 = 7$ can be solved just by looking at it to determine the value of x that will make $x + 2$ equal to 7. The solution is 5, since when x is replaced by 5 the two sides of the equation will be the same. Both sides will equal 7.

Although you can sometimes "see" the solution, as in $x + 2 = 7$, very often this is not the case, especially in more complicated equations appearing later in the chapter. Accordingly, a mechanical method is desirable. The method presented next is based on the following idea.

> If the same number is added to both sides of an equation, the equality is maintained.

For example, in the simple equation $35 = 35$, if 6 is added to each side, the equality is maintained.

$$\begin{array}{r} 35 = \quad 35 \\ +6 \quad +6 \\ \hline 41 = \quad 41 \end{array}$$

Of course, if you add a number (other than zero) to only one side of an equation, the equality will not be maintained.

The equation $x + 2 = 7$ can be solved by adding -2 to both sides, as shown next.

$$\begin{array}{r} x + 2 = \quad 7 \\ -2 \quad -2 \\ \hline x + 0 = \quad 5 \end{array}$$

or

$$x = \quad 5 \ \checkmark$$

It was decided to add -2 to both sides because -2 is the opposite of the $+2$ that appeared next to (added to) the x. Because $(+2) + (-2) = 0$, the $+2$ no longer appears on the left side, thus x is left by itself.

> To eliminate a term from one side of an equation, always add the opposite of the term to both sides of the equation.

Example 1. *Solve for m:* $m - 13 = 6$.

Add $+13$ to both sides in order to leave m by itself. We chose $+13$ because $+13$ is the opposite of -13; that is, $+13$ and -13 add to zero, thereby leaving m alone.

$$
\begin{array}{rcl}
m - 13 = & & 6 \\
+13 & & +13 \\
\hline
m & = & 19 \checkmark
\end{array}
$$

The solution is 19.

EXERCISES 12.2

Answers to starred exercises are given in the back of the book.

*1. Solve each linear equation.

(a) $x + 1 = 6$ (b) $x + 5 = 8$ (c) $x - 7 = 2$
(d) $x - 2 = 15$ (e) $m + 5 = 2$ (f) $y + 8 = 1$
(g) $r + 2 = -7$ (h) $x - 2 = 158$ (i) $x - 1 = -1$
(j) $m - 7 = -5$ (k) $c + 6 = 0$ (l) $x - 3 = -9$
(m) $x - 9 = 0$ (n) $x + 9 = -17$ (o) $y + 4 = 4$
(p) $y - 5 = 0$

2. Solve each linear equation.

*(a) $x - \dfrac{1}{8} = \dfrac{1}{4}$ *(b) $y + \dfrac{2}{3} = \dfrac{3}{4}$

*(c) $x + 9.6 = 15.2$ *(d) $t - 7.4 = -5.8$

*(e) $x - \dfrac{3}{8} = \dfrac{3}{2}$ (f) $x - \dfrac{2}{3} = \dfrac{11}{12}$

*(g) $x + \dfrac{1}{3} = \dfrac{1}{3}$ (h) $x + \dfrac{5}{8} = \dfrac{5}{8}$

*(i) $x - 1.4 = -2.6$ (j) $n - 9.2 = -15.7$

*(k) $y + 8.4 = -5.7$ (l) $y + 17.6 = -11.9$

12.3 DIVIDING BY THE COEFFICIENT

The equation $3x = 60$ can be solved just by looking at it to determine the value of x that will make $3x$ equal to 60. The solution is 20, since when x is replaced by 20 the two sides of the equation will be the same. Both sides will equal 60.

Although you can sometimes "see" the solution, as in $3x = 60$, very often this is not the case, especially in more complicated equations appearing later in the chapter. Accordingly, a mechanical method is desirable. The method presented next is based on the following idea.

> If both sides of an equation are divided
> by the same number, the equality is maintained.

For example, in the simple equation $40 = 40$, if both sides are divided by 8, the equality is maintatined.

$$\frac{40}{8} = \frac{40}{8}$$

$$5 = 5$$

Of course, if you divide only one side of an equation by a number (other than 1), the equality will not be maintained.

The equation $3x = 60$ can be solved by dividing both sides by 3, as shown next.

$$3x = 60$$

$$\frac{3x}{3} = \frac{60}{3}$$

$$1x = 20$$

$$x = 20$$

The idea is to change the coefficient of the unknown to 1. In this case the coefficient of x was 3, so we divided both sides by 3. Dividing $3x$ by 3 produces $1x$, or x, as desired.

> To change the coefficient of the unknown
> to 1, divide both sides of the equation
> by the coefficient of the unknown.

Example 2. *Solve for x:* $4x = 20$.

The coefficient of x is 4, so divide both sides by 4. The result is

$$4x = 20$$

$$\frac{4x}{4} = \frac{20}{4}$$

$$x = 5 \ \checkmark$$

The solution is 5.

Example 3. *Solve for x:* $-x = 16$.

$-x$ is the same as $-1x$, so the coefficient of x is -1 in this case. This means we will divide both sides of the equation by -1.

$$-x = 16$$

becomes

$$\frac{-x}{-1} = \frac{16}{-1}$$

or

$$x = -16 \quad \checkmark$$

Example 4. *Solve:* $-3y = 7$.

The coefficient of y is -3, so divide both sides by -3. The result is

$$\frac{-3y}{-3} = \frac{7}{-3}$$

$$y = -\frac{7}{3} \quad \checkmark$$

Note. In algebra it is usually acceptable to leave the solution as an improper fraction, and this will be done throughout. However, if the fraction can be reduced, it should be. For example, 8/6 should be reduced to 4/3, and 4/3 is an acceptable form.

Example 5. *Solve* $9x = 0$.

Divide both sides by 9, the coefficient of x.

$$\frac{9x}{9} = \frac{0}{9}$$

$$x = 0 \quad \checkmark$$

Example 6. *Solve* $2.6x = 88.4$.

Divide both sides of the equation by 2.6, the coefficient of x. Then

$$2.6x = 88.4$$

becomes

$$\frac{2.6x}{2.6} = \frac{88.4}{2.6}$$

or

$$x = 34 \quad \checkmark$$

EXERCISES 12.3

*1. Solve each linear equation.

(a) $7x = 21$

(b) $5x = 30$

(c) $3m = 144$

(d) $9y = 189$

(e) $4x = 17$

(f) $8x = 3$

(g) $5t = 2$

(h) $3x = 19$

(i) $7x = -35$

(j) $4p = -56$

(k) $-3x = 36$

(l) $-7x = 42$

(m) $-5c = 17$

(n) $-8x = 5$

(o) $-6x = -60$

(p) $-2n = -184$

(q) $-5x = -9$

(r) $-6x = -1$

(s) $-x = 19$

(t) $-y = -7$

(u) $7x = 0$

(v) $-4w = 0$

(w) $1.9x = 38$

(x) $3.2t = 72.32$

(y) $-2.1y = 1050$

(z) $-9.1x = -59.15$

2. Solve each linear equation.

*(a) $4x = 65$ (b) $3h = -17$ *(c) $-x = 16$

(d) $19x = 0$ *(e) $8x = -63$ (f) $-5x = -120$

*(g) $-65y = -17$ (h) $-m = -93$ *(i) $8x = 0$

(j) $10t = -160$

3. Solve each linear equation.

*(a) $x + 13 = -7$ (b) $13x = -7$

*(c) $-4t = 1$ (d) $m - 5 = -1$

*(e) $2m = 9$ (f) $y - 8 = 0$

*(g) $y - 3 = -3$ (h) $2y = 0$

*(i) $-6p = -37$ (j) $p - 6 = -37$

12.4 MULTIPLYING BY RECIPROCALS

As shown in the preceding section, equations such as $3x = 19$, $1.2x = 72$, and $-5x = -16$ are solved by dividing both sides by the coefficient of x. However, when the coefficient of the unknown is a fraction or if the constant on the other side is a fraction, then another approach is recommended.

reciprocal

To prepare for the alternative approach, we need to know about reciprocals. The *reciprocal* of a fraction is the fraction created by interchanging the numerator and denominator. Here is a table showing several examples.

Number	Reciprocal
$\dfrac{3}{7}$	$\dfrac{7}{3}$
$\dfrac{7}{3}$	$\dfrac{3}{7}$
$\dfrac{1}{9}$	9
4	$\dfrac{1}{4}$
-2	$-\dfrac{1}{2}$

The result of multiplying a number by its reciprocal is always 1. Here are some examples.

$$\frac{3}{7} \cdot \frac{7}{3} = 1$$

$$\frac{1}{9} \cdot 9 = 1$$

$$(-2)\left(-\frac{1}{2}\right) = 1$$

In general:

$$\boxed{\text{(number)} \cdot \text{(its reciprocal)} = 1}$$

The method presented next is based on the use of reciprocals and the following idea.

$$\boxed{\begin{array}{l} \text{If both sides of an equation are multiplied} \\ \text{by the same number, the equality is maintained.} \end{array}}$$

Specifically, we will multiply both sides of the equation by the reciprocal of the coefficient of the unknown. As a result, the coefficient of the unknown will become 1.

$$\boxed{\begin{array}{l} \text{When fractions are present, to change the} \\ \text{coefficient of the unknown to 1, multiply} \\ \text{both sides of the equation by the reciprocal} \\ \text{of the coefficient of the unknown.} \end{array}}$$

Example 7. *Solve* $\frac{1}{3}x = 7$.

Multiply both sides of the equation by 3, the reciprocal of the coefficient of x. This produces

$$3 \cdot \frac{1}{3}x = 3 \cdot 7$$

$$x = 21 \quad \checkmark$$

In some cases it is more difficult to determine the coefficient of the unknown. Here are some examples.

Term	Coefficient of x
$\frac{3x}{8}$	$\frac{3}{8}$ $\left(\text{since } \frac{3x}{8} \text{ is } \frac{3}{8}x\right)$
$\frac{x}{4}$	$\frac{1}{4}$ $\left(\text{since } \frac{x}{4} \text{ is } \frac{1x}{4} \text{ or } \frac{1}{4}x\right)$
$-\frac{7x}{3}$	$-\frac{7}{3}$

Example 8. *Solve* $-\dfrac{x}{5} = 13$.

The coefficient of the unknown is $-\frac{1}{5}$, so multiply both sides of the equation by -5, the reciprocal of the coefficient. The result is

$$(-5)\left(-\frac{x}{5}\right) = (-5)(13)$$

or $\qquad\qquad\qquad\qquad x = -65 \quad \checkmark$

Example 9. *Solve* $\dfrac{4x}{7} = 9$.

Multiply both sides by $\frac{7}{4}$, the reciprocal of the coefficient of x. This produces

$$\frac{7}{4} \cdot \frac{4x}{7} = \frac{7}{4} \cdot 9$$

or $\qquad\qquad\qquad\qquad x = \dfrac{63}{4} \quad \checkmark$

Example 10. *Solve* $3y = \dfrac{4}{11}$.

Since there is a fraction in the equation, multiply both sides by the reciprocal of the coefficient of y, namely, $\frac{1}{3}$.

$$\frac{1}{3} \cdot 3y = \frac{1}{3} \cdot \frac{4}{11}$$

$$y = \frac{4}{33} \quad \checkmark$$

EXERCISES 12.4

*1. Solve each linear equation.

(a) $\dfrac{1}{2}x = 7$ $\;(14)$

(b) $\dfrac{3}{4}y = 7$

(c) $\dfrac{n}{3} = 5$

(d) $-\dfrac{x}{6} = 4$ $\;\;-24$

(e) $-\dfrac{5}{7}x = \dfrac{9}{14}$

(f) $\dfrac{5x}{4} = 2$

(g) $\dfrac{2m}{7} = -8$ $\qquad$ (h) $-\dfrac{2m}{9} = -\dfrac{11}{5}$ $\qquad$ (i) $3p = \dfrac{4}{5}$

(j) $-7t = -\dfrac{11}{13}$

2. Solve each linear equation.

*(a) $\dfrac{2x}{9} = 5$ $\qquad$ (b) $7t = \dfrac{11}{4}$ $\qquad$ *(c) $-\dfrac{m}{7} = -2$

(d) $\dfrac{5y}{9} = \dfrac{3}{2}$ $\qquad$ *(e) $6y = -\dfrac{13}{7}$ $\qquad$ (f) $\dfrac{x}{2} = -9$

*(g) $-\dfrac{3}{5}x = \dfrac{7}{4}$ $\qquad$ (h) $-x = \dfrac{2}{7}$

3. Solve each linear equation.

*(a) $7x = -3$ $\qquad$ (b) $-3x = 17$ $\qquad$ *(c) $\dfrac{m}{9} = -6$

(d) $5x = 0$ $\qquad$ *(e) $-4x = \dfrac{27}{11}$ $\qquad$ (f) $\dfrac{3y}{14} = 5$

*(g) $-t = \dfrac{6}{5}$ $\qquad$ (h) $-f = -13$ $\qquad$ *(i) $-\dfrac{1}{3}x = \dfrac{9}{7}$

(j) $-\dfrac{y}{3} = -17$

12.5 COMBINING THE METHODS

To solve the equation $3x - 1 = -8$ requires addition of an opposite (add $+1$ to both sides) and then division by the coefficient of x (divide both sides by 3).

$$
\begin{aligned}
3x - 1 &= -8 \\
+1 \quad &\quad +1 \\
\hline
3x \quad &= -7
\end{aligned}
$$
$\left\{\begin{array}{l}\text{First, add } +1 \\ \text{to both sides.}\end{array}\right.$

$$\dfrac{3x}{3} = \dfrac{-7}{3}$$
$\left\{\begin{array}{l}\text{Second, divide} \\ \text{both sides by 3.}\end{array}\right.$

$$x = -\dfrac{7}{3} \ \checkmark$$

In general,

> To solve an equation in which you must both *add an opposite* and *divide by the coefficient*, it is advisable to add the opposite *first* and then divide by the coefficient.

By the way, if you happen to divide first (before adding the opposite) you will usually introduce fractions immediately and then have to add fractions before obtaining a solution. Consequently, we will follow the boxed suggestion above.

Example 11. *Solve $5n + 3 = 12$.*

$$
\begin{array}{rl}
5n + 3 = 12 & \\
\underline{-3 \quad\; -3} & \left\{ \begin{array}{l} \text{adding } -3 \text{ to both sides} \\ \text{to get } 5n = \cdots \end{array} \right. \\
5n \quad\;\; = \;\; 9 &
\end{array}
$$

$$
\frac{5n}{5} = \frac{9}{5} \qquad \left\{ \begin{array}{l} \text{dividing both sides} \\ \text{by 5 to get } n = \cdots \end{array} \right.
$$

$$
n = \frac{9}{5} \;\; \checkmark
$$

Example 12. *Solve $-2x + 9 = -5$.*

$$
\begin{array}{rl}
-2x + 9 = -5 & \\
\underline{-9 \quad\; -9} & \left\{ \begin{array}{l} \text{adding } -9 \text{ to both sides} \\ \text{to get } -2x = \cdots \end{array} \right. \\
-2x \quad\;\; = -14 &
\end{array}
$$

$$
\frac{-2x}{-2} = \frac{-14}{-2} \qquad \left\{ \begin{array}{l} \text{dividing both sides by} \\ -2 \text{ to get } x = \cdots \end{array} \right.
$$

$$
x = 7 \;\; \checkmark
$$

Example 13. *Solve $5 - 3y = 0$.*

First, add -5 to both sides, in order to get the y term alone.

$$
\begin{array}{rl}
5 - 3y = & 0 \\
\underline{-5 \qquad\;\; -5} & \\
-3y = & -5
\end{array}
$$

Next, divide both sides by -3, the coefficient of y.

$$
\frac{-3y}{-3} = \frac{-5}{-3}
$$

$$
y = \frac{5}{3} \;\; \checkmark
$$

Remember that when fractions are present, the coefficient of the unknown can be changed to 1 by multiplying both sides by the reciprocal of the coefficient. This means that with an equation such as $\frac{4}{3}t - 1 = -12$ we first add $+1$ to both sides and then multiply both sides by $\frac{3}{4}$, the reciprocal of $\frac{4}{3}$.

Example 14. *Solve* $\frac{4}{3}t - 1 = -12$.

$$\frac{4}{3}t - 1 = -12$$
$$\underline{+1 = +1} \qquad \text{adding } +1 \text{ to both sides}$$
$$\frac{4}{3}t \qquad = -11$$

$$\frac{3}{4} \cdot \frac{4}{3}t = \frac{3}{4}(-11) \qquad \text{multiplying both sides by } \frac{3}{4}$$

$$t = -\frac{33}{4} \checkmark$$

In a linear equation, the unknown can appear on the right side rather than on the left side of the equals sign. While you may prefer to have the unknown on the left side, there is nothing wrong with it being on the right side instead. In reality it does happen, so you should get used to it. Here are some examples of equations where the unknown is on the right side.

$$19 = 2x + 3$$
$$15 = 4x$$
$$8 = x - 5$$

As an example, $19 = 2x + 3$ is solved next.

Example 15. *Solve* $19 = 2x + 3$.

First, add -3 to both sides to get the x term by itself.

$$19 = 2x + 3$$
$$\underline{-3 \qquad\quad -3}$$
$$16 = 2x$$

Next, divide both sides by 2, the coefficient of x.

$$\frac{16}{2} = \frac{2x}{2}$$
$$8 = x \checkmark$$

EXERCISES 12.5

*1. Solve each linear equation.

(a) $3x - 4 = 6$

(b) $2x + 1 = 5$

(c) $7m - 3 = -18$

(d) $5x + 2 = -3$

(e) $7x + 3 = 0$

(f) $4y + 1 = 1$

(g) $-2x - 7 = -5$

(h) $-7x - 3 = 41$

(i) $\frac{1}{2}x + 6 = 9$ (j) $-\frac{x}{5} + 3 = 18$

(k) $5 + 2x = 17$ (l) $9 - 7x = 4$

(m) $\frac{17}{2}n + 1 = 35$ (n) $\frac{3m}{5} - 2 = 1$

(o) $-\frac{3}{5}x - 5 = 16$ (p) $-\frac{2x}{3} + 6 = -19$

*2. Solve each linear equation.

  (a) $25 = 3x$ (b) $13 = t - 7$ (c) $15 = 3 + 7x$

  (d) $-8 = 3x - 10$ (e) $0 = 1 - 15y$ (f) $7 = 8x + 7$

3. Solve each linear equation.

  *(a) $\frac{x}{3} = 13$ (b) $5x = -\frac{1}{7}$

  *(c) $5 - 2m = 17$ (d) $9 + 6x = -14$

  *(e) $3 = 1 - \frac{1}{2}x$ (f) $\frac{4}{5}x + 3 = -22$

  *(g) $6y - \frac{1}{3} = \frac{1}{6}$ (h) $5w + \frac{1}{2} = \frac{3}{4}$

  *(i) $1.6x + 3 = 9.4$ (j) $1.3y - 8.6 = -3.4$

  *(k) $-7.8p + 9.5 = -6.1$ (l) $-2.3t + 1.9 = -2.7$

12.6 OTHER LINEAR EQUATIONS

This section presents several examples of more complicated linear equations along with step-by-step solutions. In each instance the equations are simplified to resemble those of previous sections. Here are guidelines we will use in the seven examples presented in this section.

To solve complicated linear equations, follow these steps—in sequence.

1. If the unknown is within parentheses, multiply out to free the unknown. (See Examples 20–22)
2. Combine like terms. (Examples 16, 17, 19, 21, 22)
3. If the unknown appears on both sides of the equation, use an opposite to eliminate it from one of the sides. (Examples 18, 19, 21)
4. At this point the equation has been simplified to resemble those solved in earlier sections. So proceed now as you have in previous sections. (Examples 16–22)

Example 16. *Solve* $3x + 4x - 5x = 6 + 2 - 3$.

There are no parentheses, so begin by combining like terms. The terms on the left combine to $2x$; the terms on the right combine to 5. Thus, the equation becomes

$$2x = 5$$

Next,

$$x = \frac{5}{2} \checkmark$$

after dividing both sides by 2.

Example 17. *Solve* $7n + 2 - 3n + 4 = 6 - 4$.

Begin by combining like terms. The result is

$$4n + 6 = 2$$

This type of equation was solved in Section 12.5. First, add -6 to both sides.

$$
\begin{array}{rcr}
4n + 6 =& & 2 \\
-6 & & -6 \\
\hline
4n & =& -4
\end{array}
$$

Next, divide both sides by 4, the coefficient of the unknown.

$$\frac{4n}{4} = \frac{-4}{4}$$

or

$$n = -1 \checkmark$$

Example 18. *Solve* $8x + 1 = x - 4$.

There are no parentheses and there are no like terms to be combined. However, the unknown does appear on both sides. So (step 3) add $-x$ to both sides to eliminate the x term from the right side.

$$
\begin{array}{rcr}
8x + 1 =& & x - 4 \\
-x & & -x \\
\hline
7x + 1 =& & -4
\end{array}
$$

Continuing, we add -1 to both sides to get the x term alone.

$$
\begin{array}{rcr}
7x + 1 =& -4 \\
-1 & -1 \\
\hline
7x & =& -5
\end{array}
$$

Finally, after dividing both sides by 7,

$$x = -\frac{5}{7} \checkmark$$

Example 19. *Solve* $8y + 2 - 3 + y = 5 + 15y - 4$.

First, combine like terms, so that

$$8y + 2 - 3 + y = 5 + 15y - 4$$

becomes

$$9y - 1 = 15y + 1$$

Notice that the unknown, y, appears on both sides of the equation. So apply step 3 and eliminate the $15y$ from the right side by adding $-15y$ to both sides.

$$
\begin{array}{rcr}
9y - 1 = & & 15y + 1 \\
-15y & & -15y \\
\hline
-6y - 1 = & & +1
\end{array}
$$

Add $+1$ to both sides to get the y term alone.

$$
\begin{array}{rcr}
-6y - 1 = & +1 \\
+1 & +1 \\
\hline
-6y & = & 2
\end{array}
$$

Finally, divide both sides by -6.

$$\frac{-6y}{-6} = \frac{2}{-6}$$

$$y = -\frac{2}{6} = -\frac{1}{3} \quad \checkmark$$

Example 20. *Solve* $2(t + 1) = 17$.

The unknown, t, is within parentheses, so multiply out the $2(t + 1)$ in order to free the unknown. Thus

$$2(t + 1) = 17$$

becomes

$$2t + 2 = 17$$

and then

$$2t = 15 \qquad \text{by adding } -2 \text{ to both sides}$$

Finally,

$$t = \frac{15}{2} \quad \checkmark$$

Example 21. *Solve* $5(3 - x) + 6 = 2 + 3x$.

The unknown, x, is within parentheses, so multiply out the $5(3 - x)$ in order to free the unknown.

$$5(3 - x) + 6 = 2 + 3x$$

becomes

$$15 - 5x + 6 = 2 + 3x$$

Now combine the like terms 15 and 6 that appear on the left side.

$$21 - 5x = 2 + 3x$$

Next, add $-3x$ to both sides to eliminate the x term from the right side.

$$21 - 8x = 2$$

Add -21 to both sides to get the x term alone on the left.

$$-8x = -19$$

Divide both sides by -8 to obtain

$$x = \frac{19}{8} \quad \checkmark$$

Example 22. *Solve* $5x - (3 + 2x) = 7$.

Remove the x from parentheses by multiplying out $-(3 + 2x)$ to get $-3 - 2x$. So

$$5x - (3 + 2x) = 7$$

first becomes

$$5x - 3 - 2x = 7$$

On combining the x terms,

$$3x - 3 = 7$$

Adding $+3$ to both sides yields

$$3x = 10$$

Finally,

$$x = \frac{10}{3} \quad \checkmark$$

EXERCISES 12.6

*1. Solve each linear equation.
  (a) $5x + 7x = 12 + 8$
  (b) $3x - 7x + x = -3 - 5 + 19$
  (c) $7p + 5 - 3p + 1 = 0$
  (d) $n + 3n = 2n + 8$
  (e) $3(m + 2) = 13$
  (f) $y - (2y + 1) = 7$

(g) $15x - 17 = 2 - 3x$
(h) $8 - 3x = 5 - 4x + 1$
(i) $2(x - 1) + 5x = 3$
(j) $-7(x + 2) = 4x$

2. Solve each linear equation.
   *(a) $5(2 + 3x) = 13$
   (b) $n + n + 1 = n + 9$
   *(c) $7x + 8 = 4x + 3$
   (d) $3n + 7 - n = 4n + 3$
   *(e) $7(x + 2) - 4(x - 1) = 0$
   (f) $7(x + 5) - 3(x + 1) = 0$
   *(g) $5(1 - 9x) + 3(15x + 4) = x$
   (h) $8 - 3y = 5 - (4y + 1)$
   *(i) $4 - 2(2 - y) + 3 = 10$
   (j) $2m + 3 = m + 3$
   *(k) $2(x + 1) = 3(1 - x)$
   (l) $2x + 3x + 4x + 2 = 5 - x$
   *(m) $5(t + 1) = 5(2t + 9)$
   (n) $7x - 1 = 4(2x + 1)$
   *(o) $13 - 3(1 - p) + 2 = 17$
   (p) $2(3m + 1) + 5 = 4 + m$

*3. Solve each linear equation.

(a) $\frac{1}{4}x + 6 = 5 - \frac{3}{4}x$ (b) $\frac{3x}{4} + 7 = \frac{5x}{8} + 3$

(c) $\frac{2x}{3} - 1 = 5 - \frac{3x}{4}$ (d) $7.4x + 5.2 = 1.3x + 11.3$

12.7 CHECKING YOUR SOLUTION

To solve an equation like $5x + 2 = 17$ is to determine the value of x that "works" in the equation—in other words, determine the number that can be substituted for x to make the left-hand side equal to the right-hand side.

If you solve $5x + 2 = 17$, you get 3 as the solution. This means that if x is replaced by 3 in the original equation, the two sides will have the same value.

Take the equation

$$5x + 2 = 17$$

and substitute 3 for x.

$$5(\underline{3}) + 2 \overset{?}{=} 17$$
$$15 + 2 \overset{?}{=} 17$$
$$17 = 17 \ \checkmark$$

The two sides have the same value, 17.

As you can see, the foregoing process is a *check*. We can check "solutions" to equations to determine whether they are indeed solutions. Only a solution will check. For instance, if we "solved" $5x + 2 = 17$ and obtained 2 as a "solution," the check would fail, meaning that 2 is not a solution. This situation is shown below.

$$5\underline{x} + 2 = 17$$
$$5(\underline{2}) + 2 \overset{?}{=} 17 \qquad \text{substituting 2 for } x$$
$$10 + 2 \overset{?}{=} 17$$
$$12 \neq 17 \qquad \text{The symbol } \neq \text{ means } \textit{not equal to.} \qquad \neq$$

The check fails, since the two sides have different values.

Example 23. *Is 4 a solution of* $2(3 - x) + 5 = x - 1$?

$$2(3 - \underline{x}) + 5 = \underline{x} - 1$$
$$2(3 - \underline{4}) + 5 \overset{?}{=} \underline{4} - 1 \qquad \left\{ \begin{array}{l} \text{substituting 4 for } x \\ \text{throughout the equation} \end{array} \right.$$
$$2(-1) + 5 \overset{?}{=} 3$$
$$-2 + 5 \overset{?}{=} 3$$
$$3 = 3 \ \checkmark$$

Yes, 4 is a solution of given equation.

Example 24. *Is 1 a solution of* $5y + 2y - 3 = 5(y + 1)$?

$$5y + 2y - 3 = 5(y + 1)$$
$$5(\underline{1}) + 2(\underline{1}) - 3 \overset{?}{=} 5(\underline{1} + 1) \qquad \left\{ \begin{array}{l} \text{substituting 1 for } y \\ \text{throughout the equation} \end{array} \right.$$
$$5 + 2 - 3 \overset{?}{=} 5(2)$$
$$4 \neq 10 \qquad \text{The check fails.}$$

No, 1 fails to check. It is not a solution of the given equation.

Example 25. *Solve and check* $3x + 1 = 9$.

Solution:
$$\begin{array}{rcr} 3x + 1 = & & 9 \\ -1 & & -1 \\ \hline 3x & = & 8 \end{array}$$
$$x = \frac{8}{3} \ \checkmark$$

Check:

$$3\left(\frac{8}{3}\right) + 1 \overset{?}{=} 9$$

$$8 + 1 \overset{?}{=} 9$$

$$9 = 9 \checkmark$$

EXERCISES 12.7

*1. For which of the following equations is -2 a solution?
   (a) $3x = -6$ (b) $5x - 3x = 4$
   (c) $2x + 3 = 1$ (d) $2(x + 1) + 2 = 0$
   (e) $x + 9 = 2x + 11$ (f) $x + 2 = 0$
   (g) $3 - (x + 1) = 8$ (h) $2(x + 3) + 5 = 7 - (1 - x)$
   (i) $7x + 10 + x = x - 4$ (j) $5(1 - x) - 11 = x + 6$

2. For which of the following equations is -3 a solution?
   (a) $4x = 12$ (b) $4x = -12$
   (c) $3(x + 1) + 6 = 0$ (d) $-x = 3$
   (e) $1 - 4(1 - x) = -15$ (f) $2(x + 3) + x = -3$
   (g) $x + 2 - 4x = 3x$ (h) $5(x + 4) = 4(x + 2)$

3. Solve and check each equation.
   *(a) $5x - 2 = 18$ *(b) $3x + 1 = 10$
   *(c) $x + 2 = 2$ *(d) $5m = 12$
   *(e) $3(1 - x) + 5 = -1$ *(f) $7x + 2 + 5x = 2x + 12$
   (g) $4x + 1 = 6$ (h) $6x - 2 + x = 4 - x$
   *(i) $3(1 + 5x) - 1 = x$ (j) $7(2x - 1) + x = 3$

12.8 HOW MANY SOLUTIONS?

Thus far each linear equation in one unknown has had just one solution. We've solved dozens of such equations, and always there has been exactly one solution. However, not all linear equations have one solution as you will see.

Example 26. *Solve $x + 3 = x + 2$.*

First, add -3 to both sides.

$$\begin{array}{r} x + 3 = x + 2 \\ \underline{-3 \qquad\quad -3} \\ x \quad\;\; = x - 1 \end{array}$$

Next, add $-x$ to both sides.

$$\begin{array}{rcr} x = & & x - 1 \\ -x & & -x \\ \hline 0 = & & -1 \end{array}$$

But 0 is not equal to -1. Moreover, the x has disappeared. The inconsistent result $0 = -1$ means that the equation has *no solution*. Look back at the original equation $x + 3 = x + 2$ and think. We seek a number x such that $x + 2$ and $x + 3$ are equal. This cannot be because $x + 3$ is always 1 larger than $x + 2$.

Example 27. *Solve* $5(2x + 3) + 8 = 10x + 23$.

First, multiply out the $5(2x + 3)$.

$$5(2x + 3) + 8 = 10x + 23$$

becomes

$$10x + 15 + 8 = 10x + 23$$

Next, combine the $+15$ and the $+8$.

$$10x + 23 = 10x + 23$$

At this point you can see that the two sides of the equation are identical. In other words, $10x + 23$ is equal to $10x + 23$ regardless of the value of x. This means that *any number is a solution* of the equation $10x + 23 = 10x + 23$ or of the original equation $5(2x + 3) + 8 = 10x + 23$. The equation has an *infinite number of solutions*. Had we attempted to combine like terms and proceed from $10x + 23 = 10x + 23$, the result would have been the equation $0 = 0$, which is true regardless of the value of x.

EXERCISES 12.8

1. Determine the number of solutions of each equation: none, one, infinite.
   *(a) $x + 9 = x + 5$
   *(b) $3x + 7 = 3x + 7$
   *(c) $2x + 10 = x + 3$
   (d) $4(x + 8) = 2x - 9$
   *(e) $5x - 3 + x = 5 + 6x$
   (f) $5x + 5 + x = 5 + 6x$
   *(g) $3(1 - x) + 5x = 1 + 2(1 + x)$
   (h) $3(x + 2) - x = 2(3x + 7)$

Chapter 12.   REVIEW EXERCISES

1. Solve each linear equation, if possible.
   *(a) $m - 19 = 24$ *(b) $-5t = -17$
   *(c) $-6y = 19$ *(d) $7r = 4$

*(e) $\frac{1}{2}x = -16$

(f) $\frac{1}{4}x = 11$

*(g) $9x - 2 = -13$

(h) $6x + 5 = 22$

*(i) $17 - 4x = 28$

(j) $8 + 4y = -11$

*(k) $9y - 12 = 51$

(l) $\frac{3}{7}x + 8 = -9$

*(m) $\frac{4}{5}x + 3 = 10$

*(n) $\frac{5}{3}x + 1 = -2$

*(o) $8x + 5 - x = 4 + 7x$

(p) $14 - 5x + 2 = -3x - 17$

*(q) $5(2x - 3) + 4 = 3x + 1$

(r) $8(x + 4) - 50 = 18 - x$

*(s) $3(1 - 2t) - 2 = 16 + 3t$

(t) $3 - 2(x + 1) = 9x + 4$

*(u) $7 - (5 - x) = x - (2 - x)$

(v) $1 - 3(x - 2) = 7 - 3x$

Word Problems

13.1 INTRODUCTION

This chapter consists of a variety of examples intended to show how algebra is used in solving problems that are stated in words. In the first section you'll see how to change word phrases to algebraic expressions. The other two sections present entire problems stated in sentences. Such sentences must be examined in order to set up equations that can be solved.

Many students have difficulty with word problems because such problems have an abstract nature. As you begin to study word phrases in the next section, you will find arithmetic examples presented first. This is done to point out that many of those problems are easy to solve when you think in terms of arithmetic first. For example, if you know how to work with 4 dimes, then x dimes should not be impossible to work with. Or, if you must work with y nickels and you find this difficult, consider something more concrete, like 7 nickels, in order to see what's going on.

13.2 WORD PHRASES

Here are several examples to introduce you to the concept of changing words to algebra.

Example 1. *How many cents are there in* 4 *dimes?*

There are 40 cents in 4 dimes. Since each dime is worth 10 cents, if you have 4 dimes, you have $10 \cdot 4 = 40$ cents.

Example 2. *How many cents are there in x dimes?*

Since each dime is worth 10 cents, there are $10 \cdot x$ or $10x$ cents—that is, 10 cents each *times* the number of dimes, x.

Example 3. *How many feet are there in 72 inches?*

Since there are 12 inches in every foot, there are $72/12 = 6$ feet in 72 inches. We divide the number of inches by 12 in order to get the number of feet.

Example 4. *How many feet are there in 5m inches?*

Change inches to feet by dividing by 12 as in the last example. Thus

$$5m \text{ inches} = \frac{5m}{12} \text{ feet} \quad \checkmark$$

Note. Some students get confused about when to multiply and when to divide. When changing dimes to cents (Examples 1 and 2) we multiplied because there must be more cents than dimes in the same amount of money. When changing inches to feet (Examples 3 and 4) we divided because there must be fewer feet than inches in the same measurement.

Example 5. *If Bob is x years old today, how old will he be in 7 years?*

Today he is x years old. In 7 years he will have increased his age by 7 years. He will be $x + 7$ years old then, seven years older.

Example 6. *Write expressions to represent*

    (a) *seven more than n*
    (b) *nine less than twice x*
    (c) *the next even number after the even number x*

(a) Seven more than n is simply $n + 7$, or $7 + n$ if you prefer. Both represent a number that is seven more than n.

(b) Twice x is $2x$—that is, two times x. So nine less than twice x is nine less than $2x$—namely, $2x - 9$.

(c) Even numbers are $0, 2, 4, 6, 8, 10, 12, \ldots$. You add two to any even number to get the next one in sequence. If x is an even number, then the next one must be $x + 2$, two more than x.

Example 7. *Write an algebraic expression to represent the total number of cents in n nickels and q quarters.*

Each nickel is worth 5 cents, and so n nickels are worth $5n$ cents (5 cents each for each of n nickels). Similarly, each quarter is worth 25 cents; so q quarters are worth $25q$ cents. Thus the total cents value of n nickels and q quarters is $5n + 25q$.

The next example makes use of the formulas for perimeter and area of a rectangle. Using the letters P for perimeter, A for area, l for length, and w for width, we have the formulas shown below.

$$P = 2l + 2w$$
$$A = l \cdot w$$

Example 8. *Suppose the width of a rectangle is x units and the length is 3 more than the width. Determine (a) the perimeter and (b) the area.*

(a) You are given that the width is x. Since the length is 3 more than the width, the length is $x + 3$. The perimeter is two lengths plus two widths, so the perimeter is

$$2(x + 3) + 2(x)$$

which simplifies to

$$2x + 6 + 2x$$

and finally becomes

$$4x + 6 \checkmark$$

(b) Since area is length times width, the area is

$$(x + 3)x$$

which, when multiplied out, is

$$x^2 + 3x \checkmark$$

The next example makes use of the formula for the simple interest (i) on an investment of p dollars at r percent per year for t years. The percent r must be expressed as a fraction or decimal for calculation purposes.

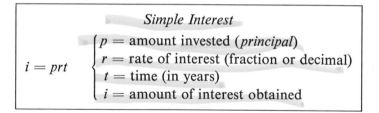

$$i = prt \qquad \begin{cases} p = \text{amount invested (\textit{principal})} \\ r = \text{rate of interest (fraction or decimal)} \\ t = \text{time (in years)} \\ i = \text{amount of interest obtained} \end{cases}$$

Simple Interest

Example 9. *Determine the interest on each investment.*

    (a) *$1200 at 8% per year for 2 years*
    (b) *$1000 at 6.5% per year for 10 years*
    (c) *d dollars at 7% per year for 3 years*
    (d) *$137 at m% per year for x years*

    (a) Here p is $1200, r is .08, and t is 2. So

$$i = (\$1200)(.08)(2) = \$192 \ \checkmark$$

    (b) Here p is $1000, r is .065, and t is 10. So

$$i = (\$1000)(.065)(10) = \$650 \ \checkmark$$

    (c) Here p is d, r is .07, and t is 3. So

$$i = (\$d)(.07)(3) = .21d \text{ dollars, or } \$.21d \ \checkmark$$

    (d) Here p is $137, r is $\dfrac{m}{100}$ or .01m, and t is x. So

$$i = (\$137)\left(\frac{m}{100}\right)(x) \quad \text{or} \quad (\$137)(.01m)(x)$$

$$= \frac{\$137mx}{100} \ \checkmark \quad \text{or} \quad \$1.37mx \ \checkmark$$

Either form is acceptable as a final answer. Note that percent means hundredths, and this $m\%$ can be expressed as hundredths by using either a fraction or decimal form.

EXERCISES 13.2

Answers to starred exercises are given in the back of the book.

*1. Write an algebraic expression to represent each phrase.
    (a) Four more than y.
    (b) Seventeen less than w.

    (c) Ten more than $5x$.

    (d) Half of m.

    (e) Twice t.

    (f) Three more than twice x.

    (g) Five less than three times y.

    (h) m more than n.

    (i) x less than y.

    (j) The product of x and y.

    (k) Seven times $6c$.

    (l) 23% of x.

    (m) $k\%$.

    (n) $k\%$ of x.

2. Write an algebraic expression to represent each phrase.

    *(a) The number of cents in t nickels.

    (b) The number of cents in x half dollars.

    *(c) The number of dollars in v cents.

    (d) The number of dollars in y dimes.

    *(e) The total number of cents in x nickels and y half dollars.

    (f) The total number of cents in p pennies and q quarters.

    *(g) The total cost in cents of x tickets at \$1.50 each and y tickets at \$3 each.

    (h) The total cost in cents of m tickets at \$2 each and n tickets at \$3.25 each.

3. Write an algebraic expression to represent each phrase.

    *(a) The number of shoes in p pairs.

    (b) The number of legs in $3x$ spiders. (A spider has eight legs.)

    *(c) The number of minutes in H hours.

    (d) The number of hours in D days.

    *(e) The number of days in T hours.

    (f) The number of hours in M minutes.

    *(g) The number of feet in $5y$ yards.

    (h) The number of inches in $7x$ feet.

    *(i) The number of feet in p inches.

    (j) The number of yards in $2k$ feet.

    *(k) The cost in cents of m articles at x cents each.

    (l) The cost in francs of n items at y francs each.

    *(m) The total cost in lira of a items at x lira each and b items at y lira each.

    (n) The number of miles in f feet. (1 mile $= 5280$ feet.)

    *(o) The profit on a book that is bought for y dollars and sold for x dollars.

*(p) The amount of money that remains if x dollars out of $5000 is spent.

(q) The number of cattle remaining in a herd of h cattle if s cattle are sold.

4. Write an algebraic expression to represent each phrase.

*(a) The interest on d dollars invested at 6% per year for one year.

*(b) The interest on d dollars invested at p% per year for two years.

(c) The interest on x dollars invested at 5% per year for three years.

(d) The interest on m dollars invested at 9% per year for y years.

*(e) The interest on m dollars invested at n% per year for w years.

*(f) 23% more than x.

(g) 17% more than y.

*5. (metric) Write an algebraic expression to represent each phrase.

(a) The number of meters in x kilometers. (1 kilometer = 1000 meters.)

(b) The number of meters in y millimeters. (1 meter = 1000 millimeters.)

(c) The number of centimeters in $3x$ meters. (1 meter = 100 centimeters.)

(d) The number of millimeters in $2y$ meters. (1 meter = 1000 millimeters.)

(e) The number of liters in m milliliters. (1 liter = 1000 milliliters.)

(f) The number of grams in x kilograms. (1 kilogram = 1000 grams.)

(g) The number of kilograms in y grams. (1 kilogram = 1000 grams.)

6. Write an algebraic expression to represent each phrase.

*(a) The total distance walked by Madeline if she walks x yards from the apartment to the pool, then y yards from the pool to her car, and finally z yards from her car to the apartment. Make a drawing if you do not see the answer immediately.

*(b) The perimeter of a rectangle with length x and width y.

*(c) The area of a rectangle with length x and width y.

*(d) The perimeter of a square with sides of length x.

(e) The area of a square with sides of length x.

*(f) The area of a rectangle with length $x + 2$ and width 7.

(g) The perimeter of a rectangle with length $x + 2$ and width 7.

*(h) The area of a rectangle whose width is x and whose length is twice the width.

(i) The perimeter of a rectangle whose width is x and whose length is twice the width.

*(j) The perimeter of a rectangle whose width is x and whose length is four more than its width.

(k) The area of a rectangle whose width is x and whose length is four more than its width.

*(l) The perimeter of a triangle with sides of lengths a, b, and c.

13.3 WORD PROBLEMS

Example 10. *The sum of four consecutive numbers is 234. What are the numbers?*

Let x be the first (or smallest) of the four numbers. Then

$x =$ smallest number (first)
$x + 1 =$ another number (second)
$x + 2 =$ another number (third)
$x + 3 =$ another number (fourth)

⎧Consecutive numbers are
⎪those that come one right
⎨after the other in the count-
⎪ing sequence; for example,
⎩7, 8, 9, 10 or 21, 22, 23, 24.

Since the sum of the four numbers is 234, we get the equation

$$(x) + (x + 1) + (x + 2) + (x + 3) = 234$$

or
$$x + x + 1 + x + 2 + x + 3 = 234$$

which simplifies to

$$4x + 6 = 234$$

and then

$$4x = 228$$

Finally,

$$x = 57 \quad \checkmark$$
$$x + 1 = 58 \quad \checkmark$$
$$x + 2 = 59 \quad \checkmark$$
$$x + 3 = 60 \quad \checkmark$$

The numbers are 57, 58, 59, and 60. They are consecutive. Check to be sure that their sum is 234.

Example 11. *15% of what number is 27?*

Let x be the number. Then 15% of x is 27. This statement can be written as the equation shown below. Keep in mind that 15% of x means .15 times x.

$$.15x = 27$$

The equation is solved by dividing both sides by .15, the coefficient of x.

$$\frac{.15x}{.15} = \frac{27}{.15}$$

$$x = 180 \quad \checkmark$$

The number is 180. That is, 15% of 180 is 27. Check this answer by actually computing 15% of 180 in order to show that it is indeed 27.

Example 12. *Dorys and Adolfo went to a nice restaurant for their anniversary and spent $68.20, including 4% tax and a 20% tip. What was the amount of the bill before tax and tip?*

Let x be the amount of the bill before tax and tip. Then we can establish the following representation.

$$x = \text{amount of the bill before tax and tip}$$

$.20x = \text{tip}$ (a 20% tip on the amount of the bill)

$.04x = \text{tax}$ (a 4% tax on the amount of the bill)

Since the sum of these three numbers must be the total amount spent, $68.20, we have the following equation.

$$x + .20x + .04x = 68.20$$

or $$1.24x = 68.20$$

which is readily solved by dividing both sides by 1.24, the coefficient of x.

$$x = \frac{68.20}{1.24} = 55 \; \checkmark$$

Thus the amount of the bill before tax and tip was $55. To check this answer, see if $55 plus 4% of $55 plus 20% of $55 is indeed equal to $68.20.

bill:	$55.00
4% of $55:	2.20
20% of $55:	11.00
Total:	$68.20

Thus, our answer checks.

Example 13. *If the sum of two numbers is 94 and the larger number is five less than two times the smaller, what are the values of the two numbers?*

Let x represent the smaller number, since the larger one is described by comparing it with the smaller.

The larger is five less than two times the smaller. Since x is the smaller, the larger is five less than two times x. This means that the larger is five less than $2x$, which means that it is $2x - 5$. So we have the representation

$$x = \text{the smaller number}$$
$$2x - 5 = \text{the larger number}$$

The statement of the problem indicates that the sum of the smaller and the larger is 94; that is,

$$\text{smaller} + \text{larger} = 94$$

or $$(x) + (2x - 5) = 94$$

We now solve this equation.

$$x + 2x - 5 = 94$$
$$3x - 5 = 94$$
$$3x = 99$$
$$x = 33 \checkmark \quad \text{smaller}$$

Also,

$$2x - 5 = 2(33) - 5$$
$$= 66 - 5$$
$$= 61 \checkmark \quad \text{larger}$$

The smaller is 33. The larger is 61.

Check the answers by returning to the statement of the problem to determine whether your answers check with the words of the problem. *Do not* merely see if they check in your equation. Your equation could be wrong!

Check:

"If the sum of two numbers is 94"	$33 + 61 \overset{?}{=} 94$
	$94 = 94 \quad \checkmark$

"Larger number is five less than two times the smaller"	$61 \overset{?}{=} 2 \cdot 33 - 5$
	$61 \overset{?}{=} 66 - 5$
	$61 = 61 \quad \checkmark$

Example 14. *A 972-meter length of fence will be used to enclose a rectangular field that is twice as long as it is wide. Determine the length and width of the field to be enclosed.*

Let $x =$ the width.
Let $2x =$ the length (since the length is 2 times, or twice, the width).

Here is a drawing of the rectangle.

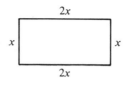

The perimeter is the sum of the four sides, $x + 2x + x + 2x$, and must be equal to 972 meters, which is the amount of fencing to be used. Thus

$$x + 2x + x + 2x = 972$$

or

$$6x = 972$$

or

$$x = \frac{972}{6} = 162 \quad \checkmark$$

and $\qquad\qquad\qquad\qquad\qquad 2x = 324$ ✓

The length of the field will be 324 meters and the width will be 162 meters.

Check:

"Twice as long as it is wide" $324 \overset{?}{=} 2(162)$
$\qquad\qquad\qquad\qquad\qquad\qquad\quad 324 = 324$ ✓

"972-meter length" $324 + 162 + 324 + 162 \overset{?}{=} 972$
$\qquad\qquad\qquad\qquad\qquad\qquad\qquad\qquad\quad 972 = 972$ ✓

Example 15. *Dan is 4 years older than Sylvia. 32 years ago he was twice as old as she was then. How old is each today?*

Today: $\begin{cases} \text{Let } x = \text{Sylvia's age} \\ x + 4 = \text{Dan's age; he's 4 years older} \end{cases}$

32 years ago: $\begin{cases} \underline{x - 32} = \text{Sylvia's age} \\ \underline{x + 4 - 32} \text{ or } x - 28 = \text{Dan's age} \end{cases}$

32 years ago (that is, when Sylvia's age was $x - 32$ and Dan's was $x - 28$), Dan's age was twice Sylvia's. This means that

$$x - 28 = 2(x - 32)$$

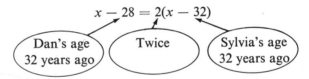

Next, solve the equation $x - 28 = 2(x - 32)$

$$\begin{array}{rcl} x - 28 &=& 2(x - 32) \\ x - 28 &=& 2x - 64 \\ \underline{+\ 64 \qquad\quad +\ 64} & & \\ x + 36 &=& 2x \\ \underline{-x \qquad\quad -x} & & \\ 36 &=& x \end{array}$$

Thus

$$x = 36 \quad ✓ \qquad \text{(Sylvia's age today)}$$
$$x + 4 = 40 \quad ✓ \qquad \text{(Dan's age today)}$$

Sylvia is 36 years old and Dan is 40.

As a check, note that Dan (age 40) is 4 years older than Sylvia (age 36). 32 years ago he was 8 and she was 4, which means that he was indeed twice her age then.

These steps should be followed, *in order*, when solving word problems.

> 1. Set up the representation of unknowns.
> 2. Form an equation.
> 3. Solve the equation.
> 4. Check the solution of the equation against the statement of the problem.

EXERCISES 13.3

Solve each word problem.

*1. The sum of three consecutive numbers is 114. Determine the numbers.

2. The sum of five consecutive numbers is 420. Determine the numbers.

*3. The sum of two numbers is 342. One number is twice the size of the other. Determine the two numbers.

4. The sum of two numbers is 252. The larger number is twice the smaller. What are the two numbers?

*5. The sum of two numbers is 153. The larger number is seven more than the smaller. What are the two numbers?

6. The sum of two numbers is 195. The larger is three more than twice the smaller. What are the two numbers?

*7. A board 17 meters long will be sawed into two pieces. One piece will be 2 meters longer than the other. Determine the length of each piece.

8. Hank and Tom have hit a total of 98 home runs. If Tom has 10 more than Hank, how many does each have?

*9. 16 percent of what number is 36?

10. 24 percent of what number is 54?

*11. 35 percent of what number is 91?

*12. What is the cost of a lamp if 3% sales tax on its cost amounts to $1.65?

13. What is the cost of a TV if 4% sales tax is $29.40?

*14. How much was the bill if a 12% tip is $5.25?

15. How much was the bill if a 15% tip is $3.60?

*16. What was the amount of the bill before tax and tip if the bill plus a 15% tip plus 5% tax totals $18.48?

17. What was the amount of the bill before tax and tip if the bill plus a 10% tip plus 5% tax totals $27.60?

*18. Charles is twice as old as Liz. If the sum of their ages is 48, how old is each of them?

19. In 16 years Linda will be three times as old as she is now. How old is she today?

*20. An 832-meter length of fence will be used to enclose a garden that is three times as long as it is wide. Determine the length and width of the field to be enclosed.

*21. A square piece of paper has a perimeter of 364 centimeters. How long is each side?

22. The length of a rectangle is 4 feet more than three times the width. If the perimeter is 64 feet, find the length and width.

23. The sum of the angles of a triangle is 180°. The largest angle is twice the smallest. The third angle is 16° more than the smallest. Determine the number of degrees in each angle.

13.4 MORE DIFFICULT WORD PROBLEMS (Optional)

Example 16. *Dennis has $4.80 in nickels and dimes. He has four times as many nickels as dimes. How many of each does he have?*

Let

$x =$ the number of dimes.

Then

$4x =$ the number of nickels, since he has four times as many nickels as dimes.

Since the total value of the nickels and dimes is $4.80, or 480 cents, our concern is with the number of *cents* in the x dimes and $4x$ nickels.

x dimes $= \underline{10} \cdot x$ or $10x$ cents, since there are $\underline{10}$ cents in each dime.
$4x$ nickels $= \underline{5} \cdot 4x$ or $20x$ cents, since there are $\underline{5}$ cents in each nickel.

Note:

$$\underbrace{\text{value of dimes (in cents)}}_{10x} + \underbrace{\text{value of nickels (in cents)}}_{20x} = \underbrace{\text{total cents}}_{480}$$

Thus we have the equation

$$10x + 20x = 480$$

which simplifies to

$$30x = 480$$

So
$$x = 16 \quad \checkmark \qquad \text{(dimes)}$$
and
$$4x = 64 \quad \checkmark \qquad \text{(nickels)}$$

There are 16 dimes and 64 nickels.

Check:

    (a) *Is the total value of 16 dimes and 64 nickels equal to $4.80?*

$$\begin{array}{r} 16 \text{ dimes} = \$1.60 \\ 64 \text{ nickels} = \$3.20 \\ \hline \text{total} = \$4.80 \quad \checkmark \end{array}$$

    (b) *Are there four times as many nickels as dimes?*
        Yes, 64 is four times 16. $\checkmark$

Example 17. *John invests a total of $9500, part at 6% (per year) and the rest at 7%. If his (one year) return on the investments is $622, how much of the $9500 was invested at 6% and how much was invested at 7%?*

If we let x be the amount invested at 6%, then how can we represent the amount invested at 7%? If he invested $2000 at 6%, it is obvious that $7500 is invested at 7%; this is simply $9500 - $2000. *Similarly*, if x dollars are invested at 6%, then $9500 - x$ dollars are left for investment at 7%. Thus

$$x = \text{amount invested at } 6\%$$
$$9500 - x = \text{amount invested at } 7\%$$

To check this representation, note that the two amounts add to $9500:

$$\begin{array}{r} x \\ 9500 - x \\ \hline 9500 \end{array}$$

We note that

$$\text{interest at } 6\% + \text{interest at } 7\% = \text{total interest (\$622)}$$

Since x dollars are invested at 6%, the interest on it is 6% of x; that is,

$$\text{interest at } 6\% = .06 \cdot x = .06x$$

Similarly, the interest on $(9500 - x)$ dollars at 7% is 7% of $(9500 - x)$; that is,

$$\text{interest at } 7\% = .07 (9500 - x)$$

So we have the equation

$$.06x + .07 (9500 - x) = 622$$

It will be easier to work with the equation if we multiply each term by 100 in order to eliminate decimals. The result is

$$6x + 7(9500 - x) = 62200$$
or $6x + 66500 - 7x = 62200$ after distributing the 7
or $-x + 66500 = 62200$ after combining $6x - 7x$

$$
\begin{array}{r}
-66500 \quad -66500 \\
\hline
-x = -4300 \\
x = 4300 \;\checkmark \\
9500 - x = 5200 \;\checkmark
\end{array}
$$

(amount at 6%)
(amount at 7%)

John invests \$4300 at 6% and \$5200 at 7%.

Check:

  (a) Does the investment return \$622?

$$.06\,(\$4300) = \$258.00$$
$$.07\,(\$5200) = \$364.00$$
$$\text{total} = \$622.00 \;\checkmark$$

  (b) Is the total investment equal to \$9500?

$$\$4300 + \$5200 = \$9500 \quad\checkmark$$

Example 18. *Two planes leave an airport at the same time and travel in opposite directions. One plane travels 50 mph faster than the other. How fast is each traveling if they are 1000 miles apart after 2 hours?*

Let's sketch the situation.

Since one plane is 50 mph faster than the other, let

$$x = \text{the speed of the slower plane}$$
and $x + 50 = \text{the speed of the faster plane}$

The planes are 1000 miles apart after 2 hours.

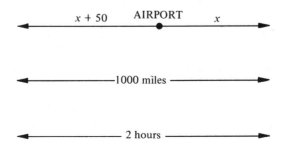

Then we use the formula

$$\boxed{\begin{array}{c} \text{distance} = \text{rate} \times \text{time} \\ d = r \cdot t \end{array}}$$

The total distance (d) is 1000 miles. The time (t) is 2 hours. The rates are x and $x + 50$. Also,

$$\binom{\text{distance of}}{\text{faster plane}} + \binom{\text{distance of}}{\text{slower plane}} = \binom{\text{total distance the}}{\text{planes are apart}}$$

or

$$\binom{\text{rate} \times \text{time}}{\text{(faster)}} + \binom{\text{rate} \times \text{time}}{\text{(slower)}} = \left(\text{distance apart}\right)$$

or

$$(x + 50)2 \quad + \quad x \cdot 2 \quad = \quad 1000$$

Note that the time (2 hours) is the same for each plane.

Since they are traveling in opposite directions, the total distance (1000) represents the sum of the distances of the two planes.

Now solve the equation formed.

$$(x + 50)2 + x \cdot 2 = 1000$$
$$2x + 100 + 2x = 1000$$
$$4x + 100 = 1000$$
$$4x = 900$$
$$x = 225 \quad \checkmark \quad \text{(slower)}$$
$$x + 50 = 275 \quad \checkmark \quad \text{(faster)}$$

The slower plane travels 225 mph; the faster travels at 275 mph.

Check:

(a) Is one plane traveling 50 mph faster than the other?
Yes, the speeds are 275 mph and 225 mph, and $275 - 225 = 50$.

(b) Will they be 1000 miles apart in 2 hours?
Yes, since $2(225) + 2(275) = 1000$.

EXERCISES 13.4

*1. Judy has $9.75 in nickels and dimes. She has twice as many dimes as nickels. How many of each does she have?

2. Carroll has $7.50 in nickels and dimes. If he has 15 more nickels than dimes, how many of each does he have?

*3. Carla has $6.05 in dimes and quarters. If she has three times as many dimes as quarters, how many of each does she have?

*4. Carol has $3.68 in pennies, nickels, and dimes. She has the same number of each. How many of each does she have?

*5. Pat buys stamps in 20¢ and 17¢ denominations. She spends a total of $7.71 and receives a total of 42 stamps. How many of each did she buy?

6. School 66 raises money by selling 102 tickets for a benefit concert. They sell $2 and $3 tickets and obtain a total of $231. How many of each did they sell?

7. The sum of two numbers is 24. If twice the smaller is three less than the larger, determine the numbers.

*8. Gil has $2.95 in nickels, dimes, and quarters. If he has twice as many dimes as quarters and four more nickels than dimes, how many of each does he have?

9. Gordon has $4 in nickels, dimes, and quarters. He has twice as many nickels as quarters. He has four more nickels than dimes. How many of each does he have?

*10. Mike has 15 coins—nickels and dimes. Their value is $1.15. How many of each does he have?

*11. Ken invests $4500, part at 5% and the rest at 6% annual interest. If his income from the two investments is $248 for the year, determine how much is invested at each rate.

12. Anita invests $2300, part at 6% and the remainder at 7%. Her total interest from the two investments is $153. How much did she invest at each rate?

*13. Arlene invests $5400 in bonds paying 8% and in stocks paying 5% interest. After one year the difference between the two returns is $68; that is, the 8% investment returns a dividend that is $68 more than that returned by the 5% investment. How much is invested at each rate?

14. Margaret invests $6700, part at 6% and the rest at 5%. If the return from the 5% investment is $60 more than that of the 6% investment, how much money was invested at each rate?

*15. The average speed of a car is 25 mph slower than that of a given train. In 3 hours and 15 minutes the train can cover the same distance that the car covers in $4\frac{1}{2}$ hours. Find the average speed of each one.

*16. Two trains leave a terminal at the same time and travel in opposite directions. One train travels 30 mph faster than the other. How fast is each train moving if they are 800 miles apart after 4 hours of traveling?

17. Jack and Wilbur are 190 miles apart. They begin to drive toward each other. Wilbur travels 15 mph faster than Jack. If they meet in 2 hours, how fast will they be going?

18. Two trains leave a terminal at the same time and travel in opposite directions. After 3 hours the trains are 465 miles apart. If one train traveled at the average rate of 70 miles per hour, how fast did the other train travel?

Chapter 13. REVIEW EXERCISES

*1. Write an algebraic expression to represent each phrase.
   (a) 15 less than the product of a and b.
   (b) The total number of cents in x dimes and $2y$ half dollars.
   (c) The number of yards in $17x$ feet.
   (d) The perimeter of a rectangle with length x and width $x - 4$.
   (e) The interest on $213 invested at $c\%$ per year for x years.
   (f) 37% more than y.

*2. Solve each word problem.
   (a) The sum of two numbers is 168. One number is three times as big as the other. Determine the two numbers.
   (b) 36% of what number is 37.8?

*3. Solve each word problem.
   (a) Flo has $4.60 in nickels and quarters. If she has two more nickels than quarters, how many of each does she have?
   (b) Joyce invests $10,000, part at 9% and the rest at 8%. The total interest from the two investments is $865. How much did she invest at each rate?

CHAPTER 14

Graphing Straight Lines

14.1 INTRODUCTION

Let us take the two number lines

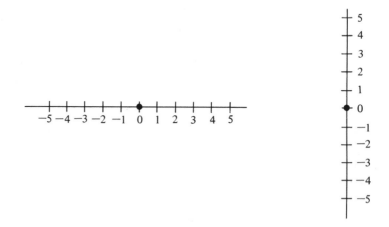

and join them so that they are perpendicular at their zeros. The result is

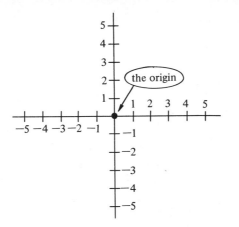

The point where the lines cross is called the *origin*. The diagram
below shows some points in this *plane*. Look at point *A*. How can we
get from the origin to the point *A*? One way is to move 5 units to the
right and then 1 unit up. There are other ways, of course.

origin

We can get to *B* (from the origin) by moving 1 unit to the right and
then up 2 units. Similarly, we get to *C* by moving right 3 and down 1.
We get to *D* by moving left 3 and up 4. *E* is left 1 and down 2.

We use the notation (5, 1) to mean 5 units to the right (of the
origin) and 1 unit up. Similarly, (−3, 4) means 3 units to the left and
4 units up.

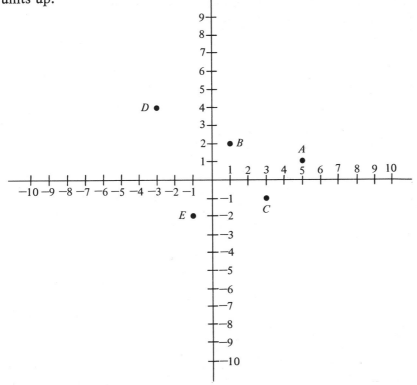

Right is positive and left is negative; up is positive and down is negative.

We tabulate the five points: *A, B, C, D, E.*

Point	Left or Right	Up or Down	Notation
A	right 5	up 1	(5, 1)
B	right 1	up 2	(1, 2)
C	right 3	down 1	(3, −1)
D	left 3	up 4	(−3, 4)
E	left 1	down 2	(−1, −2)

coordinates

We have represented a point as a pair of numbers—the *coordinates of the point.* The first coordinate of the pair indicates the left-right position of the point. The second coordinate indicates the up-down position. So you can see that the order of the numbers is important, and you can

ordered pair

also see why the pair of numbers is called an *ordered pair.*

rectangular

The system that we have drawn and described is called the *rectan-*

Cartesian

gular coordinate system. It is also called the *Cartesian coordinate system* after the French mathematician René Descartes, who developed the scheme in 1637.

Example 1. *The points (5, 1) and (1, 5) are not the same. The order of the numbers 5 and 1 is important in describing a point.*

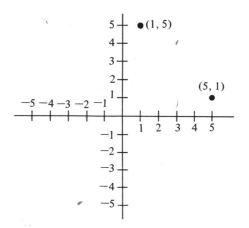

Example 2. *Plot the points (3, 4) and (−2, −5).*

To plot (3, 4), begin at the origin and move 3 to the right and up 4. To plot (−2, −5), begin at the origin and move 2 to the left and down 5.

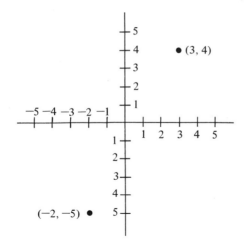

The graph on which we plot points is divided into four distinct regions. Each is called a *quadrant*. The quadrants are numbered I, II, **quadrant** III, IV, as shown in the diagram. Points on the number lines are not considered as being in a quadrant.

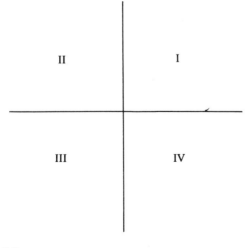

EXERCISES 14.1

Answers to starred exercises are given in the back of the book.

1. Plot each point and indicate which quadrant it is in.

(a) (3, 2)	(b) (2, 3)	(c) (4, 1)	(d) (1, 4)
(e) (1, −5)	(f) (1, 5)	(g) (3, −2)	(h) (−4, 3)
(i) (−5, −2)	(j) (0, 6)	(k) (−2, −4)	(l) (−3, 1)
(m) (3, 0)	(n) (4, 4)	(o) (5, −3)	(p) (7, −1)
(q) (0, 0)	(r) (−2, 0)	(s) (0, 2)	(t) (0, −1)

*2. Determine the coordinates of each point.

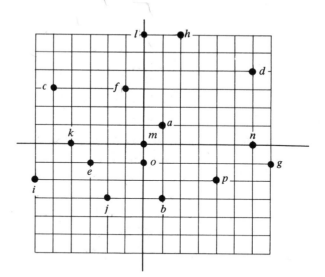

14.2 GRAPHING STRAIGHT LINES

axes   The number lines that we joined to form a plane are called *axes*, and they are usually labeled *x* and *y*. The horizontal axis is called the *x axis*. The vertical axis is called the *y axis*. The plane is then called the *xy* plane.

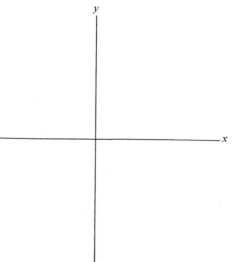

Distances left or right are values for x; distances up or down are values for y. The point (1, 2) means right 1 and up 2. It is also the point (x, y), where x is 1 and y is 2.†

Next, let's plot the points (0, 0), (1, 2), (2, 4), (3, 6), and (−1, −2) in the plane.

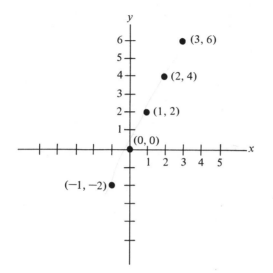

If you connect the points, you'll have a *straight line*. In other words, a straight line can be passed through the points. Also, if you chart the x coordinates and y coordinates as

		x	y
(0, 0)	⟶	0	0
(1, 2)	⟶	1	2
(2, 4)	⟶	2	4
(3, 6)	⟶	3	6
(−1, −2)	⟶	−1	−2

you might see the relationship between x and y. Each y value is *exactly twice* the corresponding x value. That is,

$$y = \text{twice } x$$

or

$$y = 2x$$

By plotting points and then passing a straight line through them,

†The x coordinate is sometimes referred to as the *abscissa* and the y coordinate as the *ordinate*.

abscissa
ordinate

graph   we would be sketching the *graph* of the relationship $y = 2x$. Later in the book we will sketch graphs that are not straight lines.

Example 3. *Sketch the graph of* $y = 3x$.

We choose some values for x and then determine the corresponding y values. This procedure will yield some points of the form (x, y). Let us choose the values $-1, 0, 2,$ and 3 for x and obtain points. Keep in mind that you can choose any values; they need not be $-1, 0, 2,$ and 3.

If $x = -1$,   then $y = 3x$ becomes $y = 3(-1) = -3$.
If $x = 0$,   then $y = 3x$ becomes $y = 3(0) = 0$.
If $x = 2$,   then $y = 3x$ becomes $y = 3(2) = 6$.
If $x = 3$,   then $y = 3x$ becomes $y = 3(3) = 9$.

So we now have the points shown in the chart below.

x	$y = 3x$	*Points*
-1	-3	$(-1, -3)$
0	0	$(0, 0)$
2	6	$(2, 6)$
3	9	$(3, 9)$

The graph is

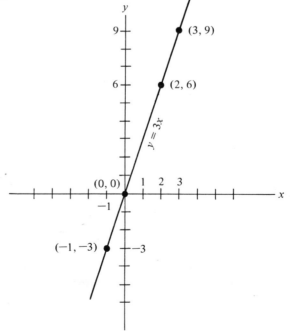

Example 4. *Sketch the graph of* $y = 2x + 1$.

If we choose x values of 0, 1, 2, 3, and 4, we get the points shown below.

x	$y = 2x + 1$	*Points*
0	1	(0, 1)
1	3	(1, 3)
2	5	(2, 5)
3	7	(3, 7)
4	9	(4, 9)

Of course, you can choose *any* values for x, but keep in mind that you are the one who must compute the y values and then plot the points. So numbers like -20 or 30 would not be good choices.

The graph is

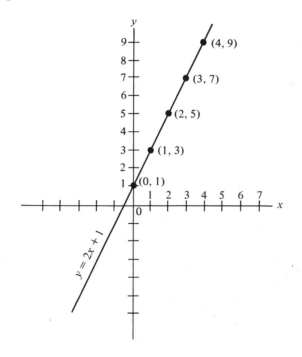

Example 5. *Sketch the graph of* $y = -3x + 2$.

Again y is given in terms of x, so select values for x and use them to determine the corresponding y values. We have chosen 0, 1, 2, and -1 as the x values. Other values could be used.

x	$y = -3x + 2$	*Points*
0	2	$(0, 2)$
1	-1	$(1, -1)$
2	-4	$(2, -4)$
-1	5	$(-1, 5)$

The graph is

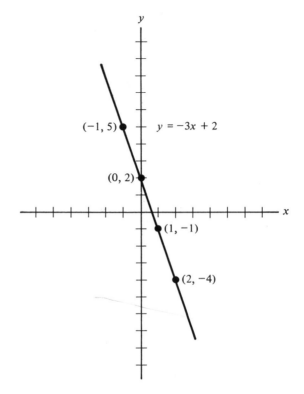

Example 6. *Sketch the graph of* $x = 2y + 1$.

Since x is given in terms of y, choose values for y and then determine the corresponding x values.

y	$x = 2y + 1$	(x, y)
0	1	$(1, 0)$
1	3	$(3, 1)$
2	5	$(5, 2)$
3	7	$(7, 3)$

The graph is

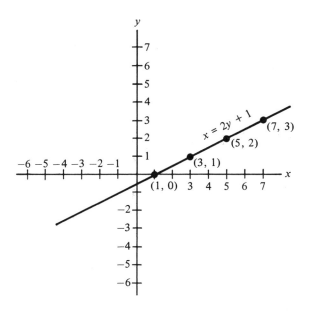

Example 7. *Sketch the graph of the line whose equation is* $3x + 2y = 6$·

The equation $3x + 2y = 6$ does not give y in terms of x or x in terms of y. The easiest way to obtain points for such a line is by letting x be 0 to get y and then letting y be 0 to get x. This will give you two points, which is technically all you need to obtain the line. But be careful, since you have no third or fourth points to serve as a check in the event you make a mistake.

First, let's use 0 for x.

$$3(0) + 2y = 6$$
$$0 + 2y = 6$$
$$2y = 6$$
$$y = 3$$

This yields the point $(0, 3)$.

Similarly, if y is 0, we have

$$3x = 6$$
$$x = 2$$

This yields the point $(2, 0)$. The graph is shown on the next page.

x	y
0	3
2	0

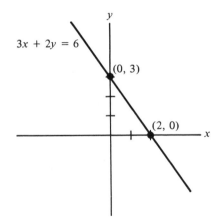

The next two examples illustrate rather unusual lines—one is the graph of an equation having no x in it, the other is the graph of an equation having no y in it.

Example 8. *Sketch the graph of the line $y = 2$.*

The equation states that y equals 2, regardless of what value x takes on. If x is 1, y is 2. If x is 3, y is 2. If x is 0, y is 2. And so on. The points: (1, 2), (3, 2), (0, 2). The graph:

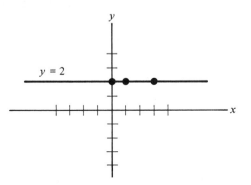

So the graph of $y = 2$ is a horizontal line 2 units above the x axis.

Example 9. *Sketch the graph of the line $x = 3$.*

The equation states that x equals 3, regardless of what value y takes on. If y is 0, x is 3. If y is 2, x is 3. If y is 3, x is 3. And so on. The points: (3, 0), (3, 2), (3, 3). The graph:

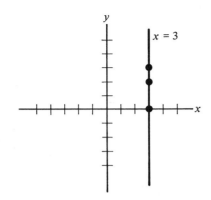

So the graph of $x = 3$ is a vertical line 3 units to the right of the y axis.

The graphs of two lines can be sketched on the same axes in order to determine their approximate point of intersection. Here is an example.

Example 10. *Sketch the graphs of $y = x + 2$ and $5x + 2y = 10$ and estimate where they meet.*

$y = x + 2$ $5x + 2y = 10$

x	y
0	2
1	3
2	4
3	5

x	y
0	5
2	0

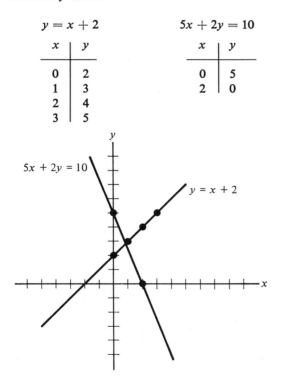

It appears that the lines meet at about (1, 3), but we cannot tell *exactly* by using a graph. Actually, these lines meet at the point ($\frac{6}{7}$, $2\frac{6}{7}$). Methods for obtaining such exact solutions are presented in the next chapter.

If two lines are parallel, then they will not meet when graphed. Here is an example.

Example 11. *Sketch the graphs of the lines whose equations are $y = 2x + 3$ and $y = 2x - 1$. You will see that the lines are parallel.*

$$y = 2x + 3$$

x	y
0	3
1	5
2	7
-1	1

$$y = 2x - 1$$

x	y
0	-1
1	1
2	3
3	5

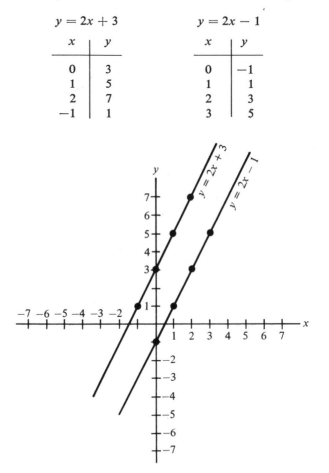

EXERCISES 14.2

1. Sketch the graph of each line.

    *(a) $y = 2x$ (b) $y = 4x$ *(c) $y = x + 3$

    (d) $y = x + 4$ *(e) $y = x - 2$ (f) $y = x - 3$

*(g) $y = x$

(h) $y = x - 1$

*(i) $y = 2x + 2$

(j) $y = 2x + 4$

*(k) $y = 3x - 2$

(l) $y = 4x - 3$

*(m) $y = -2x$

(n) $y = -3x$

*(o) $y = -3x + 4$

(p) $y = -2x + 1$

*(q) $y = -2x - 3$

(r) $y = -3x - 5$

*(s) $y = \dfrac{1}{2}x$

(t) $y = -\dfrac{1}{2}x$

*(u) $y = \dfrac{1}{2}x + 3$

(v) $y = \dfrac{1}{2}x - 1$

*(w) $y = -x$

(x) $y = -x + 2$

2. Sketch the graph of each line.

*(a) $x = y + 3$

(b) $x = y - 4$

*(c) $x = 2y - 3$

(d) $x = 3y + 1$

*(e) $x = -y$

(f) $x = -2y$

*(g) $x = -2y + 3$

(h) $x = -y + 4$

3. Sketch the graph of each line.

*(a) $x + 2y = 8$

(b) $2x + y = 6$

*(c) $x - y = 3$

(d) $x + y = 4$

*(e) $2x + 3y = 12$

(f) $2x + 5y = 10$

*(g) $2x - 3y = 12$

(h) $2x - 5y = 10$

*(i) $3y - 2x = 6$

(j) $2y - 3x = 12$

*(k) $x + y = -4$

(l) $x - y = -4$

*(m) $x + y = 0$ (*Hint*: Use values of x or y other than zero.)

(n) $x + 2y = 0$

4. Sketch the graph of each line. Use Example 8 as a guide.

*(a) $y = 3$ (b) $y = 1$ *(c) $y = -5$ (d) $y = 0$

5. Sketch the graph of each line. Use Example 9 as a guide.

*(a) $x = 2$ (b) $x = 4$ *(c) $x = -3$ (d) $x = 0$

6. Graph each pair of lines on the same axes and determine approximately where the lines meet, if they do meet.

*(a) $\begin{cases} y = x + 1 \\ y = 3x - 1 \end{cases}$

*(b) $\begin{cases} y = x - 1 \\ y = -x + 3 \end{cases}$

*(c) $\begin{cases} y = x \\ y = 2x - 4 \end{cases}$

(d) $\begin{cases} y = -4x + 2 \\ y = 2x - 5 \end{cases}$

*(e) $\begin{cases} y = x \\ y = -x \end{cases}$

(f) $\begin{cases} y = x + 4 \\ y = -x + 4 \end{cases}$

*(g) $\begin{cases} y = x + 2 \\ y = x - 2 \end{cases}$

(h) $\begin{cases} y = -x + 3 \\ y = 2x - 1 \end{cases}$

*(i) $\begin{cases} y = 0 \\ x = 0 \end{cases}$

(j) $\begin{cases} y = 2x - 3 \\ y = 2x + 2 \end{cases}$

*(k) $\begin{cases} y = 2x + 1 \\ y = 4 \end{cases}$

(l) $\begin{cases} y = x - 4 \\ x = 3 \end{cases}$

*(m) $\begin{cases} x + y = 6 \\ y = -x + 1 \end{cases}$

(n) $\begin{cases} y = 1 - 3x \\ y = 3x - 1 \end{cases}$

Chapter 14.   REVIEW EXERCISES

*1. Sketch the graph of each line.

(a) $3x + 4y = 12$ (b) $y = 2x - 5$ (c) $x = 3y - 2$

(d) $y = 4$ (e) $x - 2y = 8$ (f) $2x - y = 0$

(g) $x = -1$ (h) $x + y = 1$ (i) $y = \frac{1}{2}x + 1$

(j) $x = 3 + y$

Systems of Linear Equations

15.1 INTRODUCTION

We have solved equations that contain one unknown, such as

$$2x - 5 = 3$$

and

$$5 + 12(1 - x) = 7x - 9$$

But suppose that we must solve an equation containing two unknowns, such as

$$x + y = 10$$

There are *many solutions*; for example, $x = 7$ and $y = 3$ satisfy the equation. And here are four more solutions:

$$x = 6 \quad \text{and} \quad y = 4$$
$$x = 0 \quad \text{and} \quad y = 10$$
$$x = 10 \quad \text{and} \quad y = 0$$
$$x = 5 \quad \text{and} \quad y = 5$$

If these xy pairs were taken as points (x, y) and plotted, they would fall on a line, and the x, y coordinates of any point on the line would be a solution of the equation. As you can see then, there are an infinite number of such solutions.

On the other hand, if the problem is to solve a *pair* of equations containing two variables, such as

$$\begin{cases} x + y = 10 \\ x - y = 8 \end{cases}$$

solution   then there is only *one solution*. The *solution* is a pair of numbers (a value for x and a value for y) that satisfy *both* equations. Here the solution is $x = 9$ and $y = 1$. You might want to check that this pair satisfies both equations.

If you were to graph the two lines, the point where they meet would be the solution. However, as noted in Chapter 14, graphing gives only an *approximate* solution.

If the lines happen to be *parallel* (and don't meet), then there is *no solution*. If the two lines are the *same*, all their points are common to each other so that there are an *infinite number of solutions*.

15.2 THE ADDITION METHOD (ELIMINATION)

addition method   One way to get an exact solution is to use the *addition method* of solving two equations in two unknowns. This method is also called
elimination   *elimination*. It is explained next.

It would be easy to solve

$$\begin{cases} x + y = 10 \\ x - y = 8 \end{cases}$$

if we could eliminate either x or y, because then there would be only one unknown. Suppose that we add the two equations term for term.

$$\begin{array}{r} x + y = 10 \\ x - y = 8 \\ \hline 2x + 0 = 18 \end{array}$$
$\left\{ \begin{array}{l} \text{Since } x - y \text{ } is \text{ } equal \text{ } to \text{ } 8, \\ \text{we are adding equal quan-} \\ \text{tities to both sides of the} \\ \text{equation } x + y = 10. \end{array} \right.$

The result is an equation in one unknown, x.

$$2x = 18$$

This equation is easily solved for x.

$$x = 9$$

The first original equation indicates that $x + y = 10$. So if x is 9, then y must be 1. Therefore the solution is

$$x = 9 \quad \text{and} \quad y = 1 \; \checkmark \quad or \quad (9, 1)$$

This pair of equations was easy to solve because the y's added to zero and were thus eliminated. Usually neither of the variables can be immediately eliminated. In such cases, you must multiply either one equation by an appropriate constant or both equations by appropriate constants. This approach is demonstrated in the next two examples.

Example 1. *Solve by the addition method.*

$$\begin{cases} 2x - 3y = 4 \\ 5x + y = 27 \end{cases}$$

If we add right away, we fail.

$$\begin{array}{r} 2x - 3y = 4 \\ 5x + y = 27 \\ \hline 7x - 2y = 31 \quad ??? \end{array}$$

Instead multiply both sides of the second equation by 3 so that the coefficients of y will be equal in magnitude (both 3 in this case) and opposite in sign (one $-$, one $+$). Then the y's will add to zero.

$$\begin{cases} 2x - 3y = 4 \xrightarrow{\text{same}} 2x - 3y = 4 \\ 5x + y = 27 \xrightarrow{\times 3} 15x + 3y = 81 \end{cases}$$
$$\begin{array}{r} \hline 17x = 85 \\ x = 5 \end{array}$$

We can solve for y now by using the equation $5x + y = 27$. Since x is 5, we get $5(5) + y = 27$, or $y = 2$. The solution is $x = 5$, $y = 2$, or $(5, 2)$. Note that it does not matter which of the original equations is used to obtain the value of the second variable.

In the next example you will see a system in which both equations must be multiplied in order to eliminate one of the unknowns.

Example 2. *Solve by the addition method.*

$$\begin{cases} 2x + 7y = 3 \\ 3x + 13y = 7 \end{cases}$$

We might multiply the top equation by -3 and the bottom one by 2. As a result, the coefficients of x will be equal in magnitude and opposite in sign. Then the x terms will add to zero.

$$\begin{array}{r} 2x + 7y = 3 \xrightarrow{\times -3} -6x - 21y = -9 \\ 3x + 13y = 7 \xrightarrow{\times 2} 6x + 26y = 14 \\ \hline 5y = 5 \\ y = 1 \ \checkmark \end{array}$$

Next, use one of the original equations, say $3x + 13y = 7$, and substitute 1 for y. This step will give us the corresponding value of x.

$$3x + 13y = 7$$
$$3x + 13(1) = 7$$
$$3x + 13 = 7$$
$$3x = -6$$
$$x = -2 \;\checkmark$$

The solution is $x = -2$, $y = 1$, or $(-2, 1)$.

Example 3. (optional) *Solve by the addition method.*

$$\begin{cases} 6x + 13y = 5 \\ y = 2x - 7 \end{cases}$$

As you can see, this system is not actually set up (lined up) to be solved by the addition method. In fact, in the next section we'll solve it very easily by a method called substitution. However, if we were to insist on solving the system by the addition method, we would proceed as follows.

If $-2x$ is added to both sides of the second equation, it will be changed so that it has the same form as the top one. Then they will line up for the addition method.

$$\begin{cases} 6x + 13y = 5 \\ y = 2x - 7 \end{cases}$$
$$\downarrow$$
$$\begin{cases} 6x + 13y = 5 \xrightarrow{\text{same}} 6x + 13y = 5 \\ -2x + y = -7 \xrightarrow{\times 3} -6x + 3y = -21 \end{cases}$$
$$16y = -16$$
$$y = -1 \;\checkmark$$

Next, take one of the original equations and substitute -1 for y.

$$y = 2x - 7$$
$$-1 = 2x - 7$$
$$6 = 2x$$
$$x = 3 \;\checkmark$$

The solution is $x = 3$, $y = -1$, or $(3, -1)$.

EXERCISES 15.2

Answers to starred exercises are given in the back of the book.

1. Solve each system of equations by using the addition method (elimination).

*(a) $\begin{cases} 2x + y = 19 \\ 3x - y = 11 \end{cases}$

(b) $\begin{cases} 5x + y = 20 \\ 2x - y = -6 \end{cases}$

*(c) $\begin{cases} 2x + 5y = 13 \\ -2x + 11y = 3 \end{cases}$

*(d) $\begin{cases} -3x + 4y = 2 \\ 3x - 5y = -1 \end{cases}$

*(e) $\begin{cases} 7x - 3y = 11 \\ 4x + y = 9 \end{cases}$

*(f) $\begin{cases} 7x - y = 10 \\ 3x + 4y = -9 \end{cases}$

*(g) $\begin{cases} 5x + 8y = 14 \\ x + 5y = 13 \end{cases}$

(h) $\begin{cases} 7x - 2y = 8 \\ 5x - y = 7 \end{cases}$

*(i) $\begin{cases} 6m + 7n = 30 \\ -2m + 9n = -10 \end{cases}$

(j) $\begin{cases} 3s + 2t = -12 \\ 5s + 4t = -22 \end{cases}$

2. Solve each system of equations by using the addition method (elimination).

*(a) $\begin{cases} 7x - 4y = 10 \\ 4x + 3y = 11 \end{cases}$

*(b) $\begin{cases} -3x + 5y = 9 \\ 2x + 3y = 13 \end{cases}$

*(c) $\begin{cases} 4x - 6y = 0 \\ 7x + 5y = 31 \end{cases}$

(d) $\begin{cases} -3p + 8q = 6 \\ 5p - 7q = 9 \end{cases}$

*(e) $\begin{cases} 3x + 2y = 6 \\ 7x + 5y = 13 \end{cases}$

(f) $\begin{cases} 5x + 8y = 3 \\ 4x + 9y = -8 \end{cases}$

*(g) $\begin{cases} 6m - 7n = 9 \\ 9m - 10n = 15 \end{cases}$

(h) $\begin{cases} 3x - 7y = -12 \\ 4x - 11y = -16 \end{cases}$

*(i) $\begin{cases} 8x + 3y = 13 \\ 10x - 7y = -16 \end{cases}$

(j) $\begin{cases} 7x + 4y = 24 \\ 5x + 7y = 13 \end{cases}$

15.3 SUBSTITUTION

A better way to solve the system of Example 3 is by a method called *substitution*. The system is

substitution

$$\begin{cases} 6x + 13y = 5 \\ y = 2x - 7 \end{cases}$$

The second equation says that y is the same as $2x - 7$. So let's substitute $2x - 7$ for y *in the other equation*.

$$6x + 13\underline{y} = 5$$

becomes
$$6x + 13(\underline{2x - 7}) = 5$$

The equation $6x + 13(2x - 7) = 5$ contains only *one* variable, x, and is easily solved. In steps,

$$6x + 13(2x - 7) = 5$$
$$6x + 26x - 91 = 5$$
$$32x - 91 = 5$$
$$32x = 96$$
$$x = 3 \checkmark$$

It is now a simple matter to determine y, since we have a $y = \cdots$ equation—namely, $y = 2x - 7$.

$$y = 2\underline{x} - 7$$
becomes $y = 2(\underline{3}) - 7$ when we substitute 3 for x
$$y = 6 - 7$$
$$y = -1 \checkmark$$

The solution is $x = 3$, $y = -1$.

Example 4. *Solve by the substitution method.*

$$\begin{cases} x = 5y + 8 \\ 2x - 3y = 2 \end{cases}$$

Use $x = 5y + 8$ to substitute for x in $2x - 3y = 2$.

$$2\underline{x} - 3y = 2$$
$$2(\underline{5y + 8}) - 3y = 2 \quad \text{by substitution of } 5y + 8 \text{ for } x$$
$$10y + 16 - 3y = 2$$
$$7y + 16 = 2$$
$$7y = -14$$
$$y = -2 \checkmark$$

We now find x.

$$x = 5\underline{y} + 8$$
$$x = 5(\underline{-2}) + 8 \quad \text{when we substitute } -2 \text{ for } y$$
$$x = -10 + 8$$
$$x = -2 \checkmark$$

The solution is $x = -2$, $y = -2$.

Example 5. (optional) *Solve by substitution.*

$$\begin{cases} 5x + 4y = 13 \\ 2x + 3y = 8 \end{cases}$$

It would be better (easier) to solve this system by using the addition method, since it is already aligned for addition. Nevertheless, we *can* use substitution. The work that follows shows the procedure to use if you were to insist on solving this system by substitution.

In order to be able to make a substitution, we begin by solving the second equation for x in terms of y.

$$2x + 3y = 8$$
$$2x = -3y + 8$$
$$x = \frac{-3y + 8}{2}$$
$$x = -\frac{3}{2}y + 4$$

Next, substitute $-\frac{3}{2}y + 4$ for x in $5x + 4y = 13$ to get

$$5\left(-\frac{3}{2}y + 4\right) + 4y = 13$$

or

$$-\frac{15}{2}y + 20 + 4y = 13$$

Multiply both sides of the equation by 2 to eliminate fractions and proceed.

$$-15y + 40 + 8y = 26$$
$$-7y + 40 = 26$$
$$-7y = -14$$
$$y = 2 \checkmark$$

To solve for x, substitute 2 for y in the equation $x = -\frac{3}{2}y + 4$.

$$x = -\frac{3}{2}y + 4$$
$$x = -\frac{3}{2}(2) + 4$$
$$x = -3 + 4$$
$$x = 1 \checkmark$$

The solution is $x = 1$, $y = 2$.

In Chapter 14 we saw how two simultaneous linear equations can be solved by graphing to determine *approximately* where the two lines (each an equation) meet. So we have three methods for solving simultaneous linear equations.

1. Addition (or elimination)
2. Substitution
3. Graphing

EXERCISES 15.3

1. Solve each system of equations by using substitution.

*(a) $\begin{cases} 2x + 3y = 13 \\ y = x + 1 \end{cases}$ (b) $\begin{cases} y = 3x + 13 \\ 5x + 2y = 15 \end{cases}$ *(c) $\begin{cases} x = 4y - 4 \\ 2x + 3y = 14 \end{cases}$

(d) $\begin{cases} 5x + 6y = 11 \\ x = 3y - 2 \end{cases}$ *(e) $\begin{cases} 7x - 6y = 5 \\ y = 3x + 1 \end{cases}$ (f) $\begin{cases} x - 7y = 9 \\ y = x - 3 \end{cases}$

*(g) $\begin{cases} x = y \\ 7x + 4y = 44 \end{cases}$ (h) $\begin{cases} 8x - 3y = 20 \\ y = x \end{cases}$ *(i) $\begin{cases} y = 5x + 1 \\ 4x + 3y = 10 \end{cases}$

(j) $\begin{cases} x = y + 2 \\ 6x - 8y = 6 \end{cases}$ *(k) $\begin{cases} x = 3y - 3 \\ y = 2x - 4 \end{cases}$ (l) $\begin{cases} x = 2y - 16 \\ y = 3x + 18 \end{cases}$

*(m) $\begin{cases} a = 1 - b \\ 5a + 4b = 2 \end{cases}$ (n) $\begin{cases} 4x - 3y = 9 \\ y = 7 - 2x \end{cases}$

2. Solve each system of equations. Use the addition method (elimination) if the system is aligned for addition. Use substitution if one of the equations is already solved for x in terms of y or y in terms of x.

*(a) $\begin{cases} y = 4x - 1 \\ 10x - 2y = 2 \end{cases}$ (b) $\begin{cases} 7x - 5y = 25 \\ 2x + 3y = 16 \end{cases}$

*(c) $\begin{cases} -3x + 2y = 7 \\ 4x + 6y = -5 \end{cases}$ (d) $\begin{cases} 3x - 10y = 16 \\ x = 4y + 5 \end{cases}$

*(e) $\begin{cases} 4x + 5y = 12 \\ y = 1 - x \end{cases}$ (f) $\begin{cases} 10x + 3y = 14 \\ 6x + 5y = 18 \end{cases}$

3. (optional) Read Example 5. Then go back to Exercises 15.2 and use substitution to solve Exercises 1 (a), (f), (g), and 2 (a), (b), (c), (e).

15.4 CHECKING A SOLUTION

To check your solution to a pair of simultaneous linear equations, check the two values in both equations. *Both equations must balance.* From a graphing point of view, you are checking to see if the point is in fact on both lines.

Example 6. *Is $x = 5$, $y = 2$ the solution of the following equations?*

$$\begin{cases} 2x - 3y = 4 \\ 5x + y = 27 \end{cases}$$

Check 5 and 2 in the first equation.

$$2x - 3y = 4$$
$$2(5) - 3(2) \overset{?}{=} 4$$
$$10 - 6 \overset{?}{=} 4$$
$$4 = 4 \quad \checkmark \quad \text{Checks in the first equation.}$$

Now check 5 and 2 in the second equation.

$$5x + y = 27$$
$$5(5) + (2) \overset{?}{=} 27$$
$$25 + 2 \overset{?}{=} 27$$
$$27 = 27 \quad \checkmark \quad \text{Checks in the second equation, too.}$$

The solution is $x = 5$, $y = 2$; it checks in *both* equations.

Example 7. *Is $x = 2$, $y = 3$ the solution of the following equations?*

$$\begin{cases} 5x + 3y = 19 \\ 2x + 7y = 24 \end{cases}$$

Check 2 and 3 in the first equation.

$$5x + 3y = 19$$
$$5(2) + 3(3) \overset{?}{=} 19$$
$$10 + 9 \overset{?}{=} 19$$
$$19 = 19 \quad \checkmark \quad \text{Checks in the first equation.}$$

Now check 2 and 3 in the second equation.

$$2x + 7y = 24$$
$$2(2) + 7(3) \overset{?}{=} 24$$
$$4 + 21 \overset{?}{=} 24$$
$$25 \neq 24 \quad \text{The check fails in the second equation.}$$

Since the check fails in at least one of the equations, we know that $x = 2$ and $y = 3$ is not the correct x, y pair—they are not the solution.

EXERCISES 15.4

1. Solve and check each system of equations. In each instance use the method (addition or substitution) that appears easier.

*(a) $\begin{cases} 7x + 2y = 1 \\ 5x + 3y = 7 \end{cases}$ (b) $\begin{cases} x = 1 - 3y \\ y = 2x - 16 \end{cases}$ *(c) $\begin{cases} 9y - 5x = 2 \\ 4y + 3x = 27 \end{cases}$

(d) $\begin{cases} 10a - 7b = 24 \\ 4a - 3b = 10 \end{cases}$ *(e) $\begin{cases} 5x - 6y = 13 \\ x = 2y - 3 \end{cases}$ (f) $\begin{cases} 7x - 2y = 1 \\ y = 1 + 5x \end{cases}$

15.5 UNUSUAL CASES

You will not always succeed when you attempt to solve a system of two linear equations in two unknowns. There are two situations in which you will fail to obtain a solution.

If two lines are parallel, then they do not meet. So when you attempt to solve a system of equations that represents parallel lines, you will not succeed. Instead of a solution, you will get such results as $5 = 0$ or $3 = 7$—that is, "equations" consisting of two unequal constants, one on each side of the equal sign.

Although two equations may appear different, they may only be different forms of the same equation. This is the other situation in which you will fail to obtain a solution. In terms of lines, the two lines are in reality only one line. So all the points that are on the one line are also on the "other" line. Consequently, there is not one solution in this case but rather an infinite number of solutions. Instead of obtaining a solution when you try to solve the system algebraically, you obtain results like $0 = 0$ or $4 = 4$. That is, you get an equation consisting of two equal constants, one on each side of the equal sign.

EXERCISES 15.5

*1. Solve each system of equations *if possible*. Use whichever method you prefer—addition or substitution. *Beware of unusual cases.*

(a) $\begin{cases} x + 4y = 8 \\ x = 3y + 1 \end{cases}$

(b) $\begin{cases} 3x + 8y = 13 \\ 7x - 6y = 18 \end{cases}$

(c) $\begin{cases} y = 5x - 1 \\ 10x - 2y = 2 \end{cases}$

(d) $\begin{cases} x = 6 \\ 5x - 7y = 16 \end{cases}$

(e) $\begin{cases} 4x + 2y = 17 \\ y = 3 - 2x \end{cases}$

(f) $\begin{cases} 3x - 2y = 10 \\ -6x + 4y = 7 \end{cases}$

(g) $\begin{cases} y = x \\ 2x + 3y = 15 \end{cases}$

(h) $\begin{cases} -2x + 5y = 8 \\ 4x - 10y = -16 \end{cases}$

15.6 SYSTEMS OF THREE EQUATIONS (Optional)

Systems of three equations in three unknowns can be solved by reducing the system to two equations in two unknowns. This we can do by using the addition method.

Example 8. *Solve the system below.*

$$\begin{cases} 5x + 2y + z = 5 \\ 2x + 3y - z = -6 \\ 7x + 4y + 2z = 4 \end{cases}$$

If the first and second equations are added, z will be eliminated.

$$\begin{array}{r} 5x + 2y + z = 5 \\ 2x + 3y - z = -6 \\ \hline 7x + 5y \quad\quad = -1 \end{array}$$

Returning to the original equations, note that if twice the second equation is added to the third equation, z will be eliminated.

$$\begin{array}{r} 2x + 3y - z = -6 \xrightarrow{2} 4x + 6y - 2z = -12 \\ 7x + 4y + 2z = 4 \longrightarrow 7x + 4y + 2z = 4 \\ \hline 11x + 10y \quad\quad = -8 \end{array}$$

We now have two new equations *in the same two unknowns*, x and y: $7x + 5y = -1$ and $11x + 10y = -8$. They can be combined as one system and easily solved for x and y. Specifically, multiply $7x + 5y = -1$ by -2 in order to eliminate y.

$$\begin{cases} 7x + 5y = -1 \xrightarrow{-2} -14x - 10y = 2 \\ 11x + 10y = -8 \longrightarrow 11x + 10y = -8 \end{cases}$$
$$\begin{array}{r} \hline -3x \quad\quad = -6 \\ x \quad\quad = 2 \checkmark \end{array}$$

Next, y can be determined by using 2 for x in the equation $7x + 5y = -1$.

$$7(2) + 5y = -1$$
$$14 + 5y = -1$$
$$5y = -15$$
$$y = -3 \checkmark$$

Finally, return to the original system, select any equation, and determine z by substituting 2 for x and -3 for y. The first equation, $5x + 2y + z = 5$, appears to be the easiest to use because the coefficient of the unknown z is simply 1.

$$5x + 2y + z = 5$$
$$5(2) + 2(-3) + z = 5$$
$$10 - 6 + z = 5$$
$$4 + z = 5$$
$$z = 1 \checkmark$$

The solution, therefore, is

$$x = \quad 2$$
$$y = -3$$
$$z = \quad 1$$

Each linear equation in three variables represents a plane in three dimensions. The solution to the system then can be visualized as the point where the three planes meet. This is an (x, y, z) point in three dimensions. For the preceding example, the planes meet at $(2, -3, 1)$.

EXERCISES 15.6

*1. Solve each system of equations.

(a) $\begin{cases} x + y + z = 6 \\ 2x + 3y - z = 5 \\ 3x + 2y + z = 10 \end{cases}$
(b) $\begin{cases} 5x - y + 2z = 8 \\ 2x + y + 5z = 6 \\ 3x - y + 3z = 7 \end{cases}$

(c) $\begin{cases} 5x + 3y + z = 7 \\ x + 4y - 2z = 8 \\ 3x + 7y - 4z = 18 \end{cases}$
(d) $\begin{cases} x - 4y + 3z = 1 \\ -2x + 3y + 2z = -7 \\ 3x + 2y + 5z = 17 \end{cases}$

(e) $\begin{cases} 7x + 3y + 2z = 9 \\ 3x - y + 3z = 17 \\ 5x + 4y + 4z = 13 \end{cases}$
(f) $\begin{cases} 2x - 3y + 4z = -3 \\ 4x + 5y - 5z = 31 \\ 6x + 2y + 3z = 20 \end{cases}$

(g) $\begin{cases} 2r + 3s + 2t = 3 \\ 3r + 5s + 7t = 19 \\ 6r + 4s + 3t = 12 \end{cases}$
(h) $\begin{cases} 5m + 2n - 3p = -8 \\ 3m - 3n + 5p = 13 \\ -4m + 2n - 2p = -6 \end{cases}$

Chapter 15. REVIEW EXERCISES

*1. Solve each system of equations by the addition method. Check your solution.

(a) $\begin{cases} 6x - 5y = 6 \\ 5x - 3y = -2 \end{cases}$
(b) $\begin{cases} 5a + 7b = 13 \\ 7a + 4b = 24 \end{cases}$

*2. Solve each system of equations by substitution. Check your solution.

(a) $\begin{cases} y = 5 - 2x \\ x = 2 - y \end{cases}$
(b) $\begin{cases} x = 19 - 3y \\ 7x - 4y = 8 \end{cases}$

*3. Solve each system of equations. Use whichever method you prefer— addition or substitution.

(a) $\begin{cases} 7m + 9n = 2 \\ n = 2m + 3 \end{cases}$
(b) $\begin{cases} 4x - 3y = 7 \\ 12x + 5y = -7 \end{cases}$

CHAPTER 16

Inequalities

16.1 INTRODUCTION

Much of the algebra presented thus far has dealt with linear equations—solving a linear equation in one unknown, sketching the graph of a linear equation in two variables, and solving systems of linear equations. The concern has been to find numbers that make one expression equal to another.

In this chapter, we digress to study *inequalities*, statements in which one quantity is greater than or less than another. The inequalities will be linear, just as previous equations have been linear.

We use the symbol $>$ to mean *greater than* and $<$ to mean *less than*.

$5 > 3$ says that five is greater than three.
$3 < 5$ says that three is less than five.

In general, the statements $a > b$ and $b < a$ are equivalent; they mean the same thing.

The symbol $\geq$ means *greater than or equal to*; $\leq$ means *less than or equal to*. Thus

$$5 \geq 3 \quad \text{and} \quad 3 \leq 5$$
$$5 \geq 5 \quad \text{and} \quad 5 \leq 5$$

Margin notes: inequality; greater than; less than

223

The inequality $x > 5$ refers to all numbers x that are greater than 5. It says that x is greater than 5. This can be drawn as

The statement $x \geq 5$ refers to all numbers x that are either greater than or equal to 5. It says that x is greater than or equal to 5. This can be represented as

A rounded parenthesis is used to indicate that the endpoint is not included and a squared bracket indicates that the endpoint is included. Another convention is ∘ and • for endpoint not included and included, respectively.

All numbers between 2 and 6, including 2 and 6, would be all numbers x such that

$$x \geq 2 \quad \text{and} \quad x \leq 6$$

These two inequalities can be combined as $2 \leq x \leq 6$. You can see in it the separate statements

$$2 \leq x \qquad \text{same as } x \geq 2$$

and

$$x \leq 6$$

The statement $2 \leq x \leq 6$ can be read as either

—all x between (and including) 2 and 6, or
—all x greater than or equal to 2 and less than or equal to 6.

This is graphed as

interval notation   The *interval notation* [2, 6] is sometimes used to mean $2 \leq x \leq 6$. If we want to represent $2 < x \leq 6$, we can write (2, 6] or draw

Similarly, $2 < x < 6$ is (2, 6) or

Finally, $2 \leq x < 6$ is [2, 6) or

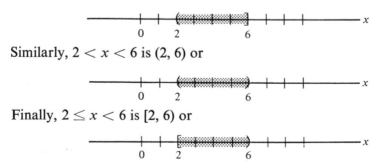

Example 1. *Draw* $-5 < m \leq 3$.

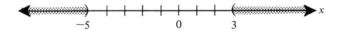

Example 2. *Draw x satisfying* $x > 3$ *or* $x < -5$.

We use *or* rather than *and* because a number cannot be both greater than 3 and less than -5. This situation is clarified by the drawing below.

It is tempting but quite *wrong* to write $x > 3$ or $x < -5$ as

$$3 < x < -5 \quad \text{(wrong)}$$

which reads $x > 3$ *and* $x < -5$, which cannot be.

EXERCISES 16.1

Answers to starred exercises are given in the back of the book.

*1. Write each inequality in interval notation.
    (a) $3 < x < 8$ (b) $5 \leq x \leq 6$ (c) $0 \leq t < 5$
    (d) $1 < y \leq 19$ (e) $10 \leq x \leq 100$ (f) $-6 \leq x < 0$
    (g) $-4 < w \leq -1$ (h) $-10 < x < 10$

*2. Write each interval using an inequality.
    (a) $(2, 5)$ (b) $[2, 5]$ (c) $(1, 7]$ (d) $[3, 11)$
    (e) $[-3, 0]$ (f) $(-2, 4]$ (g) $[-7, -3]$ (h) $(0, 7)$

*3. Make a drawing to represent each interval.
    (a) $4 < x < 7$ (b) $0 \leq y < 7$ (c) $0 \leq x \leq 1$
    (d) $-5 < t \leq 3$ (e) $[4, 12]$ (f) $[5, 8)$
    (g) $(9, 10)$ (h) $(-4, 1]$

16.2 SOLVING INEQUALITIES

Inequalities resemble equations except that they involve $> < \geq \leq$ instead of $=$. The following are examples of linear inequalities.

$$2x > 6$$
$$5x - 1 \leq 19$$
$$4(x - 1) + 3 \geq x - 3$$

In order to solve such inequalities, we use three basic properties. For the most part, these properties resemble those used to solve equations. The key distinction is given in the box below.

1. If the same number is added to both sides of an inequality, the direction of the inequality ($>$ or $<$) remains the same.

   Example. $5 > 2$; so $5 + \underline{8} > 2 + \underline{8}$—that is, $13 > 10$.
   Example. $8 > -4$; so $8 + (\underline{-3}) > -4 + (\underline{-3})$—that is, $5 > -7$.
   Example. $-2 < 4$; so $-2 + (\underline{-6}) < 4 + (\underline{-6})$—that is, $-8 < -2$.
   Example. $-8 < -3$; so $-8 + \underline{17} < -3 + 17$—that is, $9 < 14$.

2. If both sides of an inequality are multiplied or divided by a positive number, the direction of the inequality ($>$ or $<$) remains the same.

   Example. $5 > -2$; so $5 \cdot \underline{3} > -2 \cdot \underline{3}$—that is, $15 > -6$.
   Example. $4 < 7$; so $4 \cdot \underline{2} < 7 \cdot \underline{2}$—that is, $8 < 14$.
   Example. $10 > -4$; so $\dfrac{10}{2} > \dfrac{-4}{2}$—that is, $5 > -2$.
   Example. $6 < 9$; so $\dfrac{6}{3} < \dfrac{9}{3}$—that is, $2 < 3$.

3. *If both sides of an inequality are multiplied or divided by a negative number, the direction of the inequality is reversed. That is, $>$ becomes $<$ and $<$ becomes $>$.*

   Example. $-2 > -3$, so $(-2)(\underline{-5}) < (-3)(\underline{-5})$—that is, $10 < 15$.
   Example. $3 < 4$, so $(3)(\underline{-5}) > (4)(\underline{-5})$—that is, $-15 > -20$.
   Example. $-12 < -4$, so $\dfrac{-12}{-2} > \dfrac{-4}{-2}$—that is, $6 > 2$.
   Example. $24 > 16$, so $\dfrac{24}{-2} < \dfrac{16}{-2}$—that is, $-12 < -8$.

Only this last property can be troublesome. The others are essentially the same as those we used when solving *equations*. But note the one difference again.

> If you multiply or divide both sides of an inequality by a negative number, the direction of the inequality is reversed: $>$ becomes $<$ and $<$ becomes $>$.

As you follow the next three examples, notice how the method of solving these inequalities is similar to the method of solving linear equations.

Example 3. *Solve* $2x + 5 < 9$.

Add -5 to both sides to get the x term alone.

$$
\begin{array}{rcr}
2x + 5 < & & 9 \\
-5 & & -5 \\
\hline
2x & < & 4
\end{array}
$$

Now divide both sides by 2 to get a coefficient of 1 for x.

$$x < 2 \quad \checkmark$$

As a partial check, any number less than 2 should check in $2x + 5 < 9$. For, instance,

$$
\begin{array}{ll}
\underline{x = 0}: \quad 2 \cdot 0 + 5 \overset{?}{<} 9 & \qquad \underline{x = -2}: \quad 2(-2) + 5 \overset{?}{<} 9 \\
\qquad\qquad\quad 0 + 5 \overset{?}{<} 9 & \qquad\qquad\qquad\quad -4 + 5 \overset{?}{<} 9 \\
\qquad\qquad\qquad\quad 5 < 9 \quad \checkmark & \qquad\qquad\qquad\qquad\quad 1 < 9 \quad \checkmark
\end{array}
$$

Also, any number greater than or equal to 2 should not check in $2x + 5 < 9$. For instance,

$$
\begin{array}{l}
\underline{x = 3}: \quad 2 \cdot 3 + 5 \overset{?}{<} 9 \\
\qquad\qquad\quad 6 + 5 \overset{?}{<} 9 \\
\qquad\qquad\qquad\quad 11 \overset{?}{<} 9 \quad \textit{No!}
\end{array}
$$

Example 4. *Solve* $5x + 2 - 9x \geq 17$.

If we combine like terms on the left side, we obtain

$$-4x + 2 \geq 17$$

Add -2 to both sides to get the x term alone.

$$-4x \geq 15$$

Divide both sides by -4 in order to get a coefficient of 1 for x.

$$x \leq -\frac{15}{4} \qquad \left\{ \begin{array}{l} \text{Division by } -4, \text{ a } \textit{negative number,} \\ \text{reverses the direction of the inequality;} \\ \geq \text{ becomes } \leq. \end{array} \right.$$

Example 5. *Solve* $3(x + 1) - 7 > 8x + 9$.

Distribute the 3 to eliminate parentheses.

$$3x + 3 - 7 > 8x + 9$$

Combine the like terms on the left side.

$$3x - 4 > 8x + 9$$

To get the x term alone, first add $+4$ to both sides.

$$3x > 8x + 13$$

Then and $-8x$ to both sides.

$$-5x > 13$$

Obtain a coefficient of 1 for x.

$$x < -\frac{13}{5} \quad \begin{cases} \text{after division of both sides by } -5 \\ \text{and consequent change of } > \text{ to } < \end{cases}$$

EXERCISES 16.2

*1. Solve each inequality.

(a) $x + 1 < 6$

(b) $t + 15 > -9$

(c) $5x + 1 > 7$

(d) $3x - 4 \leq -8$

(e) $9y + 5 < 5y - 2$

(f) $-8n - 9 < 6n$

(g) $5x + 3 + 2x < -4$

(h) $2x - 7 - 9x \leq 6$

(i) $4(x - 2) - 3 > x$

(j) $4x + 2(3 - x) \geq 6x + 7$

2. Solve each inequality.

*(a) $y - 7 > -1$

(b) $3m - 6 \geq 0$

*(c) $-2x + 3 < 6$

(d) $2x - 9 > -3$

*(e) $3x + 1 - x \geq 13$

(f) $2m + 6 > 7m + 4$

*(g) $7t - 4 - 10t < 12$

(h) $3(x + 1) + 5 < 5$

*(i) $2(x + 3) - 6 \geq 9$

(j) $2(3 + 2x) - 4 > 7x$

*(k) $4x + 3 + 2x < 5 - 2x$

(l) $5x + 1 < 2(3x - 4) + 3$

16.3 GRAPHING INEQUALITIES (Optional)

Example 6. *Sketch the graph of the inequality $x + y < 4$.*

The points on the line $x + y = 4$ are points where $x + y$ *equals* 4. So perhaps $x + y$ is less than 4 either above the line or below the line but certainly not on the line. Let's graph the line $x + y = 4$ and consider the possibilities.

$$x + y = 4$$

x	y
0	4
4	0

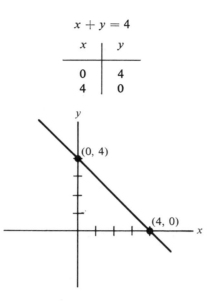

Points *above* the line, such as (6, 0), (0, 5), (4, 3), and (−1, 6), do not satisfy $x + y < 4$, since for each of these points $x + y > 4$. Observe.

x	y	x + y	x + y < 4
6	0	6	no
0	5	5	no
4	3	7	no
−1	6	5	no

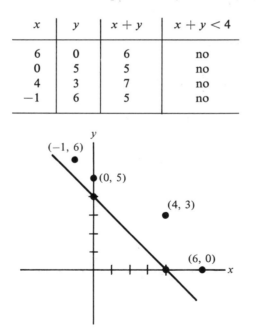

On the other hand, the following points, all of which are below the line $x + y = 4$, satisfy the inequality $x + y < 4$. Accordingly, we shade the area below the line, the desired area. The line itself is drawn dashed because its points do not satisfy $x + y < 4$. If the inequality were $x + y \leq 4$ instead, then the line would be drawn solid in order to suggest that points on the line are to be included.

x	y	$x + y$	$x + y < 4$
0	0	0	yes
1	0	1	yes
0	3	3	yes
-2	2	0	yes
0	-1	-1	yes
1	-2	-1	yes
-3	5	2	yes

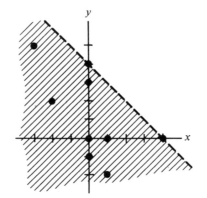

Example 7. *Find graphically the solution to the system below.*

$$\begin{cases} y \geq x + 2 \\ 3x + y \geq 6 \end{cases}$$

First, obtain two points for each line $y = x + 2$ and $3x + y = 6$. Graph them and shade each inequality. The solution, then, is all points that satisfy both inequalities—that is, all points that are shaded twice.

$y = x + 2$	
x	y
0	2
-2	0

$3x + y = 6$	
x	y
0	6
2	0

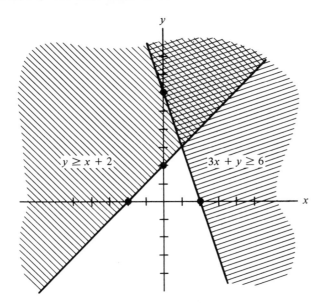

Thus the solution is all points in the V-shaped region above the lines, including the portions of the lines that make up the outline of the V-shaped region.

Example 8. *Find graphically the solution of the system below.*

$$\begin{cases} y - 2x - 5 > 0 \\ 1 + 2x \geq y \end{cases}$$

Points for the lines:

$y - 2x - 5 = 0$			$1 + 2x = y$	
x	y		x	y
0	5		0	1
−1	3		1	3

The graph is shown next. The graph shows that the two inequalities have no points in common, and thus there is no solution to this system of linear inequalities.

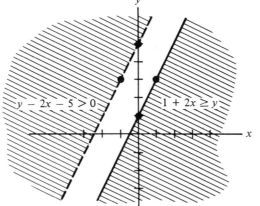

EXERCISES 16.3

1. Graph each inequality in the xy plane.
  *(a) $y > x + 1$ (b) $y > x - 5$
  *(c) $y \le 2x + 1$ (d) $y < 3x + 5$
  *(e) $x + y \le 6$ (f) $x + y > 6$
  *(g) $x - y < 4$ (h) $x - y > 7$
  *(i) $x \ge 2y - 6$ (j) $x > y + 2$
  *(k) $2x + 3y < 12$ (l) $2x - 3y \le 12$
  *(m) $x \ge 5$ (n) $x < -3$
  *(o) $y \ge 0$ (p) $y < 5$

2. Solve each system of linear inequalities by graphing.
  *(a) $\begin{cases} y > x + 1 \\ y < 2x - 2 \end{cases}$ (b) $\begin{cases} x + y \ge 7 \\ y - x \le 5 \end{cases}$

  *(c) $\begin{cases} x - y \ge 0 \\ 2x - y \le 3 \end{cases}$ (d) $\begin{cases} y \le 4x + 1 \\ 4x - y \le -4 \end{cases}$

  *(e) $\begin{cases} y \ge 4 \\ x \le -3 \end{cases}$ (f) $\begin{cases} 2x + y < 6 \\ 3x - 2y \ge 6 \end{cases}$

  *(g) $\begin{cases} y \ge 1 - 2x \\ y \le 2 \end{cases}$ (h) $\begin{cases} x \ge 0 \\ y \ge 0 \end{cases}$

  *(i) $\begin{cases} 2x + 3y \le 6 \\ 3x + 2y \ge 6 \end{cases}$ (j) $\begin{cases} y - x \le -4 \\ x - y \ge -2 \end{cases}$

3. Represent each situation by one or more inequalities. If only one variable is involved, then call it x. If two variables are needed, use x and y.
  *(a) Employees work at least 40 hours a week.
  *(b) The temperature never gets as hot as 100°.
  (c) The temperature never gets as cold as 0°.
  (d) The temperature never gets as hot as 90° or as cold as 5°.
  *(e) The supply is always greater than the demand.
  *(f) Employees work at least 40 hours but no more than 60 hours per week.
  (g) The stock number of an auto part cannot be negative.
  *(h) John works at least 20 hours and his boss works at least 40 hours. The total number of hours both work cannot exceed 75 hours.
  (i) The cost of 10 pencils and 15 pens is at least $3.50. The cost of 15 pencils and 5 pens is at most $3.

Chapter 16. REVIEW EXERCISES

*1. Write each inequality in interval notation.
  (a) $-4 < x < 6$ (b) $-5 \le t < 0$ (c) $78 \le y \le 100$
  (d) $-9 < m \le 7$

*2. Write each interval using an inequality.

   (a) $[-3, 10]$ (b) $(2, 11]$

*3. Solve each inequality.

   (a) $5x - 8 < 17$ (b) $-4x - 7 > 19$

   (c) $8x - 8 \leq 3x - 5$ (d) $9 - 7x < x + 2$

   (e) $7(1 + 4x) > -6$ (f) $8(x + 1) - 3 \geq 2x - 7$

*4. Graph each inequality in the xy plane.

   (a) $y < 3x - 2$ (b) $x < y + 5$ (c) $2x + 5y > 10$

   (d) $x \leq 7$

*5. Solve each system of linear inequalities by graphing.

   (a) $\begin{cases} y > 2x + 1 \\ x + y < 7 \end{cases}$ (b) $\begin{cases} y \geq -x + 2 \\ y \geq -5 \end{cases}$

_____ Factoring

17.1 INTRODUCTION TO FACTORING

In Chapter 11 we began the study of algebra by using the distributive property to combine like terms. Recall the first example,

$$3x + 8x = (3 + 8)x = 11x$$

Both terms, $3x$ and $8x$, contain the same factor x. Using the distributive property to take the common factor x and place it outside parentheses **factoring** is an example of the process known as *factoring*.

Example 1. *Factor $3x + mx$.*

Here x is a factor common to both $3x$ and mx. So x can be factored out according to the distributive property.

$$3x + mx = (3 + m)x \quad \checkmark$$

The factoring can be checked by applying the distributive property to multiply out the factored form. Here

$$(3 + m)x = 3 \cdot x + m \cdot x = 3x + mx$$

And $3x + mx$ is indeed the expression we started with.

Example 2. *Factor ay + by.*

Here y is common to both terms, so it can be factored out.

$$ay + by = (a + b)y \ \checkmark$$

Example 3. *Factor the expression 5mx + 10my.*

When 10 is considered as $5 \cdot 2$, you can see that $5 \cdot m$, or $5m$, is common to both terms and can therefore be factored out.

$$5mx + 10my = \underline{5} \cdot \underline{m} \cdot x + \underline{5} \cdot 2 \cdot \underline{m} \cdot y$$
$$= \underline{5m} \cdot x + \underline{5m} \cdot 2y$$
$$= \underline{5m}(x + 2y) \ \checkmark$$

Note: The direction "Factor" means *factor completely.* In Example 3, you should not factor out *only* the 5 or *only* the m. You should factor out *both* 5 and m. In this type of factoring, where a monomial factor common to all terms is factored out, the monomial factor should be the greatest factor common to all terms.

Example 4. *Factor the expression $15x^2y + 10xy^2$.*

You can see that the factor $5xy$ is present in both $15x^2y$ and $10xy^2$. Thus

$$15x^2y + 10xy^2 = 5 \cdot 3 \cdot x \cdot x \cdot y + 5 \cdot 2 \cdot x \cdot y \cdot y$$
$$= \underline{5xy} \cdot 3x + \underline{5xy} \cdot 2y$$
$$= \underline{5xy}(3x + 2y) \ \checkmark$$

Example 5. *Factor the expression $x^2y - xy$.*

Here xy is common to both terms. The factoring is

$$xy(x - 1) \ \checkmark$$

The check: $\quad xy(x - 1) = xy \cdot x - xy \cdot 1 = x^2y - xy$

Example 6. *Factor the expression $x^2 + 3x$.*

Notice that x is common to both terms, so it can be factored out. Thus

$$x^2 + 3x = x(x + 3) \ \checkmark$$

Again, you can check this result by multiplying out the factored form and seeing that it does indeed produce the original expression $x^2 + 3x$.

Example 7. *Factor the expression $m^6 + m^2$.*

The greatest factor common to both terms is m^2. Thus

$$m^6 + m^2 = \underline{m^2} \cdot m^4 + \underline{m^2} \cdot 1$$
$$= m^2(m^4 + 1) \ \checkmark$$

Example 8. *Factor the expression $12mn^2p - 16m^2n$.*

The greatest common factor is $4mn$. A common mistake is to look too quickly at the expression and see only a 2 in the 12 and 16. There is a 4 common to both 12 and 16. Thus

$$12mn^2p - 16m^2n = \underline{4mn} \cdot 3np - \underline{4mn} \cdot 4m$$
$$= 4mn(3np - 4m) \quad \checkmark$$

Example 9. *Factor the expression $4abc + 2acd + 3ace$.*

This expression consists of three terms, so factor out whatever is common to all three terms, namely, ac.

$$4abc + 2acd + 3ace = \underline{ac} \cdot 4b + \underline{ac} \cdot 2d + \underline{ac} \cdot 3e$$
$$= ac(4b + 2d + 3e) \quad \checkmark$$

Notice that there is a factor of 2 in each of the first two terms. But since there is no factor of 2 in the third term, no 2 can be factored out. A factor must appear in *every term* of an expression in order to be factored out of the expression.

Example 10. *Factor the expression $7mn + 13uv$.*

There is no factor common to both terms of this expression, so the expression cannot be factored.

EXERCISES 17.1

Answers to starred exercises are given in the back of the book.

*1. Factor each expression.

(a) $7x + nx$ (b) $mx + nx$ (c) $pr + qr$
(d) $3y - ty$ (e) $5wx + 10wy$ (f) $3xy + 12xz$
(g) $12xy - 18yz$ (h) $15mn + 20mnp$ (i) $x^2 + 6x$
(j) $y^2 - 8y$ (k) $a^2 + a$ (l) $y^2 - y$
(m) $5m - m^2$ (n) $3x^2 - x$ (o) $20m^2n + 15mn^2$
(p) $18x^2y^2 - 24xy^2$ (q) $6x + 7x^2$ (r) $17t^2 - 6t$
(s) $3x + 3$ (t) $10 - 10y$

2. Factor each expression, if possible.

*(a) $10x + 5$ (b) $6 - 3w$
*(c) $8 + 40x^2$ (d) $5x^2 + 30$
*(e) $x^5 + x^3$ (f) $y^7 + y^2$
*(g) $a^2b - 2abc^4$ (h) $16w^2x^2 - 24wx^2$
*(i) $ab + cd$ (j) $ab + bc$
*(k) $pc + qc + rc$ (l) $17xy - 33wz + 14wx$

*(m) $m^3 + m^2 + m$ (n) $x^4 + x^3 + x^2$

*(o) $xyz + 3yz + wyz$ (p) $wxy + wxz - xyz$

*(q) $100m^2n + 50m^2n^2 + 20m^2$ *(r) $5ab^2 + 10a^2b + a^2b^2$

*(s) $15uv^2 - 6u^2 + 12uv$ *(t) $mnp - npq + mqr$

*(u) $m^2x + mx^2 + mx$ *(v) $15x^2y^2 + 5x^2 + 3y^2$

17.2 FACTORING QUADRATIC EXPRESSIONS

A *quadratic* expression is one that contains the second power of the **quadratic** variable and no higher power. For example, the following expressions are quadratic in x.

$$x^2 + 5x - 4$$
$$3x^2 - 7x + 11$$
$$x^2 + 3x$$
$$x^2 - 9$$

In Chapter 11 you learned how to multiply binomials. We now need to examine a special case, the one in which each binomial is the sum of a particular variable (say x) and a constant. Here are some examples.

$$(x + 2)(x + 4)$$
$$(x - 3)(x + 5)$$
$$(x + 1)(x - 9)$$
$$(y + 6)(y + 7)$$

Observe the multiplication.

$$(x \underline{+ 2})(x \underline{+ 4}) = x^2 + 4x + 2x + 8$$
$$= x^2 \underline{+ 6x} \underline{+ 8} \quad \checkmark$$

In the final result the coefficient of x (the $+6$) is the *sum* of the $+2$ and $+4$ from the original factors $(x + 2)$ and $(x + 4)$. Note also that the constant $+8$ is the *product* of the $+2$ and $+4$ from those original factors.

Let's see another example.

$$(x \underline{+ 6})(x \underline{- 3}) = x^2 - 3x + 6x - 18$$
$$= x^2 \underline{+ 3x} \underline{- 18} \quad \checkmark$$

The sum of $+6$ and -3 is $+3$; and that is the coefficient of x. The product of $+6$ and -3 is -18; and that is the constant term.

Example 11. *Determine the product of $(x + 2)$ and $(x + 5)$ by using the shortcut suggested above.*

$$(x + 2)(x + 5) = x^2 + 7x + 10$$

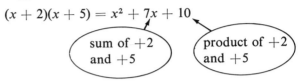

Example 12. *Find the product: $(x - 9)(x - 4)$.*

$$(x - 9)(x - 4) = x^2 - 13x + 36$$

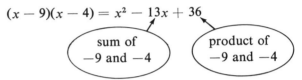

The examples of this section are all of the form

$$(\underline{x} +\quad)(\underline{x} +\quad)$$

that is,

$$(\underline{1}x +\quad)(\underline{1}x +\quad)$$

The shortcut that we have used applies only when the coefficient of x is one.

> *If the coefficient of x is not one, then the shortcut does not apply.*

A very important problem in algebra is to *reverse* this shortcut multiplication process, that is, to *factor* a quadratic expression such as $x^2 + 7x + 12$ into $(x + 3)(x + 4)$ by using what you know about sum and product from the shortcut multiplication process. If you examine $(x + 3)(x + 4)$ and $x^2 + 7x + 12$, you realize that

... x^2 comes from $x \cdot x$.
... $+7$ comes from the *sum* $+3 + 4$.
... $+12$ comes from the *product* $(+3)(+4)$.

So when given a quadratic expression such as $x^2 + 7x + 12$ to factor, the thinking might go as follows:

1. The factors are of the form $(x \underline{\quad})(x \underline{\quad})$.
2. The product is $+12$. Using integers, this could be any of the

following pairs:

$$(+12)(+1)$$
$$(-12)(-1)$$
$$(+6)(+2)$$
$$(-6)(-2)$$
$$(+4)(+3)$$
$$(-4)(-3)$$

3. *But* the sum must be $+7$, and only $(+4)$ and $(+3)$ add up to $+7$.

$$(+12)(+1)$$
$$(-12)(-1)$$
$$(+6)(+2)$$
$$(-6)(-2)$$
$$(+4)(+3) \quad \longleftarrow$$
$$(-4)(-3)$$

So the factors are $(x + 4)$ and $(x + 3)$. Therefore $x^2 + 7x + 12 = (x + 4)(x + 3)$.

As a check, multiply out the $(x + 4)(x + 3)$ to show that it does in fact equal $x^2 + 7x + 12$.

Example 13. *Factor the expression* $x^2 + 8x + 7$.

Again the form is $(x\_\_\_)(x\_\_\_)$. From the expression you can see that the product is $+7$. Using integers, this is either $(+7)(+1)$ or $(-7)(-1)$. Since the sum must be $+8$, the choice is $(+7)$ and $(+1)$. Thus

$$x^2 + 8x + 7 = (x + 7)(x + 1) \quad \checkmark$$

The factorization $(x + 7)(x + 1)$ can also be written $(x + 1)(x + 7)$.

Example 14. *Factor the expression* $x^2 - 5x + 4$.

The form is $(x\_\_\_)(x\_\_\_)$.
The product is $+4$. This could be any of the following.

$$(+4)(+1)$$
$$(-4)(-1)$$
$$(+2)(+2)$$
$$(-2)(-2)$$

The choice is $(-4)(-1)$, since the sum must be -5 in order to produce the $-5x$ term. So

$$x^2 - 5x + 4 = (x - 4)(x - 1) \quad \checkmark$$

Example 15. *Factor the expression* $x^2 - 12x + 20$.

The form is $(x\underline{\quad})(x\underline{\quad})$.
The product, $+20$, could come from any of the following.

$$(+20)(+1)$$
$$(-20)(-1)$$
$$(+10)(+2)$$
$$(-10)(-2)$$
$$(+5)(+4)$$
$$(-5)(-4)$$

The choice is (-10) and (-2), since the sum is -12. Thus

$$x^2 - 12x + 20 = (x - 10)(x - 2) \quad \checkmark$$

Example 16. *Factor the expression* $x^2 + 3x - 10$.

The form is $(x\underline{\quad})(x\underline{\quad})$. The product is -10. This could be any of the following.

$$(-10)(+1)$$
$$(+10)(-1)$$
$$(-5)(+2)$$
$$(+5)(-2)$$

The choice is the last one, $(+5)(-2)$, since the sum must be $+3$ (from $+3x$). Thus

$$x^2 + 3x - 10 = (x + 5)(x - 2) \quad \checkmark$$

Note: $(x + 5)(x - 2)$ can be written $(x - 2)(x + 5)$, using the commutative property, but it *cannot* be written $(x + 2)(x - 5)$ or $(x - 5)(x + 2)$. These last two are not commuted forms of $(x + 5)(x - 2)$; they are completely different—and they do not check. Multiply the factors to see that each multiplies to give $x^2 - 3x - 10$, which is not the original expression.

Example 17. *Factor the expression* $n^2 - n - 12$.

The form is $(n\underline{\quad})(n\underline{\quad})$. The product is -12. Consider

$$(+12)(-1)$$
$$(-12)(+1)$$
$$(-6)(+2)$$
$$(+6)(-2)$$
$$(+4)(-3)$$
$$(-4)(+3)$$

The choice is $(-4)(+3)$, since the sum must be -1 (because of the $-n$ term, where $-n$ means $-1n$). Thus

$$n^2 - n - 12 = (n - 4)(n + 3) \quad \checkmark$$

Example 18. *Factor the expression $x^2 + 6x + 12$.*

The form is $(x\underline{})(x\underline{})$. The product is 12. The choices:

$$(+12)(+1)$$
$$(-12)(-1)$$
$$(+6)(+2)$$
$$(-6)(-2)$$
$$(+4)(+3)$$
$$(-4)(-3)$$

But *none* of these pairs gives a sum of $+6$. This means the expression $x^2 + 6x + 12$ *cannot be factored* by the methods that we have studied so far.

Example 19. *Factor the expression $x^2 + 7x$.*

This expression may have caught you off guard, but it is a simple one to factor. It is of the type factored in Section 17.1, before we even began to study quadratic expressions. There are only two terms in $x^2 + 7x$, and there is an x common to both terms. So factor out the x. The result is

$$x^2 + 7x = x(x + 7) \quad \checkmark$$

Example 20. *Factor the expression $x^2 - 9$.*

The form is $(x\underline{})(x\underline{})$. The product is -9. The choices:

$$(-9)(+1)$$
$$(+9)(-1)$$
$$(+3)(-3)$$

Note that $x^2 - 9$ is the same as $x^2 + 0x - 9$. There are *zero* x's; that is why the x term does not appear in $x^2 - 9$. Thus the sum (which is the coefficient of x) is 0. So $(+3)(-3)$ is the proper choice, since $(+3) + (-3) = 0$. And therefore

$$x^2 - 9 = (x + 3)(x - 3) \quad \checkmark$$

Consider the following similar factorizations.

$$x^2 - 4 \longrightarrow (x + 2)(x - 2)$$
$$x^2 - 25 \longrightarrow (x + 5)(x - 5)$$
$$y^2 - 16 \longrightarrow (y + 4)(y - 4)$$

In each case, the factors involve square roots. Notice the pattern for such factoring. These are examples of factoring the *difference of two squares*. Consider the examples below.

$x^2 - 9$	(*x* squared, 3 squared)	$\longrightarrow (x + 3)(x - 3)$
$m^2 - 4$	(*m* squared, 2 squared)	$\longrightarrow (m + 2)(m - 2)$
$x^2 - 25$	(*x* squared, 5 squared)	$\longrightarrow (x + 5)(x - 5)$

difference of two squares

$$49 - x^2 \qquad \text{(7 squared, } x \text{ squared)} \qquad \longrightarrow (7 + x)(7 - x)$$
$$y^2 - 16 \qquad \text{(} y \text{ squared, 4 squared)} \qquad \longrightarrow (y + 4)(y - 4)$$
$$x^2 - a^2 \qquad \text{(} x \text{ squared, } a \text{ squared)} \qquad \longrightarrow (x + a)(x - a)$$
$$x^2 - 9y^2 \qquad \text{(} x \text{ squared, } 3y \text{ squared)} \qquad \longrightarrow (x + 3y)(x - 3y)$$
$$4m^2 - 25n^2 \qquad \text{(} 2m \text{ squared, } 5n \text{ squared)} \longrightarrow (2m + 5n)(2m - 5n)$$

Example 21. *Factor the expression* $16x^2 - y^2$.

The expression $16x^2 - y^2$ is the difference of two squares, since $16x^2$ is the square of $4x$, y^2 is the square of y, and the connecting sign is a minus. Thus
$$16x^2 - y^2 = (4x + y)(4x - y) \quad \checkmark$$

Example 22. *Factor the expression* $x^2 + y^2$.

Although this expression contains two squares, it is not the difference of two squares, rather it is the sum of two squares. The sum of two squares cannot be factored.

EXERCISES 17.2

1. Determine the product by using the shortcut explained in Examples 11 and 12.

   *(a) $(x + 2)(x + 3)$ *(b) $(x + 5)(x + 9)$
   *(c) $(m + 6)(m + 4)$ *(d) $(m + 1)(m + 1)$
   *(e) $(x - 5)(x - 2)$ (f) $(x - 7)(x - 1)$
   *(g) $(m - 6)(m - 4)$ (h) $(m - 2)(m - 3)$
   *(i) $(y - 3)(y + 5)$ (j) $(y - 6)(y + 2)$
   *(k) $(x - 7)(x + 7)$ *(l) $(x + 5)(x - 5)$
   *(m) $(m + 3)(m - 8)$ (n) $(m + 6)(m - 5)$
   *(o) $(y + 8)(y - 3)$ (p) $(y + 3)(y - 3)$
   *(q) $(x - 9)(x + 2)$ (r) $(x + 5)(x + 5)$
   *(s) $(m - 2)(m - 10)$ (t) $(m + 9)(m - 3)$

*2. Factor each expression.

   (a) $x^2 + 5x + 6$ (b) $x^2 + 6x + 5$
   (c) $x^2 + 4x + 4$ (d) $x^2 - 6x + 5$
   (e) $t^2 - 6t - 7$ (f) $x^2 - 12x + 20$
   (g) $m^2 - 5m - 6$ (h) $x^2 + 3x$
   (i) $x^2 + x - 12$ (j) $a^2 - 9a$
   (k) $y^2 - 4y - 12$ (l) $m^2 + m - 30$
   (m) $x^2 + x - 2$ (n) $y^2 + 5y - 24$

3. Factor each expression or indicate that it cannot be factored.

*(a) $x^2 - 9x + 20$ (b) $x^2 - 7x - 18$

*(c) $m^2 - 5m + 16$ (d) $y^2 - 14y + 24$

*(e) $x^2 - 2x + 1$ (f) $x^2 - 15x - 24$

*(g) $x^2 - 36$ (h) $n^2 - 64$

*(i) $x^2 - 81y^2$ (j) $x^2 - m^2$

*(k) $m^2 + 2m - 18$ (l) $m^2 - 5m + 14$

*(m) $x^2 + 10x + 25$ (n) $y^2 + 7y + 10$

*(o) $25a^2 + 4b^2$ (p) $16 - m^2$

*(q) $25 - 4t^2$ (r) $9x^2 - 25y^2$

*(s) $x^2 - 9x$ (t) $y^2 - 25y$

*(u) $x^2 - 14x + 45$ (v) $x^2 + 17x + 16$

*(w) $m^2 - 8m$ (x) $m^2 + 13m - 14$

*(y) $100y^2 - 49x^2$ (z) $x^2 - 13x + 36$

17.3 MORE QUADRATIC FACTORING

Sometimes the methods of Sections 17.1 and 17.2 must be combined to completely factor a quadratic expression. When applicable,

---

First: Factor out the greatest common monomial factor (as in Section 17.1).

Second: Factor the remaining quadratic expression (as in Section 17.2).

Example 23. *Factor the expression $2t^2 - 4t - 16$.*

Before attempting to apply the quadratic factoring techniques of the preceding section, you should examine the expression for common monomial factors. Notice the factor of 2 in each term of $2t^2 - 4t - 16$. As it happens, 2 is the greatest common monomial factor. Thus

$$2t^2 - 4t - 16 = 2(t^2 - 2t - 8)$$

Now proceed to factor the quadratic expression $t^2 - 2t - 8$. The result is

$$= 2(t - 4)(t + 2) \quad \checkmark$$

Example 24. *Factor the expression $3x^2 + 9x - 12$.*

$$3x^2 + 9x - 12 = 3(x^2 + 3x - 4)$$
$$= 3(x + 4)(x - 1) \quad \checkmark$$

Example 25. *Factor the expression* $5y^2 + 35y + 10$.

$$5y^2 + 35y + 10 = 5(y^2 + 7y + 2)$$

The quadratic expression $y^2 + 7y + 2$ cannot be factored, so the factored form $5(y^2 + 7y + 2)$ is the final result.

A more involved problem arises when the coefficient of the x^2 term is not 1 *and* it cannot be factored out. Here are a few examples of quadratic expressions of that type.

$$2x^2 + 5x + 3$$
$$3x^2 - 11x + 6$$
$$4x^2 - 13x + 9$$

Consider the expression $2x^2 + 5x + 3$. The factors take the form

$$(2x \quad)(x \quad)$$

since you need $2x \cdot x$ to get the $2x^2$ term of $2x^2 + 5x + 3$. The product is 3, and it can only come from $1 \cdot 3$. The sum is $+5$ (or $+5x$), and this seems impossible to obtain using 3 and 1 until you examine the form

$$(2x \ \_)(x \ \_)$$

and realize that part of your sum is *twice* (from the $2x$) the number placed after x. There are two possibilities:

$$(2x + 1)(x + 3) \quad \text{and} \quad (2x + 3)(x + 1)$$

The sum from $(2x + 1)(x + 3)$ is $7x$.

$$x + 6x$$

The sum from $(2x + 3)(x + 1)$ is $5x$, as needed.

$$3x + 2x$$

So $(2x + 3)(x + 1)$ is the correct factorization.

Example 26. *Factor the expression* $3y^2 - 11y + 6$.

The $3y^2$ comes from $3y \cdot y$, so we will proceed with the form $(3y\_\_\_)(y\_\_\_)$. The product $+6$ suggests $(+6)(+1)$, $(-6)(-1)$, $(+3)(+2)$, and $(-3)(-2)$. Considering that the coefficient of y is a *negative* number, there is no way that $(+6)(+1)$ or $(+3)(+2)$ could work. So the possible factors of $3y^2 - 11y + 6$ are:

$$(3y - 6)(y - 1)$$
$$(3y - 1)(y - 6)$$
$$(3y - 3)(y - 2)$$
$$(3y - 2)(y - 3)$$

Only the last one multiplies out to yield $3y^2 - 11y + 6$. Thus

$$3y^2 - 11y + 6 = (3y - 2)(y - 3) \quad \checkmark$$

Example 27. *Factor the expression* $6x^2 - x - 12$.

The $6x^2$ could have come from either $6x \cdot x$ *or* $3x \cdot 2x$. This means the form of the factoring could be either $(6x\_\_\_)(x\_\_\_)$ or $(3x\_\_\_)(2x\_\_\_)$. Also, the product -12 could come from any of these: $(-12)(+1)$, $(+12)(-1)$, $(-6)(+2)$, $(+6)(-2)$, $(-3)(+4)$, or $(+3)(-4)$. So there are a lot of cases to try, unless you get lucky and find the right one soon. (Actually, there are 24 cases here.) After several tries, the correct factoring is produced, namely,

$$6x^2 - x - 12 = (2x - 3)(3x + 4) \quad \checkmark$$

EXERCISES 17.3

1. Factor each quadratic expression completely. Begin by factoring out the greatest common monomial factor.
   *(a) $3m^2 + 15m + 18$
   (b) $2x^2 + 6x + 4$
   *(c) $2x^2 - 18x - 20$
   (d) $2x^2 - 14x + 24$
   *(e) $3x^2 - 3$
   (f) $4x^2 - 144$
   *(g) $5x^2 - 30x + 45$
   (h) $3x^2 - 24x + 45$
   *(i) $7x^2 - 21x$
   (j) $5x^2 + 140x$

2. Factor each quadratic expression.
   *(a) $2x^2 + 13x + 15$
   *(b) $3x^2 + 5x + 2$
   *(c) $3x^2 - 5x - 2$
   *(d) $5y^2 - 17y + 6$
   *(e) $5n^2 + 13n - 6$
   (f) $7x^2 + 3x - 4$
   *(g) $6x^2 - 5x - 6$
   (h) $8x^2 - 2x - 3$

3. Factor each quadratic expression, if possible.
   *(a) $5y^2 + 22y + 8$
   *(b) $2m^2 + 5m - 12$
   *(c) $6x^2 + 61x + 10$
   *(d) $5x^2 + 2x + 1$
   *(e) $2y^2 - 12y + 18$
   (f) $12x^2 - 10x - 8$
   *(g) $3m^2 - 75$
   (h) $4x^2 + 8x + 3$
   *(i) $6x^2 - 16x + 10$
   (j) $3x^2 - 15x - 42$
   *(k) $4x^2 - 13x + 9$
   (l) $6x^2 + 9x - 6$
   *(m) $2x^2 + 2$
   (n) $5x^2 - 2x - 4$

17.4 SOLVING QUADRATIC EQUATIONS

Let m and n represent numbers. If $m \cdot n = 0$, then at least one of the numbers (m or n) must be zero. Also, if either of the numbers (m or n) is zero, then the product $m \cdot n$ is zero.

Suppose we know that $(x + 3)(x + 2) = 0$. The product is zero if either $(x + 3)$ or $(x + 2)$ is zero. Furthermore, if $(x + 3) = 0$, then $x = -3$, as shown next.

$$
\begin{array}{rcc}
x + 3 = & 0 \\
-3 & -3 \\
\hline
x & = -3 \checkmark
\end{array}
$$

root   Similarly, if $(x + 2) = 0$, then $x = -2$. Therefore -3 and -2 are the solutions (or *roots*) of $(x + 3)(x + 2) = 0$, since substitution of either -3 or -2 for x makes $(x + 3)(x + 2)$ equal to zero. You can check this statement by actually making the substitution. First, substitute -3 for x in $(x + 3)(x + 2)$.

$$
\begin{aligned}
(\underline{x} + 3)(\underline{x} + 2) &= 0 \\
(\underline{-3} + 3)(\underline{-3} + 2) &\overset{?}{=} 0 \\
(0)(-1) &\overset{?}{=} 0 \\
0 &= 0 \quad \checkmark
\end{aligned}
$$

Next, substitute -2 for x in $(x + 3)(x + 2) = 0$.

$$
\begin{aligned}
(\underline{x} + 3)(\underline{x} + 2) &= 0 \\
(\underline{-2} + 3)(\underline{-2} + 2) &\overset{?}{=} 0 \\
(1)(0) &\overset{?}{=} 0 \\
0 &= 0 \quad \checkmark
\end{aligned}
$$

Both solutions check.

Example 28. *Solve the equation* $(n - 5)(n + 1) = 0$.

Setting $(n - 5)$ equal to zero gives one solution. Setting $(n + 1)$ equal to zero gives the other.

$$
\begin{array}{c|c}
n - 5 = 0 & n + 1 = 0 \\
\text{yields } n = 5 \ \checkmark & \text{yields } n = -1 \ \checkmark
\end{array}
$$

Example 29. *Solve the equation* $x^2 - 7x + 12 = 0$.

$x^2 - 7x + 12$ is easily factored. So $x^2 - 7x + 12 = 0$ can be written as $(x - 3)(x - 4) = 0$. Thus

$$x - 3 = 0 \quad \text{or} \quad x = 3 \quad \checkmark$$

and
$$x - 4 = 0 \quad \text{or} \quad x = 4 \quad \checkmark$$

The solutions are 3 and 4.

The solutions should be checked in the original, $x^2 - 7x + 12 = 0$, rather than in the factored form, since an error could have been made in the factoring. To check 3 as a solution:

$$x^2 - 7x + 12 = 0$$
$$(3)^2 - 7(\underline{3}) + 12 \overset{?}{=} 0$$
$$9 - 21 + 12 \overset{?}{=} 0$$
$$0 = 0 \quad \checkmark$$

It is left for you to check 4 as a solution.

Example 30. *Solve the equation* $x^2 - 5x = 0$.

Because there is no constant term, we can factor out an x.

$$x^2 - 5x = 0$$

becomes
$$(x)(x - 5) = 0$$

Setting each factor equal to 0 yields

$$x = 0 \quad \checkmark \quad \bigg| \quad x - 5 = 0$$
$$x = 5 \quad \checkmark$$

The solutions are 0 and 5.

The examples we have seen suggest that *quadratic equations always have two solutions.* You should realize that there will always be two factors for the quadratic expression and hence two solutions. It is possible, however, for both solutions to be the same. This situation will occur whenever both factors are the same. Here is an example.

Example 31. *Solve the equation* $m^2 - 6m + 9 = 0$.

$$m^2 - 6m + 9 = 0$$
$$(m - 3)(m - 3) = 0$$
$$m - 3 = 0 \quad \bigg| \quad m - 3 = 0$$
$$m = 3 \quad \checkmark \quad \bigg| \quad m = 3 \quad \checkmark$$

The solutions are $m = 3$ and $m = 3$. We can say that 3 is a "double root" of the equation $m^2 - 6m + 9 = 0$.

Example 32. *Solve the equation $y^2 + 5y + 1 = 0$.*

We cannot factor $y^2 + 5y + 1$; so we cannot solve the equation $y^2 + 5y + 1 = 0$ by the methods that we have studied. More advanced methods are examined in Chapter 22.

Example 33. *Solve the equation $6y^2 + 17y + 5 = 0$.*

When the expression is factored, the equation $6y^2 + 17y + 5 = 0$ becomes

$$(3y + 1)(2y + 5) = 0$$

The solutions follow by setting the factors equal to 0 and solving the resulting equations.

$$
\begin{array}{c|c}
3y + 1 = 0 & 2y + 5 = 0 \\
3y = -1 & 2y = -5 \\
y = -\dfrac{1}{3} \ \checkmark & y = -\dfrac{5}{2} \ \checkmark
\end{array}
$$

Example 34. *Solve the equation $3x^2 - 12 = 0$.*

A 3 can be factored out of the expression $3x^2 - 12$. So

$$3x^2 - 12 = 0$$

becomes

$$3(x^2 - 4) = 0$$

The 3 can be eliminated from the equation by dividing both sides by 3. The result is

$$x^2 - 4 = 0$$

The expression on the left side is the difference of two squares and is readily factored. The equation is then easily solved.

$$(x + 2)(x - 2) = 0$$

$$
\begin{array}{c|c}
x + 2 = 0 & x - 2 = 0 \\
x = -2 \ \checkmark & x = 2 \ \checkmark
\end{array}
$$

Example 35. *Solve the equation $x^2 + 3x = 28$.*

In order to factor the quadratic expression and consider what makes it equal to zero, the equation must be in an $= 0$ form. So add -28 to both sides of the equation $x^2 + 3x = 28$ in order to obtain the desired form.

$$
\begin{array}{r}
x^2 + 3x = 28 \\
-28 = -28 \\
\hline
x^2 + 3x - 28 = 0
\end{array}
$$

Now we can proceed as in previous examples.

$$x^2 + 3x - 28 = 0$$
$$(x - 4)(x + 7) = 0$$

$x - 4 = 0$	$x + 7 = 0$
$x = 4$ ✓	$x = -7$ ✓

Example 36. *Solve the equation $x^2 = 36 - 9x$.*

In order to solve a quadratic equation by factoring, the expression must be equal to 0. So add $9x$ and -36 to both sides of this equation in order to obtain the desired form. First, add $9x$ to both sides.

$$x^2 = 36 - 9x$$
$$\underline{+9x \qquad\quad + 9x}$$
$$x^2 + 9x = 36$$

Then add -36 to both sides.

$$x^2 + 9x \qquad\quad = \quad 36$$
$$\underline{\qquad - 36 \quad -36}$$
$$x^2 + 9x - 36 = 0$$

Now we can proceed to factor and solve.

$$x^2 + 9x - 36 = 0$$
$$(x + 12)(x - 3) = 0$$

$x + 12 = 0$	$x - 3 = 0$
$x = -12$ ✓	$x = 3$ ✓

Example 37. *Solve the equation $-x^2 + x + 12 = 0$.*

The equation $-x^2 + x + 12 = 0$ appears strange because we are accustomed to working with quadratic expressions that begin x^2, *not* $-x^2$. If the equation is multiplied by -1, then the $-x^2$ term will become x^2.

$$\underline{-1}(-x^2 + x + 12) = \underline{-1}(0)$$

becomes

$$x^2 - x - 12 = 0$$

which is easily factored and solved.

$$(x - 4)(x + 3) = 0$$

$x - 4 = 0$	$x + 3 = 0$
$x = 4$ ✓	$x = -3$ ✓

EXERCISES 17.4

*1. Solve each quadratic equation by factoring.

(a) $x^2 + 7x + 6 = 0$ (b) $y^2 + 4y + 3 = 0$

(c) $x^2 - 6x + 5 = 0$ (d) $y^2 - 7y + 10 = 0$

(e) $m^2 - m - 6 = 0$ (f) $x^2 + 7x - 8 = 0$

(g) $n^2 + 2n + 1 = 0$ (h) $t^2 - 2t - 3 = 0$

(i) $x^2 - 16 = 0$ (j) $b^2 - 49 = 0$

(k) $x^2 + 7x = 0$ (l) $y^2 - 8y = 0$

2. Solve each quadratic equation by factoring, if the expression can be factored.

*(a) $y^2 - y - 20 = 0$ *(b) $x^2 - x = 0$

*(c) $x^2 - 1 = 0$ *(d) $x^2 - 15x - 14 = 0$

*(e) $x^2 + 25 = 0$ (f) $x^2 + 8x + 16 = 0$

*(g) $2x^2 - 18x + 16 = 0$ (h) $3x^2 + 3x - 60 = 0$

*(i) $2x^2 + 11x + 5 = 0$ (j) $3x^2 + 13x + 12 = 0$

3. Solve each quadratic equation by factoring, if possible.

*(a) $5x^2 + 2x - 3 = 0$ *(b) $3w^2 + 8w + 4 = 0$

*(c) $4x^2 + 8x - 5 = 0$ (d) $x^2 + 25x = 0$

*(e) $6y^2 - 54 = 0$ (f) $7p^2 - 175 = 0$

*(g) $-x^2 + 10x - 24 = 0$ (h) $-x^2 - 5x + 36 = 0$

*(i) $6n^2 + 17n + 12 = 0$ (j) $2y^2 + 9y + 7 = 0$

*(k) $x^2 + 16x + 1 = 0$ (l) $m^2 - 4m - 6 = 0$

*(m) $4x^2 + 8x - 60 = 0$ *(n) $5x^2 - 40x = 0$

4. Solve each quadratic equation by factoring, if possible.

*(a) $x^2 + 8x = 16$ (b) $y^2 + 3y = 4$

*(c) $n^2 - 8n = -7$ (d) $x^2 + 9x = 10$

*(e) $3x^2 + x = 10$ *(f) $3x^2 + 36x = -96$

*(g) $2x^2 = 13x + 7$ (h) $m^2 = 2m + 8$

*(i) $4x^2 + 6 = 11x$ (j) $2m^2 = 11m - 12$

*(k) $10x + 4 = -6x^2$ (l) $x^2 = 7x$

*(m) $b^2 = 9$ (n) $a^2 = 49$

*(o) $x^2 - 7x + 2 = 20$ (p) $x^2 - 6x + 1 = -8$

5. For each exercise, read the words carefully, establish a quadratic equation, and solve it in order to answer the question.

*(a) The length of a rectangle is three times the width. The area is 147 square meters. What are the length and width of the rectangle?

(b) The length of a rectangle is five times the width. The area is 320 square yards. What are the length and width of the rectangle?

*(c) The width of a rectangle is four feet less than the length. The area is 96 square feet. What are the length and width of the rectangle?

(d) The length of a rectangle is three centimeters more than the width. The area is 88 square centimeters. What are the length and width of the rectangle?

6. For each exercise, read the words carefully, let x represent the number you seek, establish a quadratic equation, solve it, and check the solutions in the statement of the exercise.

*(a) If you add the square of a certain number and nine times the number, the sum is 36. What is the number(s)?

*(b) If you square a number and then subtract 57, the result is 24. Determine the number(s).

*(c) If you add the square of a number and 20 times the number, the result is zero. What is the number(s)?

(d) The square of a number is added to seven times the number. That sum is 18. What is the number(s)?

(e) Seven times a number is subtracted from the square of the number. That difference is -10. What is the number(s)?

(f) If you add the square of a number and ten times the number, the result is zero. What is the number(s)?

Chapter 17.   REVIEW EXERCISES

*1. Factor each expression, if possible.

(a) $mx + nx + px$

(b) $a^2b^2c + ab^2c^2$

(c) $26ab + 39bc$

(d) $x^2 - 10xy$

(e) $10 - 30y$

(f) $x^2 + y^2$

(g) $x^2 + 3x - 10$

(h) $x^2 - 14x + 13$

(i) $x^2 - 21x$

(j) $m^2 + 30m$

(k) $n^2 - 100$

(l) $y^2 - 1$

(m) $4x^2 + 8x + 3$

(n) $2x^2 - 10x - 28$

(o) $4x^2 - 5x - 9$

(p) $3x^2 - 48$

*2. Solve each quadratic equation, if the quadratic expression can be factored.

(a) $x^2 + 13x + 12 = 0$

(b) $x^2 - 2x + 1 = 0$

(c) $t^2 - 10t + 25 = 0$

(d) $y^2 - 5y - 14 = 0$

(e) $x^2 - 9x + 12 = 0$

(f) $t^2 + 10t + 16 = 0$

(g) $m^2 + 14m + 24 = 0$

(h) $x^2 + 9x + 14 = 0$

(i) $3x^2 - 3x - 18 = 0$

(j) $2n^2 - 8n - 32 = 0$

(k) $2m^2 - 16m + 30 = 0$

(l) $2x^2 - 3x - 14 = 0$

(m) $5x^2 + 14x - 3 = 0$

(n) $m^2 - 4m = 5$

(o) $m^2 = 49$

(p) $x^2 = 12x$

REVIEW PROBLEMS FOR PART 2

1. Simplify each expression.
   *(a) $(-8)(+2)(-5) + (-3)(+2)$
   (b) $-9 + 2 - 6 + (-3)^2 - 1 + 9(-5)$
   *(c) $(-3)(+4 - 2 - 7) - 8(-9 + 15)$
   (d) $-8^2 + 6^2 + 9(-1)^{15} - 1^{11}$
   *(e) $\dfrac{(-6)(+4)}{(-2)^3} + \dfrac{(+1)}{(-1)} - 1$

2. Simplify each expression.
   *(a) $4x + 9x - 3x + x - 12x$ (b) $5x - 3 + 2x - 4 - x + 2y$
   *(c) $5x + 4(1 - 3x) + 2$ (d) $6 - 4[5 + (2 + x)7] + 3x$

3. Solve each equation.
   *(a) $3x + 9 = -2$ (b) $\frac{2}{3}x - 18 = 7$
   *(c) $5x - 9 = 7x + 18$ (d) $8(1 - 3x) = 15x + 1$
   *(e) $6 - (x - 7) = 3x + 4$

*4. Simplify each expression.
   (a) $x^5 x^9$ (b) $y^4 y^{34}$ (c) $m^6 m^7 m^8$
   (d) $\dfrac{x^{10}}{x^6}$ (e) $\dfrac{y^{15}}{y^7}$ (f) $\dfrac{m^9}{m^3}$

*5. Perform the long division.
   (a) $x - 3 \overline{)\, x^3 - 2x^2 + 4x - 18}$ (b) $x + 2 \overline{)\, x^3 - 4x^2 - 9x + 6}$

*6. Write an algebraic expression to represent each phrase.
   (a) Ten more than z.
   (b) Nine less than five times x.
   (c) The number of feet in $7x$ inches.
   (d) The interest on q dollars invested at 7% per year for r years.

7. Sketch the graph of each straight line.
   *(a) $y = 3x - 2$ (b) $x = 2y + 3$
   *(c) $x - y = 1$ (d) $3x + y = 9$

8. Solve each system of linear equations.
   *(a) $\begin{cases} 2x + 3y = 3 \\ 5x - 2y = 17 \end{cases}$ *(b) $\begin{cases} 5x + 2y = -3 \\ y = x + 2 \end{cases}$

   (c) $\begin{cases} 5x + 7y = 20 \\ 4x + 2y = 16 \end{cases}$ (d) $\begin{cases} x + 3y + 4z = 8 \\ 2x - 7y - 5z = 3 \\ -3x + 4y + 7z = 1 \end{cases}$

9. Solve each inequality.
   (a) $-x + 6 > 19$ *(b) $5x - 3 \leq -9$
   (c) $3(2 + 5x) + 4 < 1$

10. Solve each system of linear inequalities by graphing.

*(a) $\begin{cases} y > 2x + 1 \\ y \le 5 - x \end{cases}$ (b) $\begin{cases} x + y \le 6 \\ y > \frac{1}{2}x \end{cases}$

11. Factor each expression, if possible.

*(a) $x^2 + x$ (b) $7y + ny$
*(c) $x^2 - 9x - 18$ (d) $x^2 + 10x - 24$
*(e) $x^2 - y^2$ (f) $5x^2 + 10x$
*(g) $3x^2 + 17x - 28$ (h) $2x^2 + 3x - 20$

12. Solve each equation.

*(a) $x^2 - 12x + 36 = 0$ (b) $y^2 + 16y - 36 = 0$
*(c) $n^2 - 2n = 15$ (d) $t^2 + 8t + 15 = 0$
*(e) $m^2 - 11m + 18 = 0$ (f) $x^2 = x$
*(g) $t^2 = 49$ (h) $x^2 = 7x - 10$

PART 3

Fractions Containing Variables

18.1 MULTIPLICATION OF FRACTIONS

Multiplication of fractions in algebra is similar to multiplication of fractions in arithmetic (which was presented in Chapter 6). Multiply numerator times numerator and denominator times denominator.

$$\frac{3}{5} \quad \longleftarrow \text{numerator}$$
$$\phantom{\frac{3}{5}} \longleftarrow \text{denominator}$$

Example 1. *Multiply* $\frac{3}{5} \cdot \frac{4}{7}$.

$$\frac{3}{5} \cdot \frac{4}{7} = \frac{3 \cdot 4}{5 \cdot 7}$$

$$= \frac{12}{35} \quad \checkmark$$

Example 2. *Multiply* $\frac{a}{b} \cdot \frac{x}{y}$.

$$\frac{a}{b} \cdot \frac{x}{y} = \frac{a \cdot x}{b \cdot y}$$

$$= \frac{ax}{by} \quad \checkmark$$

Example 3. *Multiply* $\dfrac{x}{2} \cdot \dfrac{x+3}{y}$.

$$\frac{x}{2} \cdot \frac{x+3}{y} = \frac{x(x+3)}{2y} \quad \checkmark \quad \text{or} \quad \frac{x^2+3x}{2y} \quad \checkmark$$

Example 4. *Multiply* $\dfrac{x+4}{x} \cdot \dfrac{x+3}{7x}$.

$$\frac{x+4}{x} \cdot \frac{x+3}{7x} = \frac{(x+4)(x+3)}{x \cdot 7x}$$

$$= \frac{(x+4)(x+3)}{7x^2} \quad \checkmark \quad \text{or} \quad \frac{x^2+7x+12}{7x^2} \quad \checkmark$$

EXERCISES 18.1

Answers to starred exercises are given in the back of the book.

*1. Multiply.

(a) $\dfrac{2}{3} \cdot \dfrac{5}{7}$ (b) $\dfrac{4}{9} \cdot \dfrac{2}{5}$ (c) $\dfrac{x}{3} \cdot \dfrac{y}{5}$

(d) $\dfrac{7}{t} \cdot \dfrac{b}{6}$ (e) $\dfrac{5}{7} \cdot \dfrac{x+1}{2}$ (f) $\dfrac{2}{x} \cdot \dfrac{x-4}{5}$

(g) $\dfrac{x+5}{x} \cdot \dfrac{4}{x}$ (h) $\dfrac{x+4}{x} \cdot \dfrac{x+1}{x}$ (i) $\dfrac{mn}{5} \cdot \dfrac{3mn}{t}$

(j) $\dfrac{xy}{4w} \cdot \dfrac{xy}{4w}$

2. Multiply.

*(a) $\dfrac{3xy}{7} \cdot \dfrac{2x}{5}$ (b) $\dfrac{7wx}{3} \cdot \dfrac{4w}{5}$ *(c) $\dfrac{5r^2w}{3x} \cdot \dfrac{2rw}{x}$

(d) $\dfrac{8a^3b}{7c} \cdot \dfrac{5a^2b^6}{3c}$ *(e) $\dfrac{7}{y^2} \cdot \dfrac{5x}{y^3}$ (f) $\dfrac{4x}{y^6} \cdot \dfrac{3x}{y^2}$

*(g) $\dfrac{5xy^2}{7z} \cdot \dfrac{-3x}{2}$ (h) $\dfrac{-6m^2n^2}{5k} \cdot \dfrac{7mn^3}{13}$

3. Multiply.

*(a) $\dfrac{x^4+1}{3x} \cdot \dfrac{-7}{4x}$ (b) $\dfrac{x^2+3}{5x} \cdot \dfrac{6}{x}$ *(c) $\dfrac{x+1}{x-4} \cdot \dfrac{x+3}{x-4}$

(d) $\dfrac{x+9}{x+1} \cdot \dfrac{x-9}{x-1}$ *(e) $\dfrac{5x-3}{3x} \cdot \dfrac{x+7}{x}$ (f) $\dfrac{7x+2}{x+1} \cdot \dfrac{6}{x}$

18.2 REDUCTION OF FRACTIONS

Reduction of fractions in algebra is similar to reduction of fractions in arithmetic (presented in Chapter 6). *Factors* that appear in both numerator and denominator can be separated and divided out. We use the idea that a number divided by itself equals 1.

$$\frac{\text{number}}{\text{itself}} = 1$$

Example 5. *Reduce* $\dfrac{20}{25}$.

Reduce $\frac{20}{25}$ by finding a common factor, 5, in both numerator and denominator; then divide out the common factor.

$$\frac{20}{25} = \frac{5 \cdot 4}{5 \cdot 5} = \frac{5}{5} \cdot \frac{4}{5} = 1 \cdot \frac{4}{5} = \frac{4}{5} \quad \checkmark$$

We are allowed to reduce because we can write the fraction as *one*, $\frac{5}{5}$, times $\frac{4}{5}$. Many people prefer to show fewer steps and do the reduction as follows.

$$\frac{20}{25} = \frac{\cancel{5} \cdot 4}{\cancel{5} \cdot 5} = \frac{4}{5}$$

Example 6. *Reduce* $\dfrac{5x}{7x}$.

Here the factor x is common to both numerator and denominator.

$$\frac{5x}{7x} = \frac{5 \cdot x}{7 \cdot x} = \frac{5}{7} \cdot \frac{x}{x} = \frac{5}{7} \cdot 1 = \frac{5}{7} \quad \checkmark$$

Example 7. *Reduce* $\dfrac{12x^2y}{30y^2}$.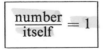

Here are the steps. Additional explanation follows.

$$\frac{12x^2y}{30y^2} = \frac{2 \cdot 6 \cdot x^2 \cdot y}{5 \cdot 6 \cdot y \cdot y}$$

$$= \frac{2 \cdot \cancel{6} \cdot x^2 \cdot \cancel{y}}{5 \cdot \cancel{6} \cdot y \cdot \cancel{y}}$$

$$= \frac{2x^2}{5y} \quad \checkmark$$

We factored 12 into $2 \cdot 6$, 30 into $5 \cdot 6$, and y^2 into $y \cdot y$ because we saw what would divide out. If you don't "see it," then factor everything into prime factors and you should see it then.

$$\frac{12x^2y}{30y^2} = \frac{2 \cdot 2 \cdot 3 \cdot x \cdot x \cdot y}{2 \cdot 3 \cdot 5 \cdot y \cdot y} = \frac{2 \cdot x \cdot x}{5 \cdot y} = \frac{2x^2}{5y}$$

As we proceed, you will see the role of factoring in the reduction of fractions. In fact,

> Fractions are reduced
> *only* by factoring.

Example 8. *Reduce* $\dfrac{c}{ac - c^2}$.

Clearly there are c's in both the numerator and denominator.

$$\frac{c}{ac - c^2} = \frac{c \cdot 1}{a \cdot c - c \cdot c}$$

The c in the numerator is a *factor*. There is also a *factor* of c in the denominator. Factor it out so that we can then divide it out with the c in the numerator.

$$= \frac{c \cdot 1}{c(a - c)}$$

$$= \frac{c \cdot 1}{c(a - c)}$$

$$= \frac{1}{a - c} \quad \checkmark$$

Example 9. *Reduce* $\dfrac{x^2 - x - 12}{x^2 + x - 20}$.

Both numerator and denominator are quadratic expressions. They should be factored if possible

$$\frac{x^2 - x - 12}{x^2 + x - 20} = \frac{(x + 3)(x - 4)}{(x + 5)(x - 4)}$$

Now divide out any factors common to both numerator and denominator.

$$= \frac{(x + 3)(x - 4)}{(x + 5)(x - 4)}$$

$$= \frac{x + 3}{x + 5} \quad \checkmark$$

And you cannot reduce $\dfrac{x + 3}{x + 5}$, as explained in the next example.

Example 10. *Reduce* $\dfrac{x+3}{x+5}$.

It is tempting to reduce $\dfrac{x+3}{x+5}$ as $\dfrac{\cancel{x}+3}{\cancel{x}+5}=\dfrac{3}{5}$. But doing so is *incorrect*. Remember, *you can only reduce fractions in which you can separate a fraction that is equal to one*. So, if $\dfrac{x+3}{x+5}$ could be written $\dfrac{x}{x}\cdot\underline{\quad}$, then $\dfrac{x}{x}$ could be removed without changing the value of the fraction. Otherwise not.

It can also be reasoned numerically that such a reduction is wrong. After all, suppose that x was equal to 2. Then

$$\frac{x+3}{x+5}=\frac{2+3}{2+5}=\frac{5}{7}$$

And $\dfrac{5}{7}\neq\dfrac{3}{5}$. So reducing $\dfrac{x+3}{x+5}$ to $\dfrac{3}{5}$ is incorrect.

If the fraction had been $\dfrac{3x}{5x}$ it would have reduced to $\dfrac{3}{5}$, as

$$\frac{3x}{5x}=\frac{3}{5}\cdot\frac{x}{x}=\frac{3}{5}\cdot 1=\frac{3}{5}$$

Example 11. *Reduce* $\dfrac{2x^2+12x+10}{4x+20}$.

Note first the common factor 2 in each term of the numerator and in each term of the denominator.

$$\frac{2x^2+12x+10}{4x+20}=\frac{2(x^2+6x+5)}{2(2x+10)}$$

$$=\frac{\cancel{2}(x^2+6x+5)}{\cancel{2}(2x+10)}$$

$$=\frac{x^2+6x+5}{2x+10}$$

Next, note that the quadratic expression factors. Also, 2 can be factored out of $2x+10$.

$$=\frac{(x+5)(x+1)}{2(x+5)}$$

$$=\frac{(\cancel{x+5})(x+1)}{2(\cancel{x+5})}$$

$$=\frac{x+1}{2}\quad\checkmark$$

Example 12. *Reduce* $\dfrac{x+3}{3+x}$.

The numerator and denominator are the same, since $x+3=3+x$. So the fraction reduces to 1.

$$\frac{x+3}{3+x}=1 \checkmark$$

Example 13. *Reduce* $\dfrac{x-5}{5-x}$.

Clearly $(x-5)$ and $(5-x)$ are different and yet they do resemble one another. In fact, with a little algebraic manipulation we can show that $(x-5)=-(5-x)$. In other words, $(x-5)$ and $(5-x)$ are opposites. Here's the manipulation.

$$-(5-x)=-5+x$$
$$=+x-5$$
$$=x-5$$

If $-(5-x)$ is substituted for $(x-5)$ in the original fraction, the fraction reduces to -1. Here is the work.

$$\frac{x-5}{5-x}=\frac{-(5-x)}{(5-x)}$$
$$=-1 \checkmark$$

This example points out a useful general result. A number divided by its opposite yields -1.

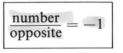

$$\frac{\text{number}}{\text{opposite}}=-1$$

Example 14. *Reduce* $\dfrac{x-y}{y-x}$.

Do you see that $(x-y)$ and $(y-x)$ are opposites?

$$x-y=-y+x=-(y-x).$$

Consequently,

$$\frac{x-y}{y-x}=-1 \checkmark$$

The last two examples were based on a "trick" that you should look for in the future. If you doubt the correctness of the procedure despite the justification presented, consider the arithmetic example given below.

$$\frac{4-9}{9-4} = \frac{-5}{+5} = -1$$

Example 15. *Reduce* $\dfrac{21-3x}{2x^2-98}$.

$$\frac{21-3x}{2x^2-98} = \frac{3(7-x)}{2(x^2-49)} \qquad \begin{cases} \text{after factoring out constant} \\ \text{factors} \end{cases}$$

$$= \frac{3(7-x)}{2(x+7)(x-7)} \qquad \begin{cases} \text{factoring the difference} \\ \text{of two squares} \end{cases}$$

$$= (-1) \cdot \frac{3(7-x)}{2(x+7)(x-7)} \qquad \text{since } \frac{7-x}{x-7} = -1 \quad \text{(opposites)}$$

$$= -\frac{3}{2(x+7)} \qquad ✓$$

EXERCISES 18.2

*1. Reduce each fraction, if possible. *Note:* "Reduce" means to reduce completely.

(a) $\dfrac{16}{80}$

(b) $\dfrac{12}{600}$

(c) $\dfrac{3x}{4x}$

(d) $\dfrac{8y}{10y}$

(e) $\dfrac{5x^2}{20x}$

(f) $\dfrac{5x}{2x^3}$

(g) $\dfrac{3x}{x^2-x}$

✗(h) $\dfrac{m}{2m-2m^2}$

(i) $\dfrac{m-n}{n-m}$

(j) $\dfrac{m-n}{m+n}$

(k) $\dfrac{x+1}{x^2+2x+1}$

(l) $\dfrac{x-5}{x^2-7x+10}$

2. Reduce each fraction, if possible.

*(a) $\dfrac{x^2+7x+12}{x^2+5x+4}$

(b) $\dfrac{x^2-9x+20}{x^2-4x-5}$

*(c) $\dfrac{x^2+9}{x^2+81}$

(d) $\dfrac{2x^2-8}{4x+8}$

*(e) $\dfrac{x^2+xy}{4x}$

✳(f) $\dfrac{3c}{bc+c^2}$

*(g) $\dfrac{8-2x}{x-4}$

(h) $\dfrac{(y-x)^2}{x^2-y^2}$

*(i) $\dfrac{4+a}{4+b}$

(j) $\dfrac{1+b}{1-b}$

*(k) $\dfrac{10x-20}{3x-6}$

(l) $\dfrac{8-4w}{6-3w}$

*(m) $\dfrac{6x-6y}{2(y-x)}$

(n) $\dfrac{12x^2-48}{2-x}$

*(o) $\dfrac{2x^2+12x+10}{4x+20}$

(p) $\dfrac{2x+8}{2x^2-14x+24}$

18.3 MORE MULTIPLICATION

Always look for reduction before you actually multiply numerator times numerator and denominator times denominator.

Example 16. *Multiply* $\dfrac{3}{5} \cdot \dfrac{5}{7}$.

The 5's divide out; that is, $\dfrac{5}{5} = 1$.

$$\frac{3}{\cancel{5}} \cdot \frac{\cancel{5}}{7} = \frac{3}{7} \quad \checkmark$$

Example 17. *Multiply* $\dfrac{3x^2y}{7x} \cdot \dfrac{5w}{6y}$.

Examination of the fractions shows that

$$\frac{3x^2y}{7x} \cdot \frac{5w}{6y} = \frac{3 \cdot x \cdot x \cdot y}{7 \cdot x} \cdot \frac{5 \cdot w}{3 \cdot 2 \cdot y}$$

$$= \frac{\cancel{3} \cdot x \cdot \cancel{x} \cdot \cancel{y}}{7 \cdot \cancel{x}} \cdot \frac{5 \cdot w}{\cancel{3} \cdot 2 \cdot \cancel{y}} \qquad (x, y, \text{ and } 3 \text{ divide out})$$

$$= \frac{x}{7} \cdot \frac{5w}{2}$$

$$= \frac{5wx}{14} \quad \checkmark$$

> When the directions say "reduce the fraction" or "multiply," the first thing you should think or do is *factor*, if possible.

Example 18. *Multiply* $\dfrac{x^2 + 2x + 1}{x^2 - 2x - 3} \cdot \dfrac{x^2 + 2x - 15}{7x + 35}$.

Factor the expressions and look for factors common to both numerator and denominator.

$$\frac{x^2 + 2x + 1}{x^2 - 2x - 3} \cdot \frac{x^2 + 2x - 15}{7x + 35} = \frac{\cancel{(x+1)}(x + 1)}{\cancel{(x+1)}\cancel{(x-3)}} \cdot \frac{\cancel{(x+5)}\cancel{(x-3)}}{7\cancel{(x+5)}}$$

$$= \frac{x + 1}{7} \quad \checkmark$$

Example 19. *Multiply* $\dfrac{x^2+6x+8}{x-5} \cdot \dfrac{x}{x^2-4}$.

$$\frac{x^2+6x+8}{x-5} \cdot \frac{x}{x^2-4} = \frac{(x+4)(x+2)}{(x-5)} \cdot \frac{x}{(x+2)(x-2)}$$

$$= \frac{(x+4)x}{(x-5)(x-2)} \checkmark$$

$$\text{or} \quad \frac{x^2+4x}{x^2-7x+10} \checkmark$$

EXERCISES 18.3

*1. Multiply. Be sure to reduce when possible.

(a) $\dfrac{3x}{5} \cdot \dfrac{10}{x}$

(b) $\dfrac{15x^2}{y} \cdot \dfrac{y^2}{60x}$

(c) $\dfrac{y+3}{5} \cdot \dfrac{15}{2y+6}$

(d) $\dfrac{20a}{3c} \cdot \dfrac{6x}{4a^3b^2c}$

(e) $\dfrac{x^2-25}{x+1} \cdot \dfrac{2}{2x+10}$

(f) $\dfrac{5}{x^2-9} \cdot \dfrac{4x-12}{7}$

(g) $\dfrac{3x+3}{x^2+7x+6} \cdot \dfrac{x+6}{6x^2}$

(h) $\dfrac{n^2-n-12}{n^2-16} \cdot \dfrac{2n-8}{n+3}$

2. Multiply. Reduce when possible.

*(a) $\dfrac{3x^2+15x+18}{x^2+2x} \cdot \dfrac{5x^3}{15x^2}$

(b) $\dfrac{2x^2-14x+20}{x^2-25} \cdot \dfrac{x^4}{2x}$

*(c) $\dfrac{2x-2y}{4x} \cdot \dfrac{3y}{3y-3x}$

(d) $\dfrac{5m-5n}{25m} \cdot \dfrac{7mn}{3n-3m}$

*(e) $\dfrac{x^2-3x-10}{x^2-25} \cdot \dfrac{x^2+9x+20}{x^2+2x}$

(f) $\dfrac{a^2-16}{a+2} \cdot \dfrac{2a}{a^2-4a}$

*(g) $\dfrac{x^2-9}{x+3} \cdot \dfrac{4x}{2x-6}$

(h) $\dfrac{b-2a}{b+2a} \cdot \dfrac{7a}{7b}$

*(i) $\dfrac{4x-4}{9x} \cdot \dfrac{3x^2}{1-x}$

(j) $\dfrac{a-b}{x-y} \cdot \dfrac{y-x}{b-a}$

18.4 DIVISION OF FRACTIONS

Division of fractions in algebra is similar to division of fractions in arithmetic (which was presented in Chapter 6). Invert the second fraction (the divisor) and multiply by it.

Example 20. *Divide* $\dfrac{3}{5} \div \dfrac{7}{4}$.

$$\dfrac{3}{5} \div \dfrac{7}{4} = \dfrac{3}{5} \cdot \dfrac{4}{7}$$

$$= \dfrac{3 \cdot 4}{5 \cdot 7}$$

$$= \dfrac{12}{35} \quad \checkmark$$

Example 21. *Divide* $\dfrac{4}{9} \div \dfrac{5}{3}$.

$$\dfrac{4}{9} \div \dfrac{5}{3} = \dfrac{4}{9} \cdot \dfrac{3}{5}$$

$$= \dfrac{4}{\cancel{3} \cdot 3} \cdot \dfrac{\cancel{3}}{5}$$

$$= \dfrac{4 \cdot 1}{3 \cdot 5}$$

$$= \dfrac{4}{15} \quad \checkmark$$

As we proceed to algebra examples, note the steps to be followed, *in order*, as division of fractions is carried out.

> *Division of Fractions*
> 1. Change to multiplication.
> 2. Factor wherever possible.
> 3. Reduce whenever possible.
> 4. Multiply what remains.

Example 22. *Divide* $\dfrac{2x + 4}{6x} \div \dfrac{x^2 - 4}{7x^3}$.

You cannot try to reduce until after you invert the second fraction and change to multiplication.

$$\dfrac{2x + 4}{6x} \div \dfrac{x^2 - 4}{7x^3} = \dfrac{2x + 4}{6x} \cdot \dfrac{7x^3}{x^2 - 4}$$

Now that we have a *multiplication* problem, each expression should be factored, if possible.

$$= \dfrac{2(x + 2)}{2 \cdot 3 \cdot x} \cdot \dfrac{7 \cdot x \cdot x^2}{(x + 2)(x - 2)}$$

Identical factors can now be divided out.

$$= \frac{\cancel{2}(\cancel{x+2})}{\cancel{2} \cdot 3 \cdot \cancel{x}} \cdot \frac{7 \cdot \cancel{x} \cdot x^2}{(\cancel{x+2})(x-2)}$$

When what remains is multiplied, the result is

$$\frac{7x^2}{3(x-2)} \quad \checkmark$$

If you prefer, multiply it out completely to get

$$\frac{7x^2}{3x-6} \quad \checkmark$$

Example 23. *Divide* $\dfrac{x^2 + 7x + 12}{x^2 - 1} \div \dfrac{x^2 + 6x + 9}{2x + 2}.$

$$\frac{x^2 + 7x + 12}{x^2 - 1} \div \frac{x^2 + 6x + 9}{2x + 2} = \frac{x^2 + 7x + 12}{x^2 - 1} \cdot \frac{2x + 2}{x^2 + 6x + 9}$$

$$= \frac{(x+3)(x+4)}{(x+1)(x-1)} \cdot \frac{2(x+1)}{(x+3)(x+3)}$$

$$= \frac{(\cancel{x+3})(x+4)}{(\cancel{x+1})(x-1)} \cdot \frac{2(\cancel{x+1})}{(\cancel{x+3})(x+3)}$$

$$= \frac{2(x+4)}{(x-1)(x+3)} \quad \checkmark$$

which you may prefer to multiply out completely and leave as

$$\frac{2x + 8}{x^2 + 2x - 3} \quad \checkmark$$

Example 24. *Divide* $\dfrac{2 - x}{x + 5} \div \dfrac{x^2 - 4}{2x + 10}.$

$$\frac{2 - x}{x + 5} \div \frac{x^2 - 4}{2x + 10} = \frac{2 - x}{x + 5} \cdot \frac{2x + 10}{x^2 - 4}$$

$$= \frac{2 - x}{x + 5} \cdot \frac{2(x + 5)}{(x + 2)(x - 2)}$$

The $(x + 5)$ factors divide out.

$$= \frac{2 - x}{\cancel{x+5}} \cdot \frac{2(\cancel{x+5})}{(x + 2)(x - 2)}$$

There is $2 - x$ in the numerator and its opposite, $x - 2$, in the denominator.

They reduce to -1. To illustrate, we'll cross them out and put a minus in front of the entire product. Compare with Example 15.

$$= -\frac{\cancel{2} \cdot x}{\cancel{x+5}} \cdot \frac{2(\cancel{x+5})}{(x+2)(\cancel{x-2})}$$

$$= -\frac{2}{x+2} \quad \checkmark$$

EXERCISES 18.4

*1. Divide. Be sure to reduce when possible.

(a) $\dfrac{3x}{7} \div \dfrac{x}{x+5}$

(b) $\dfrac{x-6}{8} \div \dfrac{2x-12}{4}$

(c) $\dfrac{5x^2}{3} \div \dfrac{x^2+3x}{9y}$

(d) $\dfrac{x^2+2x+1}{9x} \div \dfrac{x^2-1}{3}$

(e) $\dfrac{x-3}{x+3} \div \dfrac{6-2x}{3+x}$

(f) $\dfrac{m^2-9m+20}{15m} \div \dfrac{m^2-3m-10}{5m+10}$

2. Divide. Reduce when possible.

*(a) $\dfrac{9xy}{y} \div \dfrac{3x^2}{5}$

(b) $\dfrac{15abc}{2c} \div \dfrac{3a}{8}$

*(c) $\dfrac{x^2-6x+8}{3x-6} \div \dfrac{x-4}{x}$

(d) $\dfrac{m^2-9m}{6m} \div \dfrac{m-9}{3}$

*(e) $\dfrac{x^2+x+12}{5x+15} \div \dfrac{x^2-16}{2x-8}$

(f) $\dfrac{x^2+4x}{2x+8} \div \dfrac{x^2+x-12}{2x-6}$

*(g) $\dfrac{x^2+9x}{x-3} \div \dfrac{x^2+13x+36}{6-2x}$

(h) $\dfrac{6t-t^2}{9t} \div \dfrac{t^2-5t-6}{6t^2}$

*(i) $\dfrac{3x+3y}{x^2-64} \div \dfrac{3xy}{2x^2+16x+32}$

(j) $\dfrac{2y-2x}{x^2-y^2} \div \dfrac{4xy^2}{2x}$

18.5 *ADDITION AND SUBTRACTION OF FRACTIONS*

Addition of fractions in algebra is similar to addition of fractions in arithmetic. A common denominator must be determined, and all fractions being added must either have this common denominator or be changed so as to have the common denominator. Then the numerators can be added and placed over the common denominator. The presentation here assumes that you are familiar with the concept of least common multiple (LCM) and its use in arithmetic to determine the lowest common denominator. Unless your instructor reviews this concept in class or offers another approach, you should *go back to Chapter 6 and read about LCM and addition of fractions*. Once again, to add fractions:

> 1. Determine common denominator (LCM).
> 2. Change all fractions to fractions having the common denominator.
> 3. Add the numerators.
> 4. Place that sum over the common denominator.

Example 25. *Add* $\dfrac{3}{8} + \dfrac{5}{12}$.

$$8 = 2^3$$

$$12 = 2^2 \cdot 3$$

So LCM $(8, 12) = 2^3 \cdot 3 = 24$. This means that 24 is the common denominator. Next change both fractions to fractions with denominator 24.

$$\frac{3}{8} = \frac{3}{8} \cdot \frac{3}{3} = \frac{9}{24}$$

$$\frac{5}{12} = \frac{5}{12} \cdot \frac{2}{2} = \frac{10}{24}$$

Now we can add the fractions.

$$\frac{3}{8} + \frac{5}{12} = \frac{9}{24} + \frac{10}{24}$$

$$= \frac{19}{24} \quad \checkmark$$

Example 26. *Add* $\dfrac{a}{u} + \dfrac{b}{v}$.

Since the denominators are u and v, the common denominator is uv. Change each fraction to a fraction with denominator uv and then add the fractions.

$$\frac{a}{u} \cdot \frac{v}{v} = \frac{av}{uv}$$

$$\frac{b}{v} \cdot \frac{u}{u} = \frac{bu}{uv}$$

Thus

$$\frac{a}{u} + \frac{b}{v} = \frac{av}{uv} + \frac{bu}{uv} = \frac{av + bu}{uv} \quad \checkmark$$

Example 27. *Add* $\dfrac{3}{5x} + \dfrac{7}{10y}.$

Since

$$5x = 5 \cdot x$$

and $\qquad\qquad 10y = 2 \cdot 5 \cdot y$

$$\text{LCM } (5x, 10y) = 2 \cdot 5 \cdot x \cdot y = 10xy$$

$$\frac{3}{5x} + \frac{7}{10y} = \frac{3}{5x} \cdot \frac{2y}{2y} + \frac{7}{10y} \cdot \frac{x}{x}$$

The denominator $5x$ must be multiplied by $2y$ to produce $10xy$. The denominator $10y$ must be multiplied by x to produce $10xy$.

$$= \frac{6y}{10xy} + \frac{7x}{10xy}$$

$$= \frac{6y + 7x}{10xy} \quad \checkmark$$

As in arithmetic, subtraction of fractions resembles addition of fractions.

Example 28. *Combine* $\dfrac{3a}{bcd} - \dfrac{7}{5c}.$

Since LCM $(bcd, 5c)$ is $5bcd$, the common denominator is $5bcd$. Change each fraction to one with a denominator of $5bcd$ and then combine them.

$$\frac{3a}{bcd} - \frac{7}{5c} = \frac{3a}{bcd} \cdot \frac{5}{5} - \frac{7}{5c} \cdot \frac{bd}{bd}$$

$$= \frac{15a}{5bcd} - \frac{7bd}{5bcd}$$

$$= \frac{15a - 7bd}{5bcd} \quad \checkmark$$

Example 29. *Combine* $\dfrac{3w}{20x^2y} + \dfrac{25z}{24wxy^3}.$

$$\left. \begin{array}{l} 20x^2y = 2^2 \cdot 5 \cdot x^2 \cdot y \\ 24wxy^3 = 2^3 \cdot 3 \cdot w \cdot x \cdot y^3 \end{array} \right\} \qquad \text{LCM is } 2^3 \cdot 3 \cdot 5 \cdot w \cdot x^2 \cdot y^3 = 120wx^2y^3.$$

$$\frac{3w}{20x^2y} + \frac{25z}{24wxy^3} = \frac{3w}{20x^2y} \cdot \frac{6wy^2}{6wy^2} + \frac{25z}{24wxy^3} \cdot \frac{5x}{5x}$$

$$= \frac{18w^2y^2}{120wx^2y^3} + \frac{125xz}{120wx^2y^3}$$

$$= \frac{18w^2y^2 + 125xz}{120wx^2y^3} \quad \checkmark$$

Example 30. *Combine* $\dfrac{3}{x+1} + \dfrac{5}{x}$.

The denominators $x + 1$ and x have no *factors* in common. (Note that x is *not* a *factor* of $x + 1$.) This means that the common denominator for the two fractions is $(x + 1)x$. Thus

$$\frac{3}{x+1} + \frac{5}{x} = \frac{3}{x+1} \cdot \frac{x}{x} + \frac{5}{x} \cdot \frac{x+1}{x+1}$$

$$= \frac{3x}{(x+1)x} + \frac{5(x+1)}{(x+1)x}$$

$$= \frac{3x + 5(x+1)}{(x+1)x}$$

$$= \frac{3x + 5x + 5}{(x+1)x}$$

$$= \frac{8x + 5}{(x+1)x} \quad \checkmark$$

Notice that the numerator simplified. This does not always happen, but you should watch for the possibility.

Example 31. *Combine* $\dfrac{7}{2x+2} + \dfrac{x-3}{4x}$.

$$\left. \begin{array}{l} 2x + 2 = 2(x + 1) \\ 4x = 2^2 \cdot x \end{array} \right\} \quad \text{LCM is } 2^2 \cdot x \cdot (x + 1) = 4x(x + 1).$$

$$\frac{7}{2x+2} + \frac{x-3}{4x} = \frac{7}{2(x+1)} \cdot \frac{2x}{2x} + \frac{x-3}{4x} \cdot \frac{x+1}{x+1}$$

$$= \frac{14x}{4x(x+1)} + \frac{(x-3)(x+1)}{4x(x+1)}$$

$$= \frac{14x + (x-3)(x+1)}{4x(x+1)}$$

$$= \frac{14x + x^2 - 2x - 3}{4x(x+1)}$$

$$= \frac{x^2 + 12x - 3}{4x(x+1)} \quad \checkmark$$

Example 32. *Combine* $\dfrac{x-1}{x^2+5x+6} + \dfrac{6}{x^2+7x+10}$.

$$
\begin{aligned}
x^2+5x+6 &= (x+3)(x+2) \\
x^2+7x+10 &= (x+5)(x+2)
\end{aligned}
\Big\} \quad \text{LCM is } (x+3)(x+2)(x+5).
$$

$$\dfrac{x-1}{x^2+5x+6} + \dfrac{6}{x^2+7x+10}$$

$$= \dfrac{x-1}{(x+3)(x+2)} + \dfrac{6}{(x+5)(x+2)}$$

$$= \dfrac{x-1}{(x+3)(x+2)} \cdot \dfrac{(x+5)}{(x+5)} + \dfrac{6}{(x+5)(x+2)} \cdot \dfrac{(x+3)}{(x+3)}$$

$$= \dfrac{(x-1)(x+5) + 6(x+3)}{(x+3)(x+2)(x+5)}$$

$$= \dfrac{x^2+4x-5+6x+18}{(x+3)(x+2)(x+5)}$$

$$= \dfrac{x^2+10x+13}{(x+3)(x+2)(x+5)} \quad \checkmark$$

EXERCISES 18.5

1. Combine the fractions.

*(a) $\dfrac{4}{15} + \dfrac{3}{10}$

*(b) $\dfrac{5}{12} + \dfrac{1}{30}$

*(c) $\dfrac{7}{x} + \dfrac{5}{y}$

*(d) $4 + \dfrac{1}{x}$ $\left(\textit{Hint: } 4 = \dfrac{4}{1}\right)$

*(e) $3 + \dfrac{5}{m}$

*(f) $\dfrac{2}{c} - 9$

*(g) $\dfrac{5}{2x} + \dfrac{3}{x}$

(h) $\dfrac{3}{10x} - \dfrac{7}{20y}$

*(i) $\dfrac{3}{wx^2} + \dfrac{2x}{5w^2}$

(j) $\dfrac{5}{3x^2} - \dfrac{7c}{12x}$

*(k) $\dfrac{8x}{15ab} + \dfrac{3y}{20abc}$

*(l) $\dfrac{3}{a} + \dfrac{4}{b} + \dfrac{5}{c}$

(m) $\dfrac{a}{b} - \dfrac{c}{d} + \dfrac{e}{f}$

*(n) $\dfrac{a}{2b} + \dfrac{c}{3d} - \dfrac{e}{4f}$

*(o) $\dfrac{1}{3x^2y} + \dfrac{7}{10xy^2}$

(p) $\dfrac{3}{2m^2n} - \dfrac{13}{6mn^2}$

*(q) $\dfrac{x}{18a^3b} + \dfrac{y}{12a^2b^2}$

*(r) $\dfrac{5}{xyz^2} + \dfrac{4}{xy^2z} + \dfrac{3}{x^2yz}$

*(s) $\dfrac{3}{a^2bc^2} + \dfrac{4}{ac} - \dfrac{5}{abc}$

(t) $\dfrac{3}{4x^2y} + \dfrac{1}{12xy^2} - \dfrac{5}{9}$

2. Combine the fractions.

*(a) $\dfrac{5}{3x + 2} + \dfrac{7}{x}$

(b) $\dfrac{3}{x + 5} - \dfrac{8}{x}$

*(c) $2x + \dfrac{7}{x + 1}$

(d) $\dfrac{5}{x - 2} + 7x$

*(e) $\dfrac{3}{x - 2} + 2$

(f) $3 + \dfrac{5}{x + 4}$

*(g) $\dfrac{5}{x + 1} + \dfrac{3}{x + 2}$

(h) $\dfrac{1}{x - 5} - \dfrac{4}{x + 1}$

*(i) $\dfrac{1}{x - 5} - \dfrac{7}{x + 3}$

(j) $\dfrac{2}{x + 5} - \dfrac{4}{x - 1}$

*(k) $\dfrac{4}{3x - 3} + \dfrac{1}{6x}$

(l) $\dfrac{7}{5x + 10} + \dfrac{9}{20x}$

*(m) $\dfrac{5}{2x} + \dfrac{3}{2x + 6}$

(n) $\dfrac{7}{3x} + \dfrac{1}{3x - 6}$

3. Combine the fractions.

*(a) $\dfrac{8}{x^2 - 4} + \dfrac{2}{x - 2}$

(b) $\dfrac{x}{x^2 - 9} + \dfrac{5}{x + 3}$

*(c) $\dfrac{1}{x^2 + 3x + 2} + \dfrac{10}{x + 1}$

*(d) $\dfrac{5x}{x^2 - 9x + 14} - \dfrac{3}{x^2 - 7x}$

*(e) $\dfrac{7}{x^2 + 6x + 8} - \dfrac{5}{x^2 + 4x + 4}$

(f) $\dfrac{2}{x^2 - 5x + 6} + \dfrac{3}{x^2 - x - 2}$

*(g) $\dfrac{9}{x^2 + 4x} - \dfrac{21}{2x + 8} + \dfrac{8}{x^2}$

(h) $\dfrac{4}{x^2 + 5x} + \dfrac{3}{2x + 10} + \dfrac{5}{4x}$

18.6 COMPLEX FRACTIONS

Sometimes the numerator and denominator of fractions are themselves fractions. Such complicated fractions are called *complex fractions*. Although they look very complex, they are not all that difficult to simplify. You merely consider the complex fraction as the division-of-fractions problem that it represents.

complex fraction

Example 33. *Simplify* $\dfrac{\frac{2}{3}}{\frac{5}{7}}$.

$$\dfrac{\frac{2}{3}}{\frac{5}{7}} = \frac{2}{3} \div \frac{5}{7} \qquad \left\{\begin{array}{l}\text{Write the complex fraction}\\ \text{as division of fractions.}\end{array}\right.$$

$$= \frac{2}{3} \cdot \frac{7}{5} \qquad \left\{\begin{array}{l}\text{Change to multiplication}\\ \text{by inverting divisor.}\end{array}\right.$$

$$= \frac{14}{15} \quad \checkmark$$

Example 34. *Simplify* $\dfrac{\frac{m+5}{m+4}}{\frac{m-6}{m-2}}$.

$$\dfrac{\frac{m+5}{m+4}}{\frac{m-6}{m-2}} = \frac{m+5}{m+4} \div \frac{m-6}{m-2}$$

$$= \frac{m+5}{m+4} \cdot \frac{m-2}{m-6}$$

$$= \frac{(m+5)(m-2)}{(m+4)(m-6)} \quad \checkmark \quad \text{or} \quad \frac{m^2+3m-10}{m^2-2m-24} \quad \checkmark$$

Example 35. *Simplify* $\dfrac{\frac{x+3}{9}}{x}$.

First, write the denominator x as the fraction $\dfrac{x}{1}$ so that you can easily see the division of one fraction by another. Then proceed by changing the complex fraction to a division of fractions.

$$\dfrac{\frac{x+3}{9}}{x} = \dfrac{\frac{x+3}{9}}{\frac{x}{1}} = \frac{x+3}{9} \div \frac{x}{1}$$

$$= \frac{x+3}{9} \cdot \frac{1}{x}$$

$$= \frac{x+3}{9x} \quad \checkmark$$

Example 36. *Simplify* $\dfrac{\dfrac{x+1}{x}}{\dfrac{1-x}{x}}$.

$$\frac{\dfrac{x+1}{x}}{\dfrac{1-x}{x}} = \frac{x+1}{x} \div \frac{1-x}{x}$$

$$= \frac{x+1}{x} \cdot \frac{x}{1-x}$$

$$= \frac{x+1}{\cancel{x}} \cdot \frac{\cancel{x}}{1-x}$$

$$= \frac{x+1}{1-x} \quad \checkmark$$

When the numerator or denominator of a complex fraction contains more than one term, as in the next example, a special procedure should be used, namely,

1. Consider all the fractions appearing in the numerator and denominator, and determine the least common denominator of all their denominators.

2. Multiply both the numerator and denominator (all terms) by that common denominator.

Example 37. *Simplify* $\dfrac{2 - \dfrac{x}{y}}{y - \dfrac{1}{x}}$.

Notice that the numerator contains two terms and so does the denominator. Since the numerator and/or denominator contain more than one term, consider all the fractions appearing, namely $\dfrac{x}{y}$ and $\dfrac{1}{x}$. The common denominator of these two fractions is xy. So multiply both numerator and denominator (all terms) by xy.

$$\frac{2 - \dfrac{x}{y}}{y - \dfrac{1}{x}} = \frac{xy\left(2 - \dfrac{x}{y}\right)}{xy\left(y - \dfrac{1}{x}\right)}$$

$$= \frac{xy \cdot 2 - xy \cdot \dfrac{x}{y}}{xy \cdot y - xy \cdot \dfrac{1}{x}} \qquad \text{after distributing the } xy$$

At this point you can see why this special procedure will indeed simplify the complex fraction. The denominator y will divide out with the y of xy. The denominator x will divide out with the x of the xy.

$$= \frac{2xy - x^2}{xy^2 - y} \quad \checkmark \qquad \text{after simplification}$$

EXERCISES 18.6

1. Simplify each complex fraction.

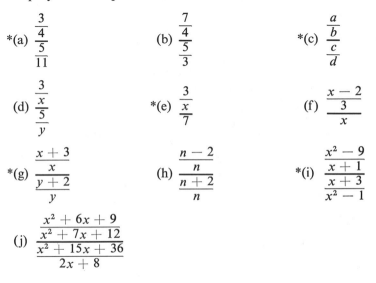

*(a) $\dfrac{\dfrac{3}{4}}{\dfrac{5}{11}}$

(b) $\dfrac{\dfrac{7}{4}}{\dfrac{5}{3}}$

*(c) $\dfrac{\dfrac{a}{b}}{\dfrac{c}{d}}$

(d) $\dfrac{\dfrac{3}{x}}{\dfrac{5}{y}}$

*(e) $\dfrac{\dfrac{3}{x}}{\dfrac{7}{}}$

(f) $\dfrac{\dfrac{x-2}{3}}{x}$

*(g) $\dfrac{\dfrac{x+3}{x}}{\dfrac{y+2}{y}}$

(h) $\dfrac{\dfrac{n-2}{n}}{\dfrac{n+2}{n}}$

*(i) $\dfrac{\dfrac{x^2-9}{x+1}}{\dfrac{x+3}{x^2-1}}$

(j) $\dfrac{\dfrac{x^2+6x+9}{x^2+7x+12}}{\dfrac{x^2+15x+36}{2x+8}}$

2. Simplify each complex fraction.

*(a) $\dfrac{1+\dfrac{1}{x}}{1-\dfrac{1}{x}}$

(b) $\dfrac{x+\dfrac{1}{2}}{y+\dfrac{1}{3}}$

*(c) $\dfrac{1+\dfrac{a}{b}}{1-\dfrac{a}{b}}$

(d) $\dfrac{\dfrac{x}{y}+2}{3+\dfrac{1}{y}}$

*(e) $\dfrac{\dfrac{3}{xy}+1}{\dfrac{2}{x}+\dfrac{3}{y}}$

(f) $\dfrac{\dfrac{5}{x}-\dfrac{2}{y}}{\dfrac{4}{y}+\dfrac{15}{x}}$

*(g) $\dfrac{\dfrac{1}{2}-\dfrac{7}{x}}{\dfrac{8}{y}+\dfrac{1}{3}}$

(h) $\dfrac{\dfrac{x}{2y}+\dfrac{y}{x}}{\dfrac{x}{y}-\dfrac{y}{x}}$

*(i) $\dfrac{m+\dfrac{x}{y}}{\dfrac{y}{x}-m}$

(j) $\dfrac{p-\dfrac{q}{r}}{p+\dfrac{r}{q}}$

*(k) $\dfrac{9}{\dfrac{a}{b}+4}$

(l) $\dfrac{7}{1-\dfrac{x}{y}}$

18.7 EQUATIONS WITH FRACTIONS

If an equation contains fractions, begin solving it by first eliminating all fractions. This procedure can be done by multiplying both sides by the lowest common denominator (LCM) of all fractions involved.

Example 38. *Solve for x:* $7 = \dfrac{2}{3x}$.

The common denominator is $3x$; so multiply both sides of the equation by $3x$ in order to eliminate fractions.

$$7 \cdot 3x = \frac{2}{3x} \cdot 3x$$

or

$$21x = 2$$

Thus

$$x = \frac{2}{21} \quad \checkmark$$

Example 39. *Solve for x:* $x + \dfrac{x+1}{5} = \dfrac{1}{2}$.

The common denominator is LCM (5, 2), which is 10. So multiply each term by 10 to eliminate all denominators.

$$x \cdot 10 + \frac{x+1}{5} \cdot 10 = \frac{1}{2} \cdot 10 \qquad \begin{cases} x \cdot 10 = 10x \\ \dfrac{x+1}{5} \cdot 10 = \dfrac{(x+1) \cdot 2 \cdot \cancel{5}}{\cancel{5}} = (x+1)2 \\ \dfrac{1}{2} \cdot 10 = 5 \end{cases}$$

or

$$10x + (x+1)2 = 5$$

$$10x + 2x + 2 = 5$$

$$12x + 2 = 5$$

$$12x = 3$$

$$x = \frac{3}{12}$$

or

$$x = \frac{1}{4} \quad \checkmark$$

Example 40. *Solve for n:* $\dfrac{5}{n+1} - \dfrac{3}{20} = \dfrac{9}{5}$.

Since LCM $(n + 1, 20, 5) = (20)(n + 1)$, multiply all terms by $(20)(n + 1)$.

$$\frac{5}{n+1} \cdot (20)(n+1) - \frac{3}{20} \cdot (20)(n+1) = \frac{9}{5} \cdot (20)(n+1)$$

or

$$\frac{(5)}{(n+1)} \cdot (20)(n+1) - \frac{(3)}{(20)} \cdot (20)(n+1) = \frac{(9)}{(5)} \cdot \overset{4}{(20)}(n+1)$$

which simplifies to

$$(5)(20) - (3)(n+1) = (9)4(n+1)$$
$$100 - 3(n+1) = 36(n+1)$$
$$100 - 3n - 3 = 36n + 36$$
$$97 - 3n = 36n + 36$$
$$97 = 39n + 36$$
$$61 = 39n$$
$$39n = 61$$
$$n = \frac{61}{39} \quad \checkmark$$

Example 41. *Solve for x:* $\dfrac{7}{3x+1} + 5 = \dfrac{5x}{x-1}$.

LCM $(3x + 1, x - 1) = (3x + 1)(x - 1)$. So multiply all terms by $(3x + 1)(x - 1)$.

$$\frac{7}{3x+1} \cdot (3x+1)(x-1) + 5 \cdot (3x+1)(x-1) = \frac{5x}{x-1} \cdot (3x+1)(x-1)$$

Notice now how the denominators divide out and that no fractions are left.

$$\frac{7}{3x+1} \cdot (3x+1)(x-1) + 5 \cdot (3x+1)(x-1) = \frac{5x}{x-1} \cdot (3x+1)(x-1)$$

The simplified equation is

$$7(x-1) + 5(3x+1)(x-1) = 5x(3x+1)$$

Then

$$7x - 7 + 15x^2 - 10x - 5 = 15x^2 + 5x$$

Although this may seem to be a quadratic equation, when $-15x^2$ is added to each side in order to combine the x^2 terms on the same side, the x^2 terms add up to zero. If they did not, we would indeed have a quadratic equation to solve. As it is, however, the equation is linear. With the $15x^2$ terms removed, the equation is

$$7x - 7 - 10x - 5 = 5x$$

or

$$-3x - 12 = 5x$$

or

$$-8x - 12 = 0$$

$$-8x = 12$$

$$x = \frac{12}{-8} = -\frac{3}{2} \quad \checkmark$$

Two notes are in order here.

Note 1. Since division by zero is not defined, *no fraction can have a denominator equal to zero* in any equation with fractions. In Example 38, x cannot be 0 or else the denominator $3x$ would be 0. In Example 41, x cannot be either 1 or $-\frac{1}{3}$ for the same reason: If $x = 1$, then the denominator $x - 1$ is 0; if $x = -\frac{1}{3}$, then the denominator $3x + 1$ is 0.

Note 2. Whenever both sides of an equation are multiplied by a variable (as in Examples 38, 40, and 41), the solution(s) obtained must be checked in the original equation. *A number can be a solution of the altered equation even though it may not be a solution of the original.*

EXERCISES 18.7

1. Solve each equation.

*(a) $\dfrac{4}{9x} = 5$

(b) $9 = \dfrac{2}{7x}$

*(c) $x + \dfrac{x + 2}{3} = 7$

*(d) $x + \dfrac{x + 3}{5} = \dfrac{1}{3}$

*(e) $\dfrac{m}{4} + \dfrac{m + 1}{3} = 5$

*(f) $\dfrac{5}{x} + \dfrac{6}{5x} = 2$

*(g) $\dfrac{x + 7}{3} + \dfrac{2}{15} = \dfrac{3}{8}$

(h) $\dfrac{16}{x} + \dfrac{5}{7} = \dfrac{2}{x}$

*(i) $\dfrac{x + 1}{2x} - \dfrac{1}{x} = \dfrac{3}{5}$

(j) $\dfrac{x + 5}{x} - \dfrac{2}{3} = \dfrac{7}{x}$

2. Solve each equation.

*(a) $\dfrac{6}{x-1} = 4$

(b) $\dfrac{3}{x+5} = 7$

*(c) $\dfrac{3}{x+10} - \dfrac{8}{11} = \dfrac{1}{2}$

(d) $\dfrac{5}{12} + \dfrac{2}{x+7} = \dfrac{1}{4}$

*(e) $\dfrac{5}{x-1} + \dfrac{3x}{x-1} + \dfrac{7}{x-1} = 0$

(f) $\dfrac{x}{x+2} + \dfrac{1}{5} + \dfrac{5}{x+2} = 0$

*3. Solve each equation.

(a) $\dfrac{7x}{x+2} + \dfrac{3}{x-5} = 7$

(b) $\dfrac{9}{x+1} + \dfrac{3}{x-2} = 0$

(c) $\dfrac{x-9}{x+2} = \dfrac{x-6}{x-3}$

(d) $\dfrac{x+5}{x-4} = \dfrac{x+1}{x+7}$

Chapter 18. REVIEW EXERCISES

*1. Perform the indicated operation. Reduce if possible.

(a) $\dfrac{5x}{8+x} \cdot \dfrac{x+8}{10y}$

(b) $\dfrac{x^2-9}{x+3} \cdot \dfrac{4x}{3-x}$

(c) $\dfrac{9xy}{10} \div \dfrac{3y}{5x}$

(d) $\dfrac{4}{2x+10} \div \dfrac{x+3}{x^2+3x-10}$

(e) $\dfrac{w}{x} + \dfrac{y}{z}$

(f) $\dfrac{3}{7x} + \dfrac{2}{5x}$

(g) $\dfrac{9}{x+1} + \dfrac{y}{x}$

(h) $\dfrac{5}{3xy} + \dfrac{3}{xz} - \dfrac{7}{5yz}$

(i) $\dfrac{x^2+6x+5}{x^2-1} \cdot \dfrac{3x-3}{18x}$

(j) $\dfrac{16-x^2}{x-4} \div \dfrac{2x+8}{x+1}$

(k) $\dfrac{16x^3y^2}{7abx} \cdot \dfrac{21ab^2}{4x}$

(l) $\dfrac{b}{a} + \dfrac{d}{c} + \dfrac{f}{e}$

(m) $\dfrac{t^2-u^2}{2t+2u} \div \dfrac{5u}{4t}$

(n) $\dfrac{4x^2+8}{4x} \cdot \dfrac{10x}{5x^2+10}$

(o) $\dfrac{5}{3xy} + \dfrac{19}{x^2} + \dfrac{1}{2y}$

(p) $\dfrac{5}{18x} + \dfrac{4}{15y}$

(q) $\dfrac{1-x}{x+y} \cdot \dfrac{y+x}{x-1}$

(r) $\dfrac{3}{x^2-x} + \dfrac{4}{5x-5}$

*2. Simplify.

(a) $\dfrac{\dfrac{x+1}{y}}{\dfrac{x-1}{x}}$

(b) $\dfrac{3-\dfrac{1}{m}}{\dfrac{2}{m}+1}$

*3. Solve each equation.

(a) $x + \dfrac{x + 5}{2} = 28$

(b) $\dfrac{x}{x + 3} - \dfrac{5}{8} = 0$

4. Reduce each fraction, if possible.

*(a) $\dfrac{2x^2 - 2}{x + 1}$

(b) $\dfrac{2x^2 + 7x + 3}{x^2 + 4x + 3}$

*(c) $\dfrac{2x^2 - 18}{2x^2 + 10x + 12}$

(d) $\dfrac{3x^2 + 7x - 6}{3x^2 - 4x - 4}$

*(e) $\dfrac{2x^2 + 8x + 8}{4x^2 - 16}$

(f) $\dfrac{2x^2 - 13x - 7}{14 - 2x}$

Properties
of Exponents

19.1 SOME BASIC CONCEPTS

The concept of exponent was introduced in Chapter 1 and was used during the arithmetic presentation. Then in Sections 11.2 and 11.4 two basic properties of exponents were explained and applied. For review, the two properties will be restated here along with some examples.

$$x^a \cdot x^b = x^{a+b}$$

Examples:

$$x^3 \cdot x^4 = x^{3+4} = x^7$$
$$y^2 \cdot y^9 = y^{2+9} = y^{11}$$
$$5x^2 \cdot x^{10} \cdot x^8 = 5 \cdot x^{2+10+8} = 5x^{20}$$
$$(-5y^3)(-7y^7) = (-5)(-7)y^3 y^7 = 35y^{10}$$

$$\frac{x^a}{x^b} = x^{a-b}$$

Examples:

$$\frac{x^{15}}{x^4} = x^{15-4} = x^{11}$$

$$\frac{m^7}{m^2} = m^{7-2} = m^5$$

Next, consider $(x^3)^5$, in which x^3 is raised to the fifth power.

$$(x^3)^5 = x^3 \cdot x^3 \cdot x^3 \cdot x^3 \cdot x^3$$

$$= \underline{xxx} \cdot \underline{xxx} \cdot \underline{xxx} \cdot \underline{xxx} \cdot \underline{xxx}$$

$$= x^{15}$$

You could have saved steps by reasoning that there would be 5 products of the form x^3, and so the resulting exponent would be five 3's or 15. In general,

$$\boxed{(x^a)^n = x^{a \cdot n}}$$

Example 1. *Simplify $(x^{10})^5$.*

$$(x^{10})^5 = x^{10 \cdot 5} = x^{50} \quad \checkmark$$

Example 2. *Simplify $(m^{40})^{20}$.*

$$(m^{40})^{20} = m^{40 \cdot 20} = m^{800} \quad \checkmark$$

Example 3. *Simplify $(x^5)^3(3x^7)$.*

$$(x^5)^3(3x^7) = x^{5 \cdot 3} \cdot 3x^7$$

$$= x^{15} \cdot 3 \cdot x^7$$

$$= 3x^{22} \quad \checkmark$$

Now consider $(xy)^5$, a product raised to a power.

$$(xy)^5 = \underline{xy} \cdot \underline{xy} \cdot \underline{xy} \cdot \underline{xy} \cdot \underline{xy}$$

which can be regrouped as

$$x \cdot x \cdot x \cdot x \cdot x \cdot y \cdot y \cdot y \cdot y \cdot y$$

and simplified to

$$x^5 y^5$$

Looking at this example for a mechanical shortcut, you see that the exponent 5 of $(xy)^5$ has been applied to each factor—to x and to y.

$$(xy)^5 = x^5 y^5$$

In general,

$$\boxed{(xy)^n = x^n y^n}$$

Example 4. *Simplify $(ab)^7$.*

$$(ab)^7 = a^7 b^7 \quad \checkmark$$

Example 5. *Simplify $(2mn)^4$.*

$$(2mn)^4 = 2^4 m^4 n^4$$
$$= 16 m^4 n^4 \quad \checkmark$$

Example 6. *Simplify $(x^2 y^3)^4$.*

$$(x^2 y^3)^4 = (x^2)^4 (y^3)^4$$
$$= x^8 y^{12} \quad \checkmark$$

This last example suggests the following general result.

$$\boxed{(x^a y^b)^n = x^{a \cdot n} y^{b \cdot n}}$$

Example 7. *Simplify $(m^9 n^3)^4$.*

$$(m^9 n^3)^4 = m^{9 \cdot 4} n^{3 \cdot 4}$$
$$= m^{36} n^{12} \quad \checkmark$$

Example 8. *Simplify $(2x^2 y^5 z)^8$.*

$$(2x^2 y^5 z)^8 = 2^8 x^{2 \cdot 8} y^{5 \cdot 8} z^8$$
$$= 2^8 x^{16} y^{40} z^8$$
$$= 256 x^{16} y^{40} z^8 \quad \checkmark$$

Powers of constants will be worked out unless the result requires too great an effort to obtain. In Example 5 we evaluated 2^4. In Example 8 we evaluated 2^8. But we would not evaluate such expressions as 2^{20}, 5^8, or 3^{10}. This decision is purely arbitrary.

Example 9. *Simplify* $(x^2 + z^5)^9$.

$$(x^2 + z^5)^9 \neq x^{18} + z^{45}$$

If the *sum* $(x^2 + z^5)$ is raised to the ninth power, it cannot be simplified by the preceding methods. On the other hand, $(x^2 \cdot z^5)^9 = x^{18}z^{45}$.

Example 10. *Simplify* $(-3mn^3)(-4mn^4)^2$.

$$\begin{aligned}
(-3mn^3)(-4mn^4)^2 &= -3mn^3 \cdot (-4)^2(m)^2(n^4)^2 \\
&= -3 \cdot m \cdot n^3 \cdot 16 \cdot m^2 \cdot n^8 \\
&= -3 \cdot 16 \cdot m \cdot m^2 \cdot n^3 \cdot n^8 \qquad \text{after rearranging} \\
&= -48m^3n^{11} \quad \checkmark
\end{aligned}$$

Example 11. *Simplify* $\left(\dfrac{x}{y}\right)^6$.

$$\begin{aligned}
\left(\frac{x}{y}\right)^6 &= \frac{x}{y} \cdot \frac{x}{y} \cdot \frac{x}{y} \cdot \frac{x}{y} \cdot \frac{x}{y} \cdot \frac{x}{y} \\
&= \frac{x \cdot x \cdot x \cdot x \cdot x \cdot x}{y \cdot y \cdot y \cdot y \cdot y \cdot y} \\
&= \frac{x^6}{y^6} \quad \checkmark
\end{aligned}$$

From this example we can generalize about a quotient raised to a power.

$$\boxed{\left(\frac{x}{y}\right)^n = \frac{x^n}{y^n}}$$

Example 12. *Simplify* $\left(\dfrac{2a}{b}\right)^4$.

$$\begin{aligned}
\left(\frac{2a}{b}\right)^4 &= \frac{(2a)^4}{b^4} \\
&= \frac{2^4 a^4}{b^4} \\
&= \frac{16a^4}{b^4} \quad \checkmark
\end{aligned}$$

Example 13. *Simplify* $\left(\dfrac{x^2 y^3}{z^4}\right)^5$.

$$\begin{aligned}
\left(\frac{x^2 y^3}{z^4}\right)^5 &= \frac{(x^2 y^3)^5}{(z^4)^5} \\
&= \frac{x^{10} y^{15}}{z^{20}} \quad \checkmark
\end{aligned}$$

EXERCISES 19.1

Answers to starred exercises are given in the back of the book.

1. Simplify if possible.
   *(a) $x^3 \cdot x^2$
   (b) $x^5 \cdot x^9$
   *(c) $d \cdot d \cdot d \cdot d \cdot d$
   *(d) $2^{15}2^{10}$
   *(e) $x^3 \cdot x^2 \cdot x \cdot x \cdot x \cdot x$
   (f) $y \cdot y^{10} \cdot y \cdot y \cdot y$
   *(g) $m^5 m^{12} m^8$
   *(h) $x^{43}5x^{154}$
   *(i) $5^{236}5^{379}$
   *(j) $x^{100} + x^{187}$
   *(k) $a^{198}a^{35}a^{17}$
   *(l) $x + x^6$
   *(m) $m^5 + m^7 + m^3$
   (n) $m^{15}m^6m^{76}$
   *(o) $5x \cdot 8x^{10}$
   *(p) $(3x^4y^2)(4x^5y^9)$
   *(q) $(-4x^2y^3)(14xy)(x^{15}y^9)$
   *(r) $3x^7 \cdot -2x \cdot x^5$
   *(s) $3a^4b^67a^23ab$
   (t) $2a^5b^45a^24a^3b^2$
   *(u) $(-6m^4n^7)(15m^5n^4)$
   (v) $(a^7bc^8)(-b^5c^9)(-a^4c)$

2. Simplify if possible.
   *(a) $(x^4)^3$
   (b) $(x^{15})^2$
   *(c) $(r^7)^4$
   (d) $(y^5)^{12}$
   *(e) $(rs)^{14}$
   (f) $(mn)^9$
   *(g) $(xy)^{50}$
   *(h) $(pqr)^{17}$
   *(i) $(x^4y^6)^5$
   (j) $(m^5n^7)^3$
   *(k) $(a^4c^7)^2$
   *(l) $(5x^3y^8)^3$
   *(m) $(x^4 + y^5)^3$
   (n) $(2m^4n^2)^5$
   *(o) $(8xy^6)^2$
   *(p) $(-3x^5y^{10})^3$
   *(q) $(-2m^3np^2)^4$
   (r) $(-4x^2y^5z^3)^4$
   *(s) $(-5a^3b^7)^3$
   (t) $(-w^6x^4y^5z^3)^{10}$

3. Simplify if possible.
   *(a) $(3mn^2)(5m^4n^7)^2$
   *(b) $(2x^5y^2)^3(3x^4y^9)^4$
   *(c) $(-2p^4q)^3(p^8q^2)^5$
   (d) $(-fg^5h^2)^2(3f^4)^3$
   *(e) $(3x^4 - y^6)^5$
   (f) $(5xy^6)^4(x^5y^{19})$
   *(g) $(6b^4c^3d)^2(-2bc^9d^2)^3$
   *(h) $3m^2n^8 + 7m^3n^5$
   *(i) $\left(\dfrac{x}{y}\right)^9$
   (j) $\left(\dfrac{r}{s}\right)^{12}$
   *(k) $\left(\dfrac{2m}{n}\right)^5$
   (l) $\left(\dfrac{ab^3}{3}\right)^4$
   *(m) $\left(\dfrac{x^4y^5}{z^3}\right)^2$
   *(n) $\left(\dfrac{b^2c^7}{a^4d^5}\right)^3$
   *(o) $\left(\dfrac{m^7n^8p^2}{5t^2}\right)^4$
   (p) $\left(\dfrac{3x^4y^6}{2z^5}\right)^3$

*4. Evaluate. Write each as one number.
   (a) 2^3
   (b) 2^6
   (c) 5^4
   (d) 3^5
   (e) $2 \cdot 3^2$
   (f) $3 \cdot 2^4$
   (g) $(2^4)^2$
   (h) $(3^2)^3$
   (i) $\left(\dfrac{2}{3}\right)^4$
   (j) $\left(\dfrac{3}{4}\right)^3$
   (k) $(2 + 3)^4$
   (l) $(4 + 1)^3$

5. *(a) Write 4^5 as a power of 2 by first substituting for 4 a power of 2.
   (b) Write 8^3 as a power of 2.
   (c) Write 9^4 as a power of 3.

19.2 NEGATIVE EXPONENTS

Consider this example, which involves a quotient of two powers of x.

$$\frac{x^7}{x^3} = x^4$$

It can also be seen as

$$x^7 \cdot \frac{1}{x^3} = x^4$$

Suppose that we want to represent the $\frac{1}{x^3}$ as x to a power. In other words, replace $\frac{1}{x^3}$ by a power of x, as

$$x^7 \cdot x^? = x^4$$

So that it will agree with the properties of exponents, the exponent that we seek must be -3, since $7 + (-3) = 4$—that is, $x^7 \cdot x^{-3} = x^4$. This example suggests the following notation.

$$\frac{1}{x^3} = x^{-3}$$

Similarly,

$$\frac{1}{m^9} = m^{-9}$$

$$\frac{1}{y^4} = y^{-4}$$

Also,

$$x^{-5} = \frac{1}{x^5}$$

In general,

$$\boxed{x^{-n} = \frac{1}{x^n}}$$

Example 14. *Evaluate 3^{-2}.*

$$3^{-2} = \frac{1}{3^2} = \frac{1}{9} \quad \checkmark$$

Notice that this expression containing a negative exponent was changed to an expression with a positive exponent so that it could be evaluated.

Example 15. *Write $7x^{-5}y^4z^{-8}$ without using negative exponents.*

$$7x^{-5}y^4z^{-8} = 7 \cdot x^{-5} \cdot y^4 \cdot z^{-8}$$

$$= \frac{7}{1} \cdot \frac{x^{-5}}{1} \cdot \frac{y^4}{1} \cdot \frac{z^{-8}}{1} \qquad \left\{ \begin{array}{l} \text{making each a fraction} \\ \text{with denominator 1} \end{array} \right.$$

$$= \frac{7}{1} \cdot \frac{1}{x^5} \cdot \frac{y^4}{1} \cdot \frac{1}{z^8} \qquad \left\{ x^{-5} = \frac{1}{x^5} \text{ and } z^{-8} = \frac{1}{z^8} \right.$$

$$= \frac{7 \cdot 1 \cdot y^4 \cdot 1}{1 \cdot x^5 \cdot 1 \cdot z^8}$$

$$= \frac{7y^4}{x^5 z^8} \quad \checkmark$$

Because a negative exponent can appear in the denominator of a fraction, we need to know the meaning of such exponents. Observe.

$$\frac{1}{x^{-n}} = \frac{1}{\frac{1}{x^n}} = x^n$$

Thus

$$\boxed{\frac{1}{x^{-n}} = x^n}$$

For instance,

$$\frac{1}{x^{-3}} = x^3$$

$$\frac{1}{m^{-5}} = m^5$$

$$\frac{1}{2^{-4}} = 2^4$$

$$n^4 = \frac{1}{n^{-4}}$$

$5n^6$

Example 16. *Write $\dfrac{5x^{-2}yz^9}{m^2 n^{-6}w^3}$ without using negative exponents.*

$\times 2$

$$\frac{5x^{-2}yz^9}{m^2 n^{-6}w^3} = \frac{5x^{-2}yz^9}{1} \cdot \frac{1}{m^2 n^{-6}w^3}$$

$$= \frac{5}{1} \cdot \frac{x^{-2}}{1} \cdot \frac{y}{1} \cdot \frac{z^9}{1} \cdot \frac{1}{m^2} \cdot \frac{1}{n^{-6}} \cdot \frac{1}{w^3}$$

Change the two negative exponents that appear.

$$= \frac{5}{1} \cdot \frac{1}{x^2} \cdot \frac{y}{1} \cdot \frac{z^9}{1} \cdot \frac{1}{m^2} \cdot \frac{n^6}{1} \cdot \frac{1}{w^3}$$

$$= \frac{5yz^9n^6}{x^2m^2w^3} \quad \text{or} \quad \frac{5n^6yz^9}{m^2w^3x^2} \quad \checkmark$$

The final form leaves the letters in alphabetical order.

Compare the final form containing only positive exponents with the original form, which contained negative exponents. Notice that a negative exponent in the numerator became a positive exponent in the denominator and that a negative exponent in the denominator became a positive exponent in the numerator. Mechanically, that is what our two boxed definitions of negative exponents accomplish.

Example 17. *Write* $\dfrac{-5x^{-3}}{y^{-4}}$ *without using negative exponents.*

$$\frac{-5x^{-3}}{y^{-4}} = \frac{-5x^{-3}}{1} \cdot \frac{1}{y^{-4}}$$

$$= \frac{-5}{1} \cdot \frac{x^{-3}}{1} \cdot \frac{1}{y^{-4}}$$

$$= \frac{-5}{1} \cdot \frac{1}{x^3} \cdot \frac{y^4}{1}$$

$$= \frac{-5y^4}{x^3} \quad \checkmark \quad \text{or} \quad -\frac{5y^4}{x^3} \quad \checkmark$$

Notice that the -5 in the numerator *does not* become a $+5$ in the denominator. It is *not an exponent* and should not be expected to behave like one.

Example 18. *Simplify* $(3x^5y^4z^{-8})^{-2}$ *and write it without using negative exponents.*

Since this is a product raised to a power, the power, -2, is distributed. The result is

$$(3x^5y^4z^{-8})^{-2} = 3^{-2}x^{-10}y^{-8}z^{16}$$

$$= \frac{1}{3^2} \cdot \frac{1}{x^{10}} \cdot \frac{1}{y^8} \cdot \frac{z^{16}}{1}$$

$$= \frac{z^{16}}{9x^{10}y^8} \quad \checkmark$$

Example 19. *Simplify $7x^{-5}x^{-3}$ and write it without using negative exponents.*

Since the bases are the same and the numbers are being multiplied, add the exponents.

$$7x^{-5}x^{-3} = 7 \cdot x^{-5-3}$$
$$= 7 \cdot x^{-8}$$
$$= 7 \cdot \frac{1}{x^8}$$
$$= \frac{7}{x^8} \quad \checkmark$$

EXERCISES 19.2

1. Evaluate. Write each as one number.
   - *(a) 2^{-3}
   - (b) 2^{-4}
   - *(c) 3^{-3}
   - (d) 5^{-2}
   - *(e) 6^{-1}
   - (f) 4^{-4}
   - *(g) $3 \cdot 5^{-1}$
   - (h) $5 \cdot 2^{-3}$
   - *(i) $7 \cdot 8^{-1}$
   - (j) $9 \cdot 3^{-2}$
   - *(k) $(2 \cdot 3)^{-2}$
   - (l) $(3 \cdot 4)^{-3}$

2. Write without using negative exponents.
   - *(a) x^{-5}
   - (b) y^{-9}
   - *(c) $4m^{-6}$
   - (d) $3x^{-2}$
   - *(e) $-7t^{-4}$
   - (f) $-5b^{-8}$
   - *(g) $(3x)^{-2}$
   - (h) $(5y)^{-3}$
   - *(i) $\dfrac{n^{-7}}{4}$
   - (j) $\dfrac{r^{-5}}{3}$
   - *(k) $\dfrac{2m^{-4}}{7}$
   - (l) $\dfrac{3x^{-9}}{2}$
   - *(m) $5x^{-7}y^{-4}$
   - (n) $6m^{-9}n^{-3}$
   - *(o) $7m^{-6}n^7$
   - (p) $19b^{-2}c^6$
   - *(q) $a^2b^{-3}c^4d^{-5}$
   - (r) $8w^6x^{-4}y^{-5}z$
   - *(s) $-8m^{-7}n^2p^{-9}$
   - (t) $-5ab^7c^{-8}d^{-2}$

3. Write without using negative exponents.
   - *(a) $\dfrac{3x^{-4}y}{z^5}$
   - (b) $\dfrac{7m^{-3}n}{p^6}$
   - *(c) $\dfrac{5a^{-4}b^{-6}}{2c}$
   - (d) $\dfrac{9wx^{-6}y^{-8}}{5z}$
   - *(e) $\dfrac{a^{-4}b^{-3}}{c^{-8}}$
   - (f) $\dfrac{x^{-9}y^{-3}}{z^{-5}}$
   - *(g) $\dfrac{7a^{-6}b^4c^{-2}}{x^5y^{-3}}$
   - (h) $\dfrac{4m^5n^{-9}}{p^{-7}r^3}$
   - *(i) $\dfrac{-5x^{-4}y^{-5}z}{v^{-3}w^2}$
   - (j) $\dfrac{-9a^{-9}b^{-7}c^{-4}}{5d^{-1}e^{-5}}$

4. Simplify each expression and write it without using negative exponents.
   - *(a) $x^{-2}x^{-7}$
   - (b) $a^{-5}a^{-7}$
   - *(c) $6t^{-4} \cdot t^{-3} \cdot t^{-6}$
   - (d) $4x^{-3} \cdot x^{-7} \cdot x^{-8}$
   - *(e) $(x^{-3}y^{-5})^{-4}$
   - (f) $(a^{-7}b^{-4})^{-2}$
   - *(g) $(2x^{-6}y^{-5})^{-3}$
   - (h) $(3x^{-2}y^{-8})^{-4}$
   - *(i) $(5a^{-3}b^{-7}c^{-5})^{-2}$
   - (j) $(4a^{-2}b^{-4}c^{-6})^{-2}$

19.3 DIVISION AND EXPONENTS

We already have the property

$$\frac{x^a}{x^b} = x^{a-b}$$

which we have used in simple cases where a was greater than b. For example,

$$\frac{x^{29}}{x^{17}} = x^{29-17} = x^{12}$$

We continue with more complicated cases.

Example 20. *Simplify* $\frac{x^5 y^9}{x^2 y^4}$.

$$\frac{x^5 y^9}{x^2 y^4} = \frac{x^5}{x^2} \cdot \frac{y^9}{y^4}$$

$$= x^{5-2} y^{9-4}$$

$$= x^3 y^5 \quad \checkmark$$

Notice that this procedure did result in the fraction being reduced.

The subtraction of exponents also applies when the power in the denominator is larger than the power in the numerator. To see this fact, notice first that

$$\frac{x^3}{x^7} = \frac{xxx}{xxxxxxx} = \frac{1}{xxxx} = \frac{1}{x^4}$$

If, however, you perform the subtraction of exponents instead, the result is

$$\frac{x^3}{x^7} = x^{3-7} = x^{-4}$$

and we know that x^{-4} is the same as $\frac{1}{x^4}$.

Example 21. *Simplify* $\frac{4l^5 m^4 n^3}{3m^9 n}$. *Do not leave negative exponents in the result.*

$$\frac{4l^5 m^4 n^3}{3m^9 n} = \frac{4}{3} \cdot \frac{l^5}{1} \cdot \frac{m^4}{m^9} \cdot \frac{n^3}{n^1}$$

$$= \frac{4}{3} l^5 m^{4-9} n^{3-1}$$

$$= \frac{4}{3} l^5 m^{-5} n^2$$

$$= \frac{4}{3} \cdot l^5 \cdot \frac{1}{m^5} \cdot n^2 = \frac{4l^5 n^2}{3m^5} \quad \checkmark$$

Example 22. *Simplify* $\dfrac{w^{-6}}{w^2}$. *Do not leave any negative exponent in the result.*

Shown below are two different ways of obtaining the final result,

$$\frac{1}{w^8} \quad \checkmark$$

1. $\dfrac{w^{-6}}{w^2} = w^{-6-2} = w^{-8} = \dfrac{1}{w^8}$

2. $\dfrac{w^{-6}}{w^2} = \dfrac{w^{-6}}{1} \cdot \dfrac{1}{w^2} = \dfrac{1}{w^6} \cdot \dfrac{1}{w^2} = \dfrac{1}{w^6 w^2} = \dfrac{1}{w^8}$

The result of Example 22 will appear as part of the next example, which is somewhat more complicated.

Example 23. *Simplify and reduce completely. Write without negative exponents.*

$$\frac{7w^{-6}x^{10}y^{-9}z^{13}}{21w^2x^{-4}z^5}$$

$$\frac{7w^{-6}x^{10}y^{-9}z^{13}}{21w^2x^{-4}z^5} = \frac{7}{21} \cdot \frac{w^{-6}}{w^2} \cdot \frac{x^{10}}{x^{-4}} \cdot \frac{y^{-9}}{1} \cdot \frac{z^{13}}{z^5}$$

$$= \frac{1}{3} \cdot \frac{1}{w^8} \cdot \frac{x^{14}}{1} \cdot \frac{1}{y^9} \cdot \frac{z^8}{1}$$

$$= \frac{x^{14}z^8}{3w^8y^9} \quad \checkmark$$

EXERCISES 19.3

1. Simplify each term, leaving positive exponents only. Be sure each fraction is completely reduced.

*(a) $\dfrac{x^{40}}{x^{10}}$ (b) $\dfrac{m^{17}}{m^2}$ *(c) $\dfrac{y^{18}}{y^{15}}$ (d) $\dfrac{n^{58}}{n^{31}}$

*(e) $\dfrac{m^6}{m^9}$ (f) $\dfrac{t^{15}}{t^{19}}$ *(g) $\dfrac{x^6 y^{10}}{x^2 y^7}$ (h) $\dfrac{m^5 n^{12}}{m^3 n^{10}}$

*(i) $\dfrac{x^5 y^7}{x y^4}$ (j) $\dfrac{x^7 y^2}{x^5 y^6}$ *(k) $\dfrac{a^3 b^8}{a^7 b^{19}}$ (l) $\dfrac{m^6 n^2}{m^{10} n^9}$

*(m) $\dfrac{p^2 q^3 r^4}{p^7 q^6 r^5}$ (n) $\dfrac{w^3 x^9 y^5}{w^{10} x y^{23}}$

2. Simplify each term, leaving positive exponents only. Be sure each fraction is completely reduced.

*(a) $\dfrac{4x^9 y^2}{x^2 y^9}$

(b) $\dfrac{-8x^2 y^{11}}{x^{24} y^{87}}$

*(c) $\dfrac{4x^9 y^2 z}{5xy^3 z^2}$

(d) $\dfrac{7a^3 b^9 c^5}{2a^4 c}$

*(e) $\dfrac{4w^{-4} x^5}{5w^3 x^{-2}}$

(f) $\dfrac{7a^8 b^{-3} c^3}{9a^{-2} b^5}$

*(g) $\dfrac{2x^2 y^{-3} z^8}{x^{-5} y z^4}$

(h) $\dfrac{25p^{-3} q^{-7} r^{-2}}{p^{-4} q^6 r}$

*(i) $\dfrac{m^{-7} n^8 p^{-9}}{m^{-9} n^9 p^{-4}}$

(j) $\dfrac{a^7 b^{-6} c^{12}}{a^{-1} b c^{-4} d^{-2}}$

*(k) $\dfrac{w^6 x^{-9} y^{-2} z^{-5}}{7w^{10} x^{-3} y^7}$

(l) $\dfrac{r^{-14} s^{-11} t^{-5}}{9r^{-2} s t^{-23}}$

19.4 ZERO EXPONENT

Zero is the only integer we had not previously used as an exponent. The zero exponent is introduced for completeness. *Any nonzero number raised to the zero power is 1*, as justified by the work shown next.

Let's attempt to simplify $\dfrac{x^5}{x^5}$ by using $\dfrac{x^m}{x^n} = x^{m-n}$.

$$\frac{x^5}{x^5} = x^{5-5} = x^0$$

We might also note that $\dfrac{x^5}{x^5} = 1$, since a number divided by itself is 1.

As a result, we define the exponent zero as follows.

$$\boxed{x^0 = 1}$$

Example 24. *Evaluate* $8^0 + 253^0$.

$$8^0 + 253^0 = 1 + 1 \qquad \text{since } 8^0 = 1 \text{ and } 253^0 = 1$$
$$= 2 \quad \checkmark$$

Example 25. *Evaluate* $(2 + 4)^0 + 3 \cdot 7^0 + (4 \cdot 2)^0$.

$$(2 + 4)^0 + 3 \cdot 7^0 + (4 \cdot 2)^0 = (6)^0 + 3 \cdot 7^0 + (8)^0$$
$$= 1 + 3 \cdot 1 + 1$$
$$= 1 + 3 + 1$$
$$= 5 \quad \checkmark$$

Example 26. *Simplify* $(3x + 2)^0$.

Once a value is supplied for x, the entire expression $3x + 2$ will be one number. (For example, if x is 5, then $3x + 2$ is 17.) Thus

$$(3x + 2)^0 = 1 \quad \checkmark$$

EXERCISES 19.4

1. Evaluate each expression.

*(a) 5^0

*(c) $(3 \cdot 4)^0$

*(e) $5x^0$

*(g) $6 \cdot 2^0$

*(i) $2^0 \cdot 7$

*(k) $-3x^0$

*(m) $(5x - 1)^0$

*(o) $9x^0 - 8 - (9x)^0$

*(b) $3 \cdot 4^0$

(d) $(abc)^0$

(f) $(5x)^0$

*(h) $4^0 + (-3)^0$

(j) $4^0 + 3^0 - 7^0$

(l) $(3m)^0 + 3m^0$

*(n) $5x - 1^0$

(p) $7 \cdot y^0 \cdot 3$

19.5 FRACTIONAL EXPONENTS

Next, we proceed to obtain an exponent notation for radicals. From the definition of square root we know that for any nonnegative number x,

$$\sqrt{x} \cdot \sqrt{x} = x$$

That is, the square root of x is a number that when squared gives x.

If we examine $\sqrt{x}\sqrt{x} = x$ in terms of exponents by letting $\sqrt{x}$ be x^m, we obtain

$x^m x^m = x$ after replacing $\sqrt{x}$ by x^m in $\sqrt{x}\sqrt{x} = x$

$x^m x^m = x^1$ since x^1 is the same as x

$x^{m+m} = x^1$ by a rule for exponents

$x^{2m} = x^1$ $m + m = 2m$

Then since $2m = 1$, m must be equal to $\frac{1}{2}$. Thus

$$x^{\frac{1}{2}} \cdot x^{\frac{1}{2}} = x^{\frac{1}{2}+\frac{1}{2}} = x^1 = x$$

So

$$\boxed{\sqrt{x} = x^{\frac{1}{2}}}$$

Example 27. *Evaluate* $25^{\frac{1}{2}}$.

$$25^{\frac{1}{2}} = \sqrt{25}$$
$$= 5 \quad \checkmark$$

Example 28. *Evaluate* $3 \cdot 16^{\frac{1}{2}}$.

$$3 \cdot 16^{\frac{1}{2}} = 3 \cdot \sqrt{16}$$
$$= 3 \cdot 4$$
$$= 12 \quad \checkmark$$

If you have forgotten the concept of cube root, note that the *cube root* of 8 is 2, since 2 cubed is 8. In symbols, $\sqrt[3]{8} = 2$, since $2^3 = 8$. Here is a brief table of some common cube roots.

$$
\begin{array}{lll}
\sqrt[3]{0} = 0 & \text{since} & 0^3 = 0 \\
\sqrt[3]{1} = 1 & \text{since} & 1^3 = 1 \\
\sqrt[3]{8} = 2 & \text{since} & 2^3 = 8 \\
\sqrt[3]{27} = 3 & \text{since} & 3^3 = 27 \\
\sqrt[3]{64} = 4 & \text{since} & 4^3 = 64 \\
\sqrt[3]{125} = 5 & \text{since} & 5^3 = 125 \\
\sqrt[3]{216} = 6 & \text{since} & 6^3 = 216 \\
\end{array}
$$

We can determine the exponent form of $\sqrt[3]{x}$ by using the same reasoning as with square root. The cube root of x is a number $\sqrt[3]{x}$ such that $\sqrt[3]{x}\sqrt[3]{x}\sqrt[3]{x} = x$. So $\sqrt[3]{x} = x^{\frac{1}{3}}$, since $x^{\frac{1}{3}} \cdot x^{\frac{1}{3}} \cdot x^{\frac{1}{3}} = x^{\frac{1}{3}+\frac{1}{3}+\frac{1}{3}} = x^1 = x$.

$$\boxed{\sqrt[3]{x} = x^{\frac{1}{3}}}$$

Example 29. *Evaluate* $64^{\frac{1}{2}} + 8^{\frac{1}{3}}$.

$$64^{\frac{1}{2}} = \sqrt{64} = 8$$
$$8^{\frac{1}{3}} = \sqrt[3]{8} = 2$$

so

$$64^{\frac{1}{2}} + 8^{\frac{1}{3}} = 8 + 2 = 10 \quad \checkmark$$

Similarly, for fourth root, $\sqrt[4]{x} = x^{\frac{1}{4}}$. Similarly, for fifth root, $\sqrt[5]{x} = x^{\frac{1}{5}}$. In general,

$$\boxed{\sqrt[n]{x} = x^{\frac{1}{n}}}$$

Next, let's consider fractional exponents in which the numerator is not 1. For instance, let's look at $x^{\frac{3}{4}}$.

$$x^{\frac{3}{4}} = x^{\frac{1}{4} \cdot 3} = (x^{\frac{1}{4}})^3 \text{ or } (x^3)^{\frac{1}{4}}$$

Similarly, for $x^{\frac{3}{2}}$,

$$x^{\frac{3}{2}} = (x^{\frac{1}{2}})^3 \text{ or } (x^3)^{\frac{1}{2}}$$

In general,

$$\boxed{x^{\frac{m}{n}} = (x^{\frac{1}{n}})^m \quad \text{or} \quad (x^m)^{\frac{1}{n}}}$$

Keep in mind that x must be nonnegative when the root is even (square root, fourth root, etc.).

Example 30. *Evaluate (a)* $8^{\frac{2}{3}}$, *(b)* $4^{\frac{5}{2}}$, *and (c)* $27^{\frac{1}{3}}$.

(a) $8^{\frac{2}{3}} = (8^{\frac{1}{3}})^2 = (\sqrt[3]{8})^2 = (2)^2 = 4$ ✓

(b) $4^{\frac{5}{2}} = (4^{\frac{1}{2}})^5 = (\sqrt[2]{4})^5 = (2)^5 = 32$ ✓

(c) $27^{\frac{1}{3}} = (27)^{\frac{1}{3}} = \sqrt[3]{27} = 3$ ✓

Generally *it is easier to simplify by taking the root first and then applying the power,* as done above, because the root is easier to determine when the number is smaller. Compare.

$$\text{Root first:} \quad 8^{\frac{5}{3}} = (8^{\frac{1}{3}})^5 = (2)^5 = 32$$
$$\text{Power first:} \quad 8^{\frac{5}{3}} = (8^5)^{\frac{1}{3}} = (32768)^{\frac{1}{3}} = 32$$

Example 31. *Evaluate* $27^{\frac{2}{3}} + 16^{\frac{5}{4}} - 4 \cdot 5^0$.

$$27^{\frac{2}{3}} + 16^{\frac{5}{4}} - 4 \cdot 5^0 = (27^{\frac{1}{3}})^2 + (16^{\frac{1}{4}})^5 - 4 \cdot 5^0$$
$$= (3)^2 + (2)^5 - 4 \cdot 1$$
$$= 9 + 32 - 4$$
$$= 37 \quad ✓$$

Example 32. *Determine the value of $9^{1.5}$.*

Begin by writing the exponent 1.5 as a mixed number, $1\frac{1}{2}$. Then change the mixed number to an improper fraction, namely, $\frac{3}{2}$.

$$9^{1.5} = 9^{1\frac{1}{2}} = 9^{\frac{3}{2}} = (9^{\frac{1}{2}})^3 = (3)^3 = 27 \quad \checkmark$$

Example 33. *Simplify $\left(\frac{49}{4}\right)^{\frac{3}{2}}$.*

$$\left(\frac{49}{4}\right)^{\frac{3}{2}} = \left[\left(\frac{49}{4}\right)^{\frac{1}{2}}\right]^3$$

$$= \left[\frac{49^{\frac{1}{2}}}{4^{\frac{1}{2}}}\right]^3$$

$$= \left(\frac{7}{2}\right)^3$$

$$= \frac{7^3}{2^3}$$

$$= \frac{343}{8} \quad \checkmark$$

Example 34. *Evaluate $9^{-\frac{1}{2}}$.*

$$9^{-\frac{1}{2}} = \frac{1}{9^{\frac{1}{2}}}$$

$$= \frac{1}{\sqrt{9}}$$

$$= \frac{1}{3} \quad \checkmark$$

Example 35. *Evaluate $8^{-\frac{5}{3}}$.*

$$8^{-\frac{5}{3}} = \frac{1}{8^{\frac{5}{3}}}$$

$$= \frac{1}{(8^{\frac{1}{3}})^5}$$

$$= \frac{1}{(2)^5}$$

$$= \frac{1}{32} \quad \checkmark$$

Example 36. *Simplify* $(x^{\frac{2}{3}})^{12}$.

$$(x^{\frac{2}{3}})^{12} = x^{\frac{2}{3} \cdot 12}$$
$$= x^8 \quad \checkmark$$

Example 37. *Simplify* $(x^2 y^{10} z)^{\frac{1}{2}}$.

$$(x^2 y^{10} z)^{\frac{1}{2}} = x^{2 \cdot \frac{1}{2}} y^{10 \cdot \frac{1}{2}} z^{\frac{1}{2}}$$
$$= x^1 \cdot y^5 \cdot z^{\frac{1}{2}}$$
$$= xy^5 z^{\frac{1}{2}} \quad \checkmark$$

Example 38. *Simplify* $x^{\frac{1}{2}} \cdot x^{\frac{1}{3}}$.

$$x^{\frac{1}{2}} \cdot x^{\frac{1}{3}} = x^{\frac{1}{2} + \frac{1}{3}}$$
$$= x^{\frac{3}{6} + \frac{2}{6}}$$
$$= x^{\frac{5}{6}} \quad \checkmark$$

Example 39. *Evaluate* $\sqrt{9 x^{10} y^6}$.

$$\sqrt{9 x^{10} y^6} = (9 x^{10} y^6)^{\frac{1}{2}}$$
$$= 9^{\frac{1}{2}} x^5 y^3$$
$$= 3 x^5 y^3 \quad \checkmark$$

Example 40. *Write* $\sqrt{x^5}$, *using fractional exponents.*

$$\sqrt{x^5} = (x^5)^{\frac{1}{2}}$$
$$= x^{\frac{5}{2}} \quad \checkmark$$

Example 41. *Write* $x^{\frac{2}{3}}$, *using a radical.*

$$x^{\frac{2}{3}} = (x^2)^{\frac{1}{3}}$$
$$= \sqrt[3]{x^2} \quad \checkmark$$

EXERCISES 19.5

1. Evaluate each expression.
   *(a) $25^{\frac{1}{2}}$ (b) $81^{\frac{1}{2}}$
   *(c) $100^{\frac{1}{2}}$ (d) $49^{\frac{1}{2}}$
   *(e) $8^{\frac{1}{3}}$ (f) $1^{\frac{1}{3}}$
   *(g) $27^{\frac{1}{3}}$ (h) $64^{\frac{1}{3}}$

*(i) $125^{\frac{1}{3}}$

*(k) $9 \cdot 49^{\frac{1}{2}}$

*(m) $5 \cdot 27^{\frac{1}{3}}$

*(o) $(64^{\frac{1}{2}} + 8^{\frac{1}{3}})^0$

*(q) $\left(\dfrac{9}{25}\right)^{\frac{1}{2}}$

*(s) $\left(\dfrac{8}{27}\right)^{\frac{1}{3}}$

*(u) $\left(\dfrac{81}{4}\right)^{\frac{1}{2}} + \left(\dfrac{27}{8}\right)^{\frac{1}{3}}$

*(j) $5 \cdot 4^{\frac{1}{2}}$

(l) $10 \cdot 144^{\frac{1}{2}}$

(n) $7 \cdot 9^{\frac{1}{2}} + 3 \cdot 125^{\frac{1}{3}}$

(p) $(27^{\frac{1}{3}} + 49^{\frac{1}{2}})^0$

(r) $\left(\dfrac{4}{9}\right)^{\frac{1}{2}}$

(t) $\left(\dfrac{64}{125}\right)^{\frac{1}{3}}$

(v) $\left(\dfrac{9}{4}\right)^{\frac{1}{2}} + \left(\dfrac{1}{8}\right)^{\frac{1}{3}}$

2. Evaluate each expression.

*(a) $4^{\frac{3}{2}}$

*(c) $16^{\frac{3}{2}}$

*(e) $125^{\frac{2}{3}}$

*(g) $5 \cdot 8^{\frac{2}{3}}$

*(i) $16^{\frac{1}{4}} + 16^{\frac{3}{4}}$

*(k) $27^{\frac{5}{3}} + 81^{\frac{3}{4}}$

(b) $8^{\frac{5}{3}}$

(d) $9^{\frac{5}{2}}$

(f) $8^{\frac{2}{3}} + 8^0$

*(h) $25^{\frac{3}{2}} + 4^{\frac{3}{2}}$

*(j) $4^{\frac{3}{2}} + 1^{\frac{1}{2}} + 3^0$

*(l) $64^{\frac{4}{3}} + 16^{\frac{3}{4}}$

3. Evaluate each expression.

*(a) $49^{-\frac{1}{2}}$

(d) $16^{-\frac{1}{4}}$

*(g) $8^{\frac{1}{3}} + 8^{-\frac{1}{3}}$

(j) $16^{-\frac{3}{4}} + 4^{-\frac{3}{2}}$

(b) $64^{-\frac{1}{2}}$

*(e) $4^{-\frac{5}{2}}$

(h) $25^{-\frac{3}{2}}$

*(c) $27^{-\frac{1}{3}}$

(f) $64^{-\frac{2}{3}}$

*(i) $1^{-\frac{5}{2}} + 9^{\frac{3}{2}}$

4. Evaluate each expression.

*(a) $4^{2.5}$

(f) $4^{3.5}$

(b) $16^{2.5}$

*(g) $9^{-1.5}$

*(c) $9^{2.5}$

(h) $4^{-2.5}$

(d) $25^{1.5}$

*(i) $1^{-2.5}$

*(e) $1^{3.5}$

(j) $16^{-1.5}$

5. Simplify.

*(a) $(x^8 y^4)^{\frac{1}{2}}$

*(d) $(-8x^3)^{\frac{1}{3}}$

*(g) $(x^{\frac{1}{4}})^8$

(j) $(27m^{12}n^6)^{\frac{4}{3}}$

(m) $\sqrt{81x^4}$

(p) $\sqrt{169n^{50}r^{20}}$

*(s) $\sqrt[4]{16x^8}$

(v) $\sqrt[3]{-m^{99}n^{36}}$

(y) $\sqrt[4]{625x^4y^{100}}$

(b) $(16x^2)^{\frac{1}{2}}$

(e) $(27b^3)^{\frac{1}{3}}$

(h) $(m^{\frac{3}{4}})^{16}$

*(k) $(121x^{10}y^{16})^{\frac{1}{2}}$

*(n) $(16x^4y^8z^{12})^{\frac{1}{4}}$

*(q) $\sqrt[3]{216m^3}$

*(t) $\sqrt[5]{-32y^5}$

*(w) $\sqrt[4]{81a^{12}b^{28}}$

(z) $\sqrt[5]{32x^{10}z^{20}}$

*(c) $(49a^2b^6)^{\frac{1}{2}}$

(f) $(64x^3y^6)^{\frac{1}{3}}$

*(i) $(-125r^3s^9)^{\frac{2}{3}}$

*(l) $\sqrt{25a^6y^{10}}$

*(o) $\sqrt{144a^{10}b^6}$

*(r) $\sqrt[3]{-64x^3y^6}$

(u) $\sqrt[3]{-8x^6y^{21}z^3}$

(x) $\sqrt[4]{256m^{84}n^{36}}$

6. Write with fractional exponents.

  *(a) $\sqrt{x^3}$ (b) $\sqrt{m^9}$ *(c) $\sqrt[3]{x^5}$ (d) $\sqrt[3]{n^4}$ *(e) $\sqrt[4]{y^7}$
  (f) $\sqrt[4]{y^3}$

7. Write using a radical.

  *(a) $x^{\frac{5}{2}}$ (b) $m^{\frac{3}{5}}$ *(c) $n^{\frac{7}{3}}$ (d) $y^{\frac{7}{2}}$ *(e) $t^{\frac{4}{5}}$
  (f) $c^{\frac{5}{4}}$

19.6 SCIENTIFIC NOTATION (Optional)

Large numbers like

$$6720000000000000000000000$$

and small numbers like

$$.0000000000000000000000053$$

can be represented with fewer digits by using *scientific notation* (also called *standard notation*).

The large number 6720000000000000000000000 is the same as 6.72×10^{22}. We use a power of 10 to eliminate the writing of a long string of zeros. Observe.

$$8400 = 840 \times 10 \quad \text{or} \quad 840 \times 10^1$$
$$= 84 \times 100 \quad \text{or} \quad 84 \times 10^2$$
$$= 8.4 \times 1000 \quad \text{or} \quad 8.4 \times 10^3$$

The most common form of numbers in scientific notation is *a number between 1 and 10, multiplied by an appropriate power of 10*. For instance,

$$8.5 \times 10^{15}$$
$$9.213 \times 10^2$$
$$1.0037 \times 10^{-12}$$
$$4.7 \times 10^{29}$$

Example 42. *Represent 765000000000000 by using scientific notation.*

The original number 765000000000000 has 14 more digits before the decimal point than does 7.65. So it is 10^{14} times 7.65. Thus, 7.65 must be multiplied by 10^{14} to be equal to 765000000000000. So we conclude that

$$765000000000000 = 7.65 \times 10^{14} \quad \checkmark$$

Multiplication by 10^{14} has the effect of moving the decimal 14 places to the right.

$$7_\wedge\underbrace{65000000000000}_{\text{14 places}}.$$

Example 43. *Represent* 89000000, *using scientific notation.*

Since 8.9 must be multiplied by 10^7 in order to move the decimal 7 places to the right, we have

$$89000000 = 8.9 \times 10^7 \quad \checkmark$$

Example 44. *Represent* .00000034, *using scientific notation.*

The number 3.4 would have to be divided by 10^7 in order to move the decimal point 7 places to the left. In other words.

$$.00000034 = \frac{3.4}{10^7}$$

Now, using

$$\frac{3.4}{10^7} = 3.4 \times 10^{-7}$$

we obtain

$$.00000034 = 3.4 \times 10^{-7} \quad \checkmark$$

In general, to move the decimal point n places to the left, multiply by 10^{-n}.

Example 45. *Represent* .00000000000001745, *using scientific notation.*

For 1.745 to be changed to .00000000000001745, the decimal point would have to be moved 14 places to the left. Thus,

$$.00000000000001745 = 1.745 \times 10^{-14} \quad \checkmark$$

Example 46. *Change* 3.6×10^7 *to a number without a power of* 10.

Multiplication by 10^7 shifts the decimal point 7 places to the right. Thus

$$3.6 \times 10^7 = 36{,}000{,}000 \quad \checkmark$$

Example 47. *Write* 3500×10^8 *in scientific notation.*

Standard form is 3.5 times the appropriate power of 10. So

$$3_\wedge500 \times 10^8 = 3.5 \times 10^3 \times 10^8$$
$$= 3.5 \times 10^{3+8}$$
$$= 3.5 \times 10^{11} \quad \checkmark$$

Example 48. *Change* $.043 \times 10^{20}$ *to standard form.*

Standard form is 4.3 times a power of 10.

$$.0\underline{4}_\wedge3 \times 10^{20} = 4.3 \times 10^{-2} \times 10^{20}$$
$$= 4.3 \times 10^{18} \quad \checkmark$$

Example 49. *Change 5270 × 10⁻⁵ to standard form.*

$$5{\scriptstyle\wedge}270 \times 10^{-5} = 5.27 \times 10^{-5+3}$$
$$= 5.27 \times 10^{-2} \ \checkmark$$

Example 50. *Change .00062 × 10⁻⁷ to standard form.*

$$.0006{\scriptstyle\wedge}2 \times 10^{-7} = 6.2 \times 10^{-7-4}$$
$$= 6.2 \times 10^{-11} \ \checkmark$$

Example 51. *Simplify the expression.*

$$\frac{30{,}000{,}000 \times 200{,}000{,}000}{.000004 \times 500{,}000}$$

Using scientific notation,

$$30{,}000{,}000 = 3 \times 10^7$$
$$200{,}000{,}000 = 2 \times 10^8$$
$$.000004 = 4 \times 10^{-6}$$
$$500{,}000 = 5 \times 10^5$$

Now the expression can be changed to scientific notation and simplified in the steps shown next.

$$\frac{30{,}000{,}000 \times 200{,}000{,}000}{.000004 \times 500{,}000} = \frac{3 \times 10^7 \times 2 \times 10^8}{4 \times 10^{-6} \times 5 \times 10^5}$$
$$= \frac{3 \times 2 \times 10^7 \times 10^8}{4 \times 5 \times 10^{-6} \times 10^5}$$
$$= \frac{6 \times 10^{15}}{20 \times 10^{-1}}$$
$$= \frac{6}{20} \times 10^{16}$$
$$= .3 \times 10^{16}$$
$$= 3.0 \times 10^{15} \ \checkmark$$

EXERCISES 19.6

1. Write each number in scientific notation.

*(a) 4700000
(b) 810000
*(c) 350000000000
(d) 2734000000000000
*(e) 453000000000000000
(f) 600000000000000000
*(g) .00023
(h) .000005
*(i) .0000007
(j) .000000000142

*(k) .0000000000013   (l) .00000000000000095
*(m) 12,000,000,000,000   (n) 5,900,000,000,000,000

2. Change each standard form to a number without a power of 10.
    *(a) 3.4×10^8 (b) 2.0×10^{15} *(c) 9.42×10^{11}
    (d) 1.829×10^{12} *(e) 1.3×10^{-9} (f) 2.81×10^{-13}
    *(g) 5.0×10^{-10} (h) 4.734×10^{-11} *(i) 7.563×10
    (j) 3.8×10^{-8}

*3. Change each number below to standard form.
    (a) 450×10^7 (b) 1200×10^{16}
    (c) $.62 \times 10^{23}$ (d) $.00731 \times 10^{15}$
    (e) 473×10^{-12} (f) 841×10^{-16}
    (g) $.621 \times 10^{19}$ (h) $.456 \times 10^{-21}$
    (i) $3,000,000,000 \times 10^{15}$ (j) $.000052 \times 10^{-17}$

*4. Simplify each expression in the manner of Example 51.
    (a) $\dfrac{.0008 \times 50,000,000,000}{20,000 \times .0000004}$

    (b) $\dfrac{.0000009 \times 2.0 \times 10^8 \times 1,000,000}{3.0 \times 10^{-3} \times 30,000,000}$

    (c) $\dfrac{800,000,000 \times 3.0 \times 10^{-5}}{12,000,000 \times 10^8}$

    (d) $\dfrac{20,000 \times 30,000 \times 50,000,000}{6,000,000 \times 5.0 \times 10^{-2}}$

Chapter 19. REVIEW EXERCISES

1. Simplify each expression.
    *(a) $x^7 \cdot x^5$ (b) $x^8 \cdot x^9$
    *(c) $(x^{12})^4$ (d) $(y^{15})^6$
    *(e) $3x^4 \cdot 5x^{12}$ (f) $6x^5 \cdot 4x^{13}$
    *(g) $(2m^4 y^5)^3$ (h) $(2a^6 b^{11})^4$
    *(i) $(5xy^3)(3x^7 y^{10})^2$ (j) $(3u^2 v)(7u^8 v^5)^2$
    *(k) $\left(\dfrac{3x^2}{y^5}\right)^4$ (l) $\left(\dfrac{2y^6}{x^3}\right)^6$
    *(m) $t^3 t^6 t^{12}$ (n) $x^5 x^{11} x^7$
    *(o) $(-2x^2 y)^5$ (p) $(-3x^7 y)^3$
    *(q) $(-5a^3 b^6)(7ab^5 c^4)$ (r) $(-6a^4 b^9)(8a^6 bc^7)$

2. Write each expression without using negative exponents.
    *(a) x^{-7} (b) c^{-3} *(c) $4m^{-2}$ (d) $6y^{-7}$

    *(e) $-7x^{-3}$ (f) $-2w^{-4}$ *(g) $\dfrac{y^{-6}}{3}$ (h) $\dfrac{m^{-3}}{8}$

3. Simplify each expression and write it without using negative exponents.

*(a) $\dfrac{x^5 y^9}{x^2 y^2}$ (b) $\dfrac{a^4 b^7}{a^3 b^5}$ *(c) $\dfrac{x^6 y^8}{x^9 y^7}$

(d) $\dfrac{a^5 b^6}{a^2 b^8}$ *(e) $(4x)^{-2}$ (f) $(2y)^{-3}$

*(g) $5x^{-3} y^4 z^{-1}$ (h) $4x^7 y^{-3} z^{-9}$ *(i) $-3x^{-7} y^{-8}$

(j) $-7a^{-4} b^{-5}$ *(k) $\dfrac{4r^{-6} s^{12}}{t^3 u^{-5}}$ (l) $\dfrac{5a^{-7} b^8}{c^{-4} d^4}$

*(m) $x^{-7} x^{-3}$ (n) $y^{-8} y^{-5}$ *(o) $3x^{-4} x^{-3} x^{-2}$

(p) $7t^{-8} t^{-2} t^{-7}$ *(q) $(b^{-4} c^{-5})^{-3}$ (r) $(x^{-9} y^{-4})^{-2}$

*(s) $\dfrac{m^5 n^{-3} p^{-7} q^6}{m^{-8} n^7 p^{-3} q^9}$ (t) $\dfrac{w^{-4} x^5 y^{-8} z^{12}}{w^7 x^{-2} y^{-1} z^9}$

*4. Evaluate each expression.

(a) $3 \cdot 2^3$ (b) 7^0 (c) $8 \cdot 3^0$ (d) $9^0 - 3$

(e) $100^{\frac{1}{2}}$ (f) $8^{\frac{1}{3}}$ (g) $16^{\frac{1}{4}}$ (h) $\left(\dfrac{49}{64}\right)^{\frac{1}{2}}$

(i) $4 \cdot 9^{\frac{1}{2}}$ (j) $4^{\frac{5}{2}}$ (k) $27^{\frac{2}{3}}$ (l) $3 \cdot 16^{\frac{3}{4}}$

(m) 2^{-4} (n) $5 \cdot 3^{-2}$ (o) $(2x)^0$ (p) $4^{3.5}$

5. Simplify each expression.

*(a) $\sqrt{49 m^{10} n^{24}}$ (b) $\sqrt{81 x^6 y^{18}}$ *(c) $(64 x^3 y^{12} z^{30})^{\frac{2}{3}}$

(d) $(27 u^6 v^3 w^{21})^{\frac{2}{3}}$

Formula
and Equation
Manipulation

20.1 INTRODUCTION

A common algebra problem is to take a formula like

$$F = \frac{9}{5}C + 32$$

and rewrite it as

$$C = \frac{5}{9}(F - 32)$$

The first formula gives a Fahrenheit (F) temperature if you supply the Celsius (C) temperature. The second formula does the opposite and can be derived from the first by carefully "solving" the first equation for C. We shall return to this problem, but first let's consider some other examples.

Example 1. *Solve $y = 3x - 2$ for x; that is, express x in terms of y in an $x = \cdots$ form.*

The problem is to eliminate the -2 from the right-hand side and then change the coefficient of x to 1. First, add $+2$ to both sides to get the x term by itself.

$$y = 3x - 2$$
$$\underline{+2 \qquad\quad +\,2}$$
$$y + 2 = 3x$$

or $3x = y + 2$ if you interchange $y + 2$ and $3x$. Next, divide both sides by 3 to change $3x = \cdots$ to $x = \cdots$.

$$\frac{3x}{3} = \frac{y+2}{3}$$

or
$$x = \frac{y+2}{3} \checkmark$$

We say that $y = 3x - 2$ has been "solved for x," since it now gives x in terms of other numbers.

Example 2. *Solve $y = mx + b$ for x.*

This problem is identical in form to the problem of the previous example. First, add $-b$ to both sides in order to get the x term by itself.

$$y = mx + b$$
$$\underline{-b = \qquad\quad -\,b}$$
$$y - b = mx$$

or $mx = y - b$ after interchanging $y - b$ and mx. Then divide both sides by m to change $mx = \cdots$ to $x = \cdots$.

$$\frac{mx}{m} = \frac{y-b}{m}$$

$$x = \frac{y-b}{m} \checkmark$$

Example 3. *Solve $C = \pi d$ for π.*

The π term is alone on the right-hand side of the equation. We need only change its coefficient to one in order to solve for π. So divide both sides by d, the coefficient of π. When this is done,

$$C = \pi d$$

becomes

$$\frac{C}{d} = \frac{\pi d}{d}$$

or
$$\pi = \frac{C}{d} \checkmark$$

In this formula C is the circumference of a circle and d is its diameter. π is the number that you know to be approximately equal to 3.14. Here you can see an interesting definition of π; it is the ratio of the circumference of a circle to its diameter.

Example 4. *Solve for b:* $a = \dfrac{b+c}{d}.$

If the denominator d were not present, it would be much easier to solve this equation for b; so multiply both sides by d, the denominator, to eliminate the fraction. Recall that to solve equations with fractions it is standard procedure to eliminate the fractions.

$$d \cdot a = \cancel{d} \cdot \frac{(b+c)}{\cancel{d}}$$

or $ad = b + c$

Next, eliminate c from the right-hand side, since we want b alone.

$$\begin{array}{r} ad = b + c \\ \underline{-c \qquad -c} \\ ad - c = b \end{array}$$

or $b = ad - c$ ✓

Example 5. *Solve for C:* $F = \frac{9}{5}C + 32.$

Begin solving for C by eliminating the fraction $\frac{9}{5}$ from $F = \frac{9}{5}C + 32$ by multiplying both sides of the equation by 5. The result is

$$5F = 9C + 160$$

Next, add -160 to both sides of this equation to obtain

$$5F - 160 = 9C$$

or $9C = 5F - 160$

Finally, divide both sides of the equation by 9 to get

$$C = \frac{5F - 160}{9} \ ✓$$

This answer is correct and the form of the answer is all right, too. However, it does not look like the form mentioned at the beginning of the chapter, namely,

$$C = \frac{5}{9}(F - 32)$$

But if you factor 5 out of the numerator, you will see that the two forms are equivalent. In steps,

$$\begin{aligned} C &= \frac{5F - 160}{9} \\ &= \frac{5(F - 32)}{9} \\ &= \frac{5}{9}(F - 32) \ ✓ \end{aligned}$$

Example 6. *Show that the proportion* $\frac{a}{b} = \frac{c}{d}$ *can be written* $ad = bc$.

The original equation

$$\frac{a}{b} = \frac{c}{d}$$

contains fractions. If both sides are multiplied by the common denominator, bd, the fractions will be eliminated.

$$bd\frac{a}{b} = bd\frac{c}{d}$$

The result is

$$ad = bc \ \checkmark$$

which happens to be the desired form.

proportion
A *proportion* is an equation in which both sides are fractions (or ratios). In the proportion

$$\frac{a}{b} = \frac{c}{d}$$

a and d are called the *extremes* of the proportion and b and c are called the *means*.

$$\frac{\text{extreme}}{\text{mean}} = \frac{\text{mean}}{\text{extreme}}$$

We have shown that $ad = bc$—that is, that the product of the extremes is equal to the product of the means. See Appendix C for additional information on ratio and proportion.

Example 7. *Solve for* y: $2x + 3y - 7 = 0$.

First, add $+7$ to both sides

$$
\begin{array}{rcl}
2x + 3y - 7 &=& 0 \\
+7 && +7 \\
\hline
2x + 3y &=& 7
\end{array}
$$

Then add $-2x$ to both sides to get the y term alone.

$$
\begin{array}{rcl}
2x + 3y &=& 7 \\
-2x && -2x \\
\hline
3y &=& -2x + 7
\end{array}
$$

Finally, divide both sides by 3 to get $y = \cdots$.

$$\frac{3y}{3} = \frac{-2x + 7}{3}$$

or

$$y = \frac{-2x + 7}{3} \ \checkmark$$

Example 8. *Solve for b:* $x = a(b + c)$.

Free the b term from within parentheses by distributing the a. Thus

$$x = a(b + c)$$

becomes

$$x = ab + ac$$

Next, get the b term, ab, by itself by adding $-ac$ to both sides.

$$
\begin{array}{r}
x = ab + ac \\
-ac = - ac \\
\hline
x - ac = ab
\end{array}
$$

Finally, get a coefficient of 1 for b by dividing both sides by a.

$$\frac{x - ac}{a} = \frac{ab}{a}$$

$$\frac{x - ac}{a} = b \; \checkmark$$

Example 9. *Solve for c:* $a = \dfrac{b}{c} + \dfrac{d}{e}$.

The lowest common denominator of the fractions is $c \cdot e$ or ce. Multiplying all terms by ce will eliminate the fractions. It will then be easier to solve for c.

$$ce \cdot a = \cancel{ce} \cdot \frac{b}{\cancel{c}} + c\cancel{e} \cdot \frac{d}{\cancel{e}}$$

or

$$cea = eb + cd$$

We can get $c = \cdots$ on the left by dividing both sides by ea. *But* there will also be a c term on the right; so we will not have solved for c. So instead of doing that, get the c terms together on the same side by adding $-cd$ to both sides.

$$
\begin{array}{r}
cea = eb + cd \\
-cd - cd \\
\hline
cea - cd = eb
\end{array}
$$

Place the coefficients of the c terms in front of c so that they are easier to work with.

$$ae\underline{c} - d\underline{c} = eb$$

Next, get *one* coefficient for c by factoring out c.

$$(ae - d)c = eb$$

Finally, divide both sides by the coefficient of c, which is $(ae - d)$.

$$\frac{(ae - d)c}{(ae - d)} = \frac{eb}{(ae - d)}$$

So

$$c = \frac{eb}{ae - d} \quad \checkmark$$

subscript   The next example includes the use of *subscripts*. The numbers x_1 and x_2 are two numbers we wish to call by similar names, both "x," so we use the subscripts 1 and 2 to distinguish them. The numbers 1 and 2 are *not* exponents. Subscripts are used frequently in statistics and calculus. You will see two examples of subscripts in the exercises—the two bases of a trapezoid will be called b_1 and b_2 and two masses will be called m_1 and m_2.

Example 10. *Solve for* x_1: $y = \dfrac{x_1 x_2}{x_1 + x_2}$.

This problem is easier to handle without fractions; so multiply both sides by the denominator $(x_1 + x_2)$. The result is

$$y(x_1 + x_2) = x_1 x_2$$

We must get the x_1 terms (arrows below) together on the same side in order to combine them.

$$\underset{\uparrow}{y(x_1 + x_2)} = \underset{\uparrow}{x_1 x_2}$$

But x_1 cannot be removed or combined until we multiply out the $y(x_1 + x_2)$. Thus

$$y(x_1 + x_2) = x_1 x_2$$

becomes

$$yx_1 + yx_2 = x_1 x_2$$

Now we can get the x_1 terms together by adding $-yx_1$ to both sides.

$$
\begin{array}{rcl}
yx_1 + yx_2 & = & x_1 x_2 \\
-yx_1 & & -yx_1 \\
\hline
yx_2 & = & x_1 x_2 - yx_1
\end{array}
$$

or $x_1 x_2 - yx_1 = yx_2$ {if we interchange the left and right sides

or $x_2 x_1 - yx_1 = yx_2$ {if we change the order of x_1 and x_2 in the first term

Next, factor x_1 out of the left-hand expression.

$$(x_2 - y)x_1 = yx_2$$

Finally, divide both sides by $x_2 - y$, the coefficient of x_1.

$$x_1 = \frac{yx_2}{x_2 - y} \ \checkmark$$

If you are unsure how to proceed in manipulating equations or formulas into the form requested, consider the guidelines given next. Be sure to *carry out the steps in the order listed here.*

Follow these steps, *in order.*

1. If there are fractions, multiply both sides (each term) by the lowest common denominator. This will eliminate fractions from the equation.

2. If the variable (or unknown) being solved for is within parentheses, multiply out to free it.

3. Use opposites to collect on one side all terms containing the desired variable and on the other side all other terms.

4. If more than one term contains the variable, factor out the variable in order to determine its coefficient.

5. Divide both sides of the equation by the coefficient of the variable.

You might want to look back over the examples of the chapter. Example 10 uses all five of the steps above, Example 9 involves all steps except number 2. Example 8 uses only steps 2, 3, and 5.

EXERCISES 20.1

Answers to starred exercises are given in the back of the book.

*1. Solve each formula for the letter indicated.

(a) $d = 2r$ for r (b) $d = rt$ for r

(c) $C = \pi d$ for d (d) $C = 2\pi r$ for r

(e) $A = lw$ for l (f) $V = lwh$ for h

(g) $i = prt$ for r (h) $V = \pi r^2 h$ for h

(i) $A = \dfrac{1}{2}bh$ for h (j) $A = \dfrac{1}{2}bh$ for b

(k) $E = \dfrac{1}{2}mv^2$ for m (l) $s = \dfrac{1}{2}at^2$ for a

2. Solve each formula for the letter indicated.
   *(a) $y = 6x$ for x *(b) $y = -5x$ for x
   *(c) $y = ax + b$ for x (d) $y = ax + b$ for b
   *(e) $y = ax + b$ for a (f) $P = 2l + 2w$ for l
   *(g) $P = 2l + 2w$ for w (h) $R = p + prt$ for t
   *(i) $V = \dfrac{1}{3}\pi r^2 h$ for h (j) $3x = 2y + 5$ for y
   *(k) $5y = 4x - 2$ for x (l) $2x + 6y = 9$ for x

3. Solve each formula for the letter indicated.
   *(a) $5x + 7y - 8 = 0$ for x (b) $5x + 7y - 8 = 0$ for y
   *(c) $ax + b = 0$ for b *(d) $ax + b = 0$ for x
   *(e) $ax + by + c = 0$ for x (f) $ax + by + c = 0$ for y
   *(g) $I = \dfrac{E}{R}$ for E *(h) $I = \dfrac{E}{R}$ for R
   (i) $v = \dfrac{s}{t}$ for s (j) $v = \dfrac{s}{t}$ for t
   *(k) $F = \dfrac{m_1 m_2}{d^2}$ for m_1 *(l) $A = \dfrac{1}{2}(b_1 + b_2)h$ for h
   *(m) $l = a + (n - 1)d$ for d *(n) $S = \dfrac{n}{2}(a + l)$ for l
   *(o) $R = p + prt$ for p (p) $R = \dfrac{1}{2\pi fc}$ for c
   *(q) $A = m(1 + p)$ for p (r) $B = \dfrac{m}{2}(a + b - c)$ for b

4. Solve each formula for the letter indicated.
   *(a) $C = \dfrac{nE}{R + nr}$ for E *(b) $C = \dfrac{nE}{R + nr}$ for R
   *(c) $C = \dfrac{nE}{R + nr}$ for r *(d) $C = \dfrac{nE}{R + nr}$ for n
   *(e) $A = \dfrac{1}{2}h(b_1 + b_2)$ for b_2 (f) $S = \dfrac{n}{2}(a + l)$ for n
   *(g) $\dfrac{1}{f} = \dfrac{1}{p} + \dfrac{1}{q}$ for p (h) $\dfrac{1}{f} = \dfrac{1}{p} + \dfrac{1}{q}$ for q
   *(i) $S = \dfrac{rl - a}{r - 1}$ for l *(j) $S = \dfrac{rl - a}{r - 1}$ for r
   *(k) $a = p(1 + rt)$ for t (l) $R = \dfrac{gs}{g + s}$ for s
   *(m) $\dfrac{a}{y} + \dfrac{b}{x} = 1$ for x (n) $\dfrac{a}{y} + \dfrac{b}{x} = 1$ for y

Chapter 20.   REVIEW EXERCISES

*1. Solve each formula for the letter indicated.

(a) $p = 3d + 4e$ for e

(b) $t = u(v + w)$ for w

(c) $m = \dfrac{a + b}{c}$ for a

(d) $a = \dfrac{1}{2}(b + c)$ for b

*2. Solve each formula for the letter indicated.

(a) $T = n(4 + rv)$ for v

(b) $x = \dfrac{3a - 2b}{c + d}$ for a

(c) $\dfrac{1}{w} = \dfrac{1}{x} + \dfrac{1}{y}$ for x

(d) $x = \dfrac{a + b}{c + d}$ for c

Radicals and Roots

21.1 REVIEW

This section offers a brief review of some of the ideas presented in Chapters 4 and 19. We'll be using these concepts throughout the chapter.

square root
The *square root* of a quantity is the number that when squared (that is, multiplied by itself) produces the original quantity. For instance, the square root of 49 is 7, since $7^2 = 49$—that is, $7 \cdot 7 = 49$. This can **radical** be written $\sqrt{49} = 7$. The symbol $\sqrt{}$ is called a *radical*. In general,

$$\sqrt{x} \cdot \sqrt{x} = x$$

perfect square
The numbers 0, 1, 4, 9, 16, 25, 36, etc. are called *perfect squares*. A whole number is a perfect square if its square root is a whole number. To illustrate, 49 is a perfect square because $\sqrt{49} = 7$, and 7 is a whole number. But 18 is not a perfect square because $\sqrt{18}$ is not equal to a whole number. Here is a short table of square roots of perfect squares.

$$\begin{array}{lll}
\sqrt{0} = 0 & \sqrt{49} = 7 & \sqrt{196} = 14 \\
\sqrt{1} = 1 & \sqrt{64} = 8 & \sqrt{225} = 15 \\
\sqrt{4} = 2 & \sqrt{81} = 9 & \sqrt{256} = 16 \\
\sqrt{9} = 3 & \sqrt{100} = 10 & \sqrt{289} = 17 \\
\sqrt{16} = 4 & \sqrt{121} = 11 & \sqrt{324} = 18 \\
\sqrt{25} = 5 & \sqrt{144} = 12 & \sqrt{361} = 19 \\
\sqrt{36} = 6 & \sqrt{169} = 13 & \sqrt{400} = 20
\end{array}$$

Numbers that can be written as the quotient of two integers are called *rational numbers.* Integers, fractions, and decimals are rational numbers. What isn't? The square root of any whole number that is not a perfect square is an example of a number that is not rational. So $\sqrt{2}, \sqrt{3}, \sqrt{5}, \sqrt{6}$, and $\sqrt{52}$ are examples of such *irrational numbers.*

rational number

irrational number

21.2 SIMPLIFICATION OF RADICALS

In many algebra problems it is acceptable to leave results that contain radicals. For instance, $5\sqrt{2}$ would be an acceptable final form in many cases. However, when we do leave an answer that contains a radical, we insist that it be *simplified.* In other words, in the end *the number left under the radical must be as small as possible.* The technique used to simplify radicals is demonstrated in the examples that follow. Too, this property is always used:

$$\boxed{\sqrt{x \cdot y} = \sqrt{x}\sqrt{y}}$$

As an example of this property, consider that $\sqrt{9 \cdot 2} = \sqrt{9}\sqrt{2}$.

Using exponent notation, we can readily show why this property is true.

$$\sqrt{xy} = (xy)^{\frac{1}{2}} = x^{\frac{1}{2}}y^{\frac{1}{2}} = \sqrt{x}\sqrt{y}$$

Example 1. *Simplify $\sqrt{18}$.*

We must write 18 as the *product* of two whole numbers, *one of which is a perfect square.* Then we can apply the property of radicals boxed above. So consider 18 as $9 \cdot 2$, because 9 is a perfect square.

$$\sqrt{18} = \sqrt{9 \cdot 2} \qquad \text{since } 9 \cdot 2 = 18$$
$$= \sqrt{9}\sqrt{2} \qquad \text{since } \sqrt{x \cdot y} = \sqrt{x}\sqrt{y}$$
$$= 3\sqrt{2} \; \checkmark \qquad \text{replacing } \sqrt{9} \text{ by } 3$$

So $\sqrt{18} = 3\sqrt{2}$. We factored 18 into $9 \cdot 2$ because 9 is a perfect square; that is, $\sqrt{9} = 3$. Had we chosen to consider 18 as $6 \cdot 3$, we could not have been able to proceed as smoothly. *When simplifying radicals, always look for the perfect square.*

Example 2. *Simplify* $\sqrt{75}$.

$$\sqrt{75} = \sqrt{25 \cdot 3} \qquad \text{Note that 25 is a perfect square.}$$
$$= \sqrt{25}\sqrt{3} \qquad \text{since } \sqrt{x \cdot y} = \sqrt{x}\sqrt{y}$$
$$= 5\sqrt{3} \; \checkmark \qquad \text{replacing } \sqrt{25} \text{ by } 5$$

You are less likely to make an error if you *place the perfect square first*, as we have been doing. In other words, write $\sqrt{25 \cdot 3}$, not $\sqrt{3 \cdot 25}$; write $\sqrt{9 \cdot 2}$, not $\sqrt{2 \cdot 9}$.

Example 3. *Simplify* $\sqrt{72}$.

$$\sqrt{72} = \sqrt{36 \cdot 2} \qquad \text{36 is a perfect square.}$$
$$= \sqrt{36}\sqrt{2} \qquad \text{since } \sqrt{x \cdot y} = \sqrt{x}\sqrt{y}$$
$$= 6\sqrt{2} \; \checkmark \qquad \text{replacing } \sqrt{36} \text{ by } 6$$

Example 4. *Suppose that in the last example you considered 72 as $9 \cdot 8$ rather than as $36 \cdot 2$. After all, 9 is also a perfect square.*

$$\sqrt{72} = \sqrt{9 \cdot 8}$$
$$= \sqrt{9}\sqrt{8}$$
$$= 3\sqrt{8}$$

Is this answer wrong? No, it's not "wrong," but it is not a completely reduced form. It can be simplified further. Remember, the number left under the radical must be made as small as possible. So continue the process.

$$3\sqrt{8} = 3 \cdot \sqrt{8}$$
$$= 3 \cdot \sqrt{4 \cdot 2} \qquad \text{by writing } \sqrt{8} \text{ as } \sqrt{4 \cdot 2}$$
$$= 3 \cdot \sqrt{4} \cdot \sqrt{2}$$
$$= 3 \cdot 2 \cdot \sqrt{2} \qquad \text{since } \sqrt{4} \text{ is } 2$$
$$= 6\sqrt{2} \; \checkmark \qquad \text{since } 3 \cdot 2 \text{ is } 6$$

As demonstrated by Examples 3 and 4, you should always try to find the *largest* perfect square factor. It simplifies the process.

Example 5. *Simplify* $\sqrt{400}$.

You may recognize 400 as a perfect square and realize that $\sqrt{400} = 20$. If so, that's it. If not, you proceed as in the other examples, perhaps considering 400 as $4 \cdot 100$.

$$\sqrt{400} = \sqrt{4 \cdot 100}$$
$$= \sqrt{4}\sqrt{100}$$
$$= 2 \cdot 10$$
$$= 20 \checkmark$$

Example 6. *Simplify* $5\sqrt{24}$.

$$5\sqrt{24} = 5 \cdot \sqrt{24}$$
$$= 5 \cdot \sqrt{4 \cdot 6}$$
$$= 5 \cdot \sqrt{4}\sqrt{6}$$
$$= 5 \cdot 2 \cdot \sqrt{6}$$
$$= 10\sqrt{6} \checkmark$$

In Example 2 we simplified $\sqrt{75}$ to $5\sqrt{3}$. One reason the form $5\sqrt{3}$ is considered "simpler" is that it can lead to simplification of a more complicated expression in which it appears. Consider the following example, which shows how a fraction can be reduced once the radical has been simplified.

Example 7. *Reduce the fraction* $\dfrac{\sqrt{75}}{5}$.

$$\frac{\sqrt{75}}{5} = \frac{5\sqrt{3}}{5} = \frac{5 \cdot \sqrt{3}}{5 \cdot 1} = \frac{5}{5} \cdot \frac{\sqrt{3}}{1} = 1 \cdot \frac{\sqrt{3}}{1} = \frac{\sqrt{3}}{1} = \sqrt{3} \checkmark$$

Sometimes it can be hard to spot a perfect square factor within the number under the radical. For example, $\sqrt{351}$ has a 9 in it and simplifies as $\sqrt{9 \cdot 39} = \sqrt{9} \cdot \sqrt{39} = 3\sqrt{39}$. Keep in mind that *it is arithmetically easy to test for the presence of a factor of 9 or 4 in a number.* You can divide the number by 9 or 4.

$$\frac{39}{9)\overline{351}} \quad \text{shows that } 351 = 9 \cdot 39$$

$$\frac{29}{4)\overline{116}} \quad \text{shows that } 116 = 4 \cdot 29$$

You can also use the divisibility tests for 9 and 4 as explained in Section 6.4. They take even less time than does the actual division process.

If you cannot see a perfect square factor within the number, and there's no factor of 4 or 9 in it, then consider dividing the number by 2 or 3 to see if that makes it easier to find a perfect square factor. For example, $\sqrt{242}$ simplifies readily when considered as $\sqrt{121 \cdot 2}$ or $\sqrt{121}\sqrt{2}$, or $11\sqrt{2}$. Dividing by 2 makes the perfect square factor 121 visible.

EXERCISES 21.2

Answers to starred exercise are given in the back of book.

*1. Simplify each radical.

(a) $\sqrt{8}$ (b) $\sqrt{12}$ (c) $\sqrt{27}$ (d) $\sqrt{36}$ (e) $\sqrt{50}$
(f) $\sqrt{60}$ (g) $\sqrt{28}$ (h) $\sqrt{44}$ (i) $\sqrt{45}$ (j) $\sqrt{100}$
(k) $\sqrt{125}$ (l) $\sqrt{63}$ (m) $\sqrt{32}$ (n) $\sqrt{48}$ (o) $\sqrt{52}$
(p) $\sqrt{56}$ (q) $\sqrt{200}$ (r) $\sqrt{900}$

2. Simplify each radical, if possible.

*(a) $\sqrt{68}$ (b) $\sqrt{84}$ *(c) $\sqrt{90}$ (d) $\sqrt{88}$
*(e) $\sqrt{83}$ (f) $\sqrt{96}$ *(g) $\sqrt{112}$ (h) $\sqrt{150}$
*(i) $\sqrt{128}$ (j) $\sqrt{115}$ *(k) $\sqrt{2500}$ (l) $\sqrt{180}$

3. Use the idea that $\sqrt{x}\sqrt{x} = x$ to simplify each product.

*(a) $\sqrt{7}\sqrt{7}$ (b) $\sqrt{c}\sqrt{c}$
*(c) $\sqrt{5}\sqrt{5}\sqrt{5}$ (d) $\sqrt{2}\sqrt{2}\sqrt{2}$
*(e) $\sqrt{3}\sqrt{3}\sqrt{3}\sqrt{3}$ (f) $\sqrt{7}\sqrt{7}\sqrt{7}\sqrt{7}$
*(g) $\sqrt{2}\sqrt{10}$ (h) $\sqrt{3}\sqrt{21}$
*(i) $\sqrt{50}\sqrt{2}$ (j) $\sqrt{12}\sqrt{3}$

4. Simplify.

*(a) $5\sqrt{8}$ *(b) $3\sqrt{25}$ (c) $4\sqrt{27}$ (d) $2\sqrt{20}$
*(e) $10\sqrt{68}$ (f) $8\sqrt{44}$ *(g) $2\sqrt{45}$ (h) $7\sqrt{32}$
*(i) $5\sqrt{300}$ (j) $3\sqrt{40}$

5. Simplify the radical, and then reduce the fraction.

*(a) $\dfrac{\sqrt{18}}{3}$ (b) $\dfrac{\sqrt{44}}{2}$ *(c) $\dfrac{\sqrt{45}}{18}$ (d) $\dfrac{\sqrt{72}}{2}$

*(e) $\dfrac{\sqrt{80}}{4}$ (f) $\dfrac{\sqrt{200}}{10}$

6. Simplify each radical, if possible. Consider using the methods suggested in the last two paragraphs of the section.

*(a) $\sqrt{117}$	(b) $\sqrt{92}$	*(c) $\sqrt{98}$	(d) $\sqrt{124}$
*(e) $\sqrt{104}$	(f) $\sqrt{108}$	*(g) $\sqrt{162}$	(h) $\sqrt{240}$
*(i) $\sqrt{143}$	(j) $\sqrt{140}$	*(k) $\sqrt{147}$	(l) $\sqrt{228}$
*(m) $\sqrt{153}$	(n) $\sqrt{221}$	*(o) $\sqrt{323}$	(p) $\sqrt{261}$
*(q) $\sqrt{279}$	(r) $\sqrt{356}$		

21.3 COMBINING LIKE RADICALS

The distributive property can be used to combine like radicals.

Example 8. *Combine* $5\sqrt{2} + 6\sqrt{2}$.

$$5\sqrt{2} + 6\sqrt{2} = (5 + 6)\sqrt{2}$$
$$= 11\sqrt{2} \checkmark$$

Example 9. *Combine* $8\sqrt{3} + 5\sqrt{2}$.

$\sqrt{3}$ and $\sqrt{2}$ are not like radicals; they cannot be combined. The distributive property does not apply.

Example 10. *Combine* $3\sqrt{8} + 7\sqrt{2}$.

$\sqrt{8}$ and $\sqrt{2}$ are not like radicals as written, but $\sqrt{8}$ can be simplied.

$$\sqrt{8} = \sqrt{4 \cdot 2} = \sqrt{4}\sqrt{2} = 2\sqrt{2}$$

So

$$3\sqrt{8} + 7\sqrt{2}$$

becomes

$$3 \cdot 2\sqrt{2} + 7\sqrt{2}$$

or

$$6\sqrt{2} + 7\sqrt{2}$$

which simplifies to

$$13\sqrt{2} \checkmark$$

Example 11. *Combine* $7\sqrt{3} + \sqrt{27} + \sqrt{3}$.

Since $\sqrt{3}$ means $1\sqrt{3}$ and $\sqrt{27} = 3\sqrt{3}$ when simplified,

$$7\sqrt{3} + \sqrt{27} + \sqrt{3} = 7\sqrt{3} + 3\sqrt{3} + 1\sqrt{3}$$
$$= (7 + 3 + 1)\sqrt{3}$$
$$= 11\sqrt{3} \checkmark$$

Example 12. *Combine* $5 + 3\sqrt{2} + \sqrt{3} + 4\sqrt{2}$.

The only like radicals are $3\sqrt{2}$ and $4\sqrt{2}$, which combine to yield $7\sqrt{2}$. The other numbers cannot be combined, so the final result is

$$5 + 7\sqrt{2} + \sqrt{3} \checkmark$$

EXERCISES 21.3

1. Combine when possible.

*(a) $4\sqrt{7} - 2\sqrt{7}$

(b) $5\sqrt{6} + \sqrt{6}$

*(c) $7\sqrt{2} + 10\sqrt{2} - 3\sqrt{2}$

(d) $8\sqrt{5} + \sqrt{7} - \sqrt{5}$

*(e) $\sqrt{8} + 9\sqrt{2}$

*(f) $\sqrt{16} + \sqrt{4}$

*(g) $\sqrt{32} + 6\sqrt{2}$

(h) $2\sqrt{3} + 5\sqrt{27}$

*(i) $\sqrt{72} - 2\sqrt{2} + \sqrt{2}$

(j) $\sqrt{54} + 3\sqrt{24}$

*(k) $5\sqrt{41} - 3\sqrt{17}$

*(l) $5\sqrt{100} + 6\sqrt{81}$

*(m) $2\sqrt{3} + 6\sqrt{75}$

(n) $\sqrt{28} + 2\sqrt{63}$

*(o) $\sqrt{45} + 8\sqrt{5}$

*(p) $7 + 6\sqrt{2} + 5\sqrt{3}$

*(q) $\sqrt{90} + \sqrt{40}$

*(r) $5\sqrt{12} + 2\sqrt{27} - \sqrt{48}$

*(s) $2\sqrt{20} - \sqrt{5} + \sqrt{10}$

(t) $\sqrt{52} + 2\sqrt{117}$

21.4 RATIONALIZING THE DENOMINATOR OR NUMERATOR

In some applications, it is undesirable to have a fraction that contains a radical in the denominator. Our problem, then, is how to remove a radical from the denominator of a fraction. In other applications, it is desirable to remove a radical from the numerator of a fraction. The examples below demonstrate how it can be done. Removing the radical from the denominator removes the irrational number from the denominator and hence *rationalizes* the denominator. Removing the radical from the numerator removes the irrational number from the numerator and hence rationalizes the numerator. The procedure is based on $\sqrt{x}\sqrt{x} = x$.

rationalize

Example 13. *Remove the radical from the denominator of* $\dfrac{5}{\sqrt{3}}$.

Multiplying the denominator $\sqrt{3}$ by $\sqrt{3}$ eliminates the radical from the denominator, since $\sqrt{3} \cdot \sqrt{3} = 3$. The numerator must also be multiplied by $\sqrt{3}$ so that the overall effect is multiplication of the fraction by 1.

$$\frac{5}{\sqrt{3}} = \frac{5}{\sqrt{3}} \cdot \frac{\sqrt{3}}{\sqrt{3}}$$

$$= \frac{5\sqrt{3}}{3} \quad \checkmark \qquad \text{since } \sqrt{3} \sqrt{3} = 3$$

Example 14. *Remove the irrational number from the numerator of* $\dfrac{\sqrt{5}}{3}$.

Multiply both numerator and denominator of $\dfrac{\sqrt{5}}{3}$ by $\sqrt{5}$. Doing so will eliminate the radical from the numerator, and yet the value of the fraction will be the same, since we are merely multiplying the fraction by an appropriate form of the number 1.

$$\frac{\sqrt{5}}{3} = \frac{\sqrt{5}}{3} \cdot \frac{\sqrt{5}}{\sqrt{5}}$$

$$= \frac{5}{3\sqrt{5}} \quad \checkmark$$

Example 15. *Rationalize the denominator of* $\dfrac{\sqrt{6}}{\sqrt{2}}$.

As before,

$$\frac{\sqrt{6}}{\sqrt{2}} = \frac{\sqrt{6}}{\sqrt{2}} \cdot \frac{\sqrt{2}}{\sqrt{2}} = \frac{\sqrt{12}}{2}$$

But $\sqrt{12}$ can be simplified, as shown next.

$$\frac{\sqrt{12}}{2} = \frac{\sqrt{4 \cdot 3}}{2} = \frac{\sqrt{4}\sqrt{3}}{2} = \frac{2\sqrt{3}}{2} = \sqrt{3} \quad \checkmark$$

Another approach:

$$\frac{\sqrt{6}}{\sqrt{2}} = \sqrt{\frac{6}{2}} = \sqrt{3} \quad \checkmark$$

The second approach showed the use of another important property of radicals. The property:

$$\boxed{\sqrt{\frac{x}{y}} = \frac{\sqrt{x}}{\sqrt{y}}}$$

We use it again in the next example.

Example 16. *Simplify* $\sqrt{\dfrac{1}{2}}$ *and leave it with a rational denominator.*

$$\sqrt{\frac{1}{2}} = \frac{\sqrt{1}}{\sqrt{2}} = \frac{1}{\sqrt{2}} \qquad \text{since } \sqrt{1} = 1$$

If we want to rationalize the denominator, we continue as

$$= \frac{1}{\sqrt{2}} \cdot \frac{\sqrt{2}}{\sqrt{2}}$$

$$= \frac{\sqrt{2}}{2} \quad \checkmark$$

Historically, people automatically rationalized the denominator of an expression like $\dfrac{1}{\sqrt{2}}$ because $\dfrac{1}{\sqrt{2}}$ was a more difficult division than was $\dfrac{\sqrt{2}}{2}$. Using 1.414 as an approximation for $\sqrt{2}$, you can see that $\dfrac{1}{1.414}$ is a more difficult division than is $\dfrac{1.414}{2}$. However, with modern calculators readily available, this no longer matters. Still, sometimes there are algebraic advantages to having a rational denominator. But there are also times when a rational *numerator* is desirable, especially in some calculus situations. So you must be able to rationalize either a denominator or a numerator.

Example 17. *Rationalize the denominator of* $\dfrac{5\sqrt{3}}{\sqrt{8}}$.

There are several ways of doing so.
One approach:

$$\frac{5\sqrt{3}}{\sqrt{8}} = \frac{5\sqrt{3}}{\sqrt{8}} \cdot \frac{\sqrt{8}}{\sqrt{8}} = \frac{5\sqrt{24}}{8} = \frac{5\sqrt{4}\sqrt{6}}{8}$$

$$= \frac{5 \cdot 2\sqrt{6}}{8} = \frac{5\sqrt{6}}{4} \quad \checkmark$$

Another:

$$\frac{5\sqrt{3}}{\sqrt{8}} = \frac{5\sqrt{3}}{2\sqrt{2}} = \frac{5\sqrt{3}}{2\sqrt{2}} \cdot \frac{\sqrt{2}}{\sqrt{2}} = \frac{5\sqrt{6}}{2 \cdot 2} = \frac{5\sqrt{6}}{4} \quad \checkmark$$

Another:

$$\frac{5\sqrt{3}}{\sqrt{8}} = \frac{5\sqrt{3}}{\sqrt{8}} \cdot \frac{\sqrt{2}}{\sqrt{2}} = \frac{5\sqrt{6}}{\sqrt{16}} = \frac{5\sqrt{6}}{4} \quad \checkmark$$

Example 18. *Reduce the fraction* $\dfrac{2 + \sqrt{24}}{2}$.

If you are tempted to simply divide out the 2's that appear in the numerator and denominator, then you had better reread Example 10 of Chapter 18. In fact, perhaps you should read the entire section on reducing fractions. In order to reduce a fraction, you must have a common *factor* in both the numerator and denominator. So to reduce this fraction, we'll need to factor a 2 out of the numerator to go with the 2 in the denominator. And this can only be done if 2 can be factored out of $\sqrt{24}$. So simplify $\sqrt{24}$ in order to find out.

$$\sqrt{24} = \sqrt{4}\,\sqrt{6}$$
$$= 2\sqrt{6}$$

Thus the original fraction will simplify, as shown in the steps below.

$$\frac{2 + \sqrt{24}}{2} = \frac{2 + 2\sqrt{6}}{2} \qquad \text{since } \sqrt{24} = 2\sqrt{6}$$

$$= \frac{2(1 + \sqrt{6})}{2} \qquad \text{after 2 is factored out of numerator}$$

$$= 1 + \sqrt{6} \ \checkmark \qquad \text{after dividing out the 2's}$$

Example 19. *Multiply* $\sqrt{3}\,\sqrt{7}$.

$$\sqrt{3}\,\sqrt{7} = \sqrt{3 \cdot 7}$$
$$= \sqrt{21} \ \checkmark$$

Example 20. *Multiply* $(2 + \sqrt{3})(5 - \sqrt{3})$.

The form of this multiplication is the same as that of multiplication of binomials (see Section 11.3).

$$(2 + \sqrt{3})(5 - \sqrt{3})$$

$$= 2 \cdot 5 - 2\sqrt{3} + \sqrt{3} \cdot 5 - \sqrt{3} \cdot \sqrt{3}$$

$$= 10 - 2\sqrt{3} + 5\sqrt{3} - 3 \qquad \begin{cases} \text{since } 2 \cdot 5 = 10 \\ \text{and } \sqrt{3} \cdot \sqrt{3} = 3 \end{cases}$$

$$= (10 - 3) - 2\sqrt{3} + 5\sqrt{3} \qquad \text{when regrouped}$$

$$= 7 + 3\sqrt{3} \ \checkmark \qquad \text{after combining like terms}$$

Example 21. *Multiply* $(5 + \sqrt{7})(2 + \sqrt{2})$.

$$(5 + \sqrt{7})(2 + \sqrt{2}) = 5 \cdot 2 + 5\sqrt{2} + \sqrt{7} \cdot 2 + \sqrt{7} \cdot \sqrt{2}$$

$$= 10 + 5\sqrt{2} + 2\sqrt{7} + \sqrt{14} \ \checkmark$$

which cannot be simplified.

Example 22. *Remove the radical from the denominator of* $\dfrac{1}{\sqrt{3}+1}$.

If the denominator $\sqrt{3}+1$ is multiplied by $\sqrt{3}-1$, the result will be the difference of two squares, $(\sqrt{3})^2 - (1)^2$, which is 2. The radical will be eliminated. Of course, the numerator must also be multiplied by $\sqrt{3}-1$. Here is the work.

$$\frac{1}{\sqrt{3}+1} = \frac{1}{\sqrt{3}+1} \cdot \frac{\sqrt{3}-1}{\sqrt{3}-1}$$

$$= \frac{\sqrt{3}-1}{3-1}$$

$$= \frac{\sqrt{3}-1}{2} \checkmark$$

EXERCISES 21.4

1. Rationalize the denominator of each fraction and simplify as much as possible.

   *(a) $\dfrac{3}{\sqrt{2}}$ (b) $\dfrac{5}{\sqrt{7}}$ *(c) $\dfrac{\sqrt{10}}{\sqrt{2}}$ (d) $\dfrac{\sqrt{21}}{\sqrt{3}}$

   *(e) $\dfrac{6\sqrt{5}}{\sqrt{8}}$ (f) $\dfrac{3\sqrt{15}}{\sqrt{32}}$ *(g) $\dfrac{5\sqrt{27}}{\sqrt{3}}$ (h) $\dfrac{\sqrt{98}}{7\sqrt{2}}$

   *(i) $\dfrac{4\sqrt{6}}{\sqrt{200}}$ (j) $\dfrac{5\sqrt{3}}{\sqrt{128}}$

2. Rationalize the numerator of each fraction and simplify as much as possible.

   *(a) $\dfrac{\sqrt{2}}{5}$ (b) $\dfrac{\sqrt{7}}{2}$ *(c) $\dfrac{\sqrt{8}}{3}$ (d) $\dfrac{\sqrt{27}}{4}$

   *(e) $\dfrac{\sqrt{24}}{12}$ (f) $\dfrac{\sqrt{54}}{3}$

3. Simplify and remove the radical from the denominator when it appears.

   *(a) $\sqrt{\dfrac{1}{3}}$ (b) $\sqrt{\dfrac{1}{5}}$ *(c) $\sqrt{\dfrac{80}{70}}$ (d) $\sqrt{\dfrac{4}{13}}$

   *(e) $\sqrt{\dfrac{1}{9}}$ (f) $\sqrt{\dfrac{27}{32}}$

4. Reduce each fraction if possible.

   *(a) $\dfrac{2+\sqrt{20}}{2}$ (b) $\dfrac{3+\sqrt{63}}{3}$

   *(c) $\dfrac{6+\sqrt{63}}{3}$ (d) $\dfrac{4-\sqrt{20}}{4}$

*(e) $\dfrac{5 + \sqrt{150}}{15}$

(f) $\dfrac{2 - \sqrt{24}}{3}$

5. Multiply.

*(a) $\sqrt{5}\,\sqrt{6}$

(b) $\sqrt{7}\,\sqrt{6}$

*(c) $(1 + \sqrt{2})(1 + \sqrt{2})$

(d) $(5 + \sqrt{3})(4 + \sqrt{3})$

*(e) $(5 + \sqrt{7})(5 - \sqrt{7})$

(f) $(3 + \sqrt{5})(2 - \sqrt{5})$

*(g) $(6 - \sqrt{13})(8 - \sqrt{3})$

(h) $(12 - \sqrt{17})(5 - \sqrt{3})$

6. Remove the radical from the denominator.

*(a) $\dfrac{1}{\sqrt{5} + 2}$

*(b) $\dfrac{3}{1 + \sqrt{2}}$

*(c) $\dfrac{6}{\sqrt{7} - 2}$

*(d) $\dfrac{7}{\sqrt{10} + 4}$

(e) $\dfrac{13}{\sqrt{17} - 4}$

(f) $\dfrac{4}{\sqrt{6} + 5}$

21.5 EQUATIONS CONTAINING RADICALS

Here is a series of examples that shows how to solve equations that contain radicals. We shall be using a new operation called *squaring both sides of an equation*. If you have an equation, then the left- and right-hand sides are equal. For instance, in the equation

squaring both sides of an equation

$$\sqrt{x} = 6$$

$\sqrt{x}$ and 6 are equal. So if we multiply each side by itself (that is, square both sides)

$$\underline{\sqrt{x}\,\sqrt{x}} = \underline{6 \cdot 6}$$

we are actually just multiplying both sides by equal quantities. This "new operation" produces the result

$$x = 36 \;\checkmark$$

since $\sqrt{x}\,\sqrt{x} = x$ and $6 \cdot 6 = 36$.

The check:

$$\sqrt{x} = 6$$

$$\sqrt{36} \overset{?}{=} 6 \qquad \text{after substituting 36 for } x$$

$$6 = 6 \;\checkmark$$

Example 23. *Solve* $\sqrt{x+1} = 11$.

Begin by squaring both sides of the equation. Squaring removes the radical, since by the definition of square root, $\sqrt{x+1}\,\sqrt{x+1} = x+1$.

$$(\sqrt{x+1})^2 = (11)^2$$
$$x+1 = 121$$
$$x = 120 \ \checkmark$$

The check:

$$\sqrt{x+1} = 11$$
$$\sqrt{120+1} \overset{?}{=} 11 \qquad \text{after substituting 120 for } x$$
$$\sqrt{121} \overset{?}{=} 11$$
$$11 = 11 \ \checkmark$$

Example 24. *Solve* $3\sqrt{7n-3} = 4$.

On squaring,

$$(3\sqrt{7n-3})^2 = (4)^2$$
$$\text{or} \qquad 3^2(\sqrt{7n-3})^2 = 4^2$$
$$9(7n-3) = 16 \qquad \text{since } (\sqrt{7n-3})^2 = 7n-3$$
$$63n - 27 = 16 \qquad \text{after distributing the 9}$$
$$63n = 43$$
$$n = \frac{43}{63} \ \checkmark$$

The check:

$$3\sqrt{7n-3} = 4$$
$$3\sqrt{7 \cdot \frac{43}{63} - 3} \overset{?}{=} 4 \qquad \text{after substituting } \frac{43}{63} \text{ for } n$$
$$3\sqrt{\frac{43}{9} - 3} \overset{?}{=} 4$$
$$3\sqrt{\frac{43}{9} - \frac{27}{9}} \overset{?}{=} 4$$
$$3\sqrt{\frac{16}{9}} \overset{?}{=} 4$$
$$3 \cdot \frac{4}{3} \overset{?}{=} 4$$
$$4 = 4 \ \checkmark$$

Example 25. *Solve* $\sqrt{6x-1} = 2\sqrt{x}$.

$$(\sqrt{6x-1})^2 = (2\sqrt{x})^2$$
$$6x - 1 = 4x$$
$$2x - 1 = 0$$
$$2x = 1$$
$$x = \frac{1}{2} \checkmark$$

The check:

$$\sqrt{6x-1} = 2\sqrt{x}$$
$$\sqrt{6 \cdot \frac{1}{2} - 1} \stackrel{?}{=} 2\sqrt{\frac{1}{2}}$$
$$\sqrt{3-1} \stackrel{?}{=} 2 \cdot \frac{\sqrt{1}}{\sqrt{2}}$$
$$\sqrt{2} \stackrel{?}{=} 2 \cdot \frac{\sqrt{1}}{\sqrt{2}} \cdot \frac{\sqrt{2}}{\sqrt{2}}$$
$$\sqrt{2} \stackrel{?}{=} 2 \cdot \frac{\sqrt{2}}{2}$$
$$\sqrt{2} = \sqrt{2} \checkmark$$

Example 26. *Solve* $\sqrt{2x} + 3 = 9$.

You *could* immediately square both sides, as we have done in the other examples. But doing so will not remove the radical. Observe.

$$(\sqrt{2x} + 3)^2 = (\sqrt{2x} + 3)(\sqrt{2x} + 3)$$
$$= 2x + 3\sqrt{2x} + 3\sqrt{2x} + 9$$
$$= 2x + 6\sqrt{2x} + 9$$

So instead of squaring immediately, simply add -3 to both sides of the original equation, $\sqrt{2x} + 3 = 9$. The result is an equation with the radical alone on one side.

$$\sqrt{2x} = 6$$

Now, on squaring both sides, we obtain a simple equation with no radical in it.

$$2x = 36$$

which yields

$$x = 18 \checkmark$$

Be sure to check 18 in the original equation.

Example 27. *Solve* $\sqrt{x} + 1 = 0$.

$$\sqrt{x} + 1 = 0$$

$$\sqrt{x} = -1 \qquad \text{after adding } -1 \text{ to both sides}$$

$$x = 1 \qquad \text{after squaring both sides of the equation}$$

But $x = 1$ *will not check in the original equation.* Try it. No number will check in the original equation; there is no solution. Squaring the -1 removed the minus and introduced an error. If you look at the original, you will notice that it suggests that the sum of two positive numbers must be equal to zero. And that cannot be true.

In the last example a solution was determined that did not check; that is, it was not a solution of the original equation. Such a "solution" **extraneous root** is called an *extraneous root*. Whenever both sides of an equation are squared, all solutions must be checked in order to determine which values are indeed solutions and which are extraneous.

As a simple example, consider the equation $x = 3$, which has only one solution, 3. Now square both sides of this equation to obtain $x^2 = 9$. This new equation has two solutions, -3 and 3, yet only one of them (3) is a solution of the original equation. In this example -3 is an extraneous root.

EXERCISES 21.5

1. Solve each equation.

*(a) $\sqrt{x} = 9$

(b) $\sqrt{m} = 11$

*(c) $\sqrt{x - 1} = 4$

(d) $\sqrt{2x + 1} = 5$

*(e) $6\sqrt{5y - 2} = 9$

(f) $10\sqrt{1 - 3x} = 12$

*(g) $\sqrt{3t - 1} = \sqrt{t + 7}$

(h) $\sqrt{2n + 7} = \sqrt{5n + 1}$

*(i) $\sqrt{5x + 10} = 4\sqrt{x}$

(j) $\sqrt{2x - 3} = 3\sqrt{x - 5}$

2. Solve each equation, if possible. Check your answer.

*(a) $\sqrt{3x - 5} + 2 = 6$

(b) $\sqrt{2n + 5} = 19$

*(c) $\sqrt{x + 1} + 4 = 0$

(d) $3 + \sqrt{x} = 1$

*(e) $6 + \sqrt{2x - 7} = 10$

(f) $8 + \sqrt{5x + 1} = 15$

*(g) $\sqrt{4n + 3} = \sqrt{6}$

*(h) $\sqrt{2x + 3} - 4 = 1$

*(i) $\sqrt{10 - 3x} = \sqrt{13}$

*(j) $\sqrt{7t - 6} - 8 = 13$

Chapter 21. REVIEW EXERCISES

1. Simplify, if possible.

   *(a) $\sqrt{24}$ (b) $\sqrt{80}$ *(c) $\sqrt{500}$ (d) $\sqrt{64}$

   *(e) $\sqrt{54}$ *(f) $\sqrt{76}$ *(g) $\sqrt{160}$ (h) $\sqrt{256}$

   *(i) $\sqrt{102}$ *(j) $\sqrt{171}$ *(k) $\sqrt{243}$

2. Simplify.

   *(a) $6\sqrt{2} + 5\sqrt{32}$ (b) $\sqrt{27} + \sqrt{75}$

3. Rationalize the denominator of each fraction.

   *(a) $\dfrac{7}{\sqrt{5}}$ (b) $\dfrac{3}{5 - \sqrt{7}}$

4. Simplify. Leave with a rational denominator.

   *(a) $\sqrt{\dfrac{1}{11}}$ (b) $\sqrt{\dfrac{9}{17}}$

5. Solve each equation.

   *(a) $\sqrt{3x + 2} = 11$ (b) $\sqrt{x + 3} - 2 = 7$

More Quadratic Equations

22.1 INTRODUCTION

In Chapter 17 you learned how to solve quadratic equations by factoring. In this chapter you will see how to solve quadratic equations which cannot be solved by factoring. The methods presented are *completing the square* (Section 22.2) and the *quadratic formula* (Section 22.3).

We proceed now with some special examples that will prepare you for the process of completing the square. The examples are based on solving quadratic equations in which the factoring for the difference of two squares can be applied.

Here are a few examples of quadratic equations that can be solved by using the factoring for the difference of two squares.

$$x^2 = 9 \longrightarrow x^2 - 9 = 0 \longrightarrow (x + 3)(x - 3) = 0 \longrightarrow x = -3, 3$$

$$x^2 = 4 \longrightarrow x^2 - 4 = 0 \longrightarrow (x + 2)(x - 2) = 0 \longrightarrow x = -2, 2$$

$$x^2 = 16 \longrightarrow x^2 - 16 = 0 \longrightarrow (x + 4)(x - 4) = 0 \longrightarrow x = -4, 4$$

plus or minus   If we use the notation $\pm$ to mean *plus or minus*, then $x = -3, 3$ can be written $x = \pm 3$; that is, $+3$ is a solution and -3 is also a solution. The preceding work can be summarized as

$$x^2 = 9 \longrightarrow x = \pm 3$$
$$x^2 = 4 \longrightarrow x = \pm 2$$
$$x^2 = 16 \longrightarrow x = \pm 4$$

Of course, it is also true that

$$\text{If } x^2 = 25, \text{ then } x = \pm 5.$$
$$\text{If } x^2 = 81, \text{ then } x = \pm 9.$$

In general,

$$\boxed{\text{If } x^2 = n, \text{ then } x = \pm \sqrt{n}.}$$

A few more examples.

$$x^2 = 49 \longrightarrow x = \pm 7$$
$$x^2 = 21 \longrightarrow x = \pm \sqrt{21}$$
$$x^2 = 3 \longrightarrow x = \pm \sqrt{3}$$

Example 1. *Solve the quadratic equation* $(x + 3)^2 = 17$.

Following the preceding pattern, if

$$(x + 3)^2 = 17$$

then $$x + 3 = \pm \sqrt{17}$$

To solve for x, add -3 to both sides. The result is

$$x = -3 \pm \sqrt{17} \; \checkmark$$

Note that $-3 \pm \sqrt{17}$ is read "minus 3 plus or minus the square root of 17." It indicates that both $-3 + \sqrt{17}$ and $-3 - \sqrt{17}$ are solutions.

Example 2. *Solve the equation* $(x - 5)^2 = 12$.

$$(x - 5)^2 = 12$$
$$x - 5 = \pm \sqrt{12}$$
$$x = 5 \pm \sqrt{12}$$

And since $\sqrt{12} = \sqrt{4 \cdot 3} = \sqrt{4} \sqrt{3} = 2\sqrt{3}$, the solutions simplify to

$$x = 5 \pm 2\sqrt{3} \; \checkmark$$

EXERCISES 22.1

Answers to starred exercises are given in the back of the book.

*1. Solve each quadratic equation.

(a) $x^2 = 64$ (b) $m^2 = 1$
(c) $y^2 = 36$ (d) $c^2 = 100$
(e) $x^2 = 7$ (f) $x^2 = 11$
(g) $a^2 = 15$ (h) $t^2 = 47$

2. Solve each quadratic equation.

*(a) $(x + 1)^2 = 11$ (b) $(x - 1)^2 = 13$
*(c) $(x - 3)^2 = 19$ (d) $(m + 5)^2 = 3$
*(e) $(y - 4)^2 = 8$ (f) $(t + 2)^2 = 24$
*(g) $(p + 1)^2 = 28$ (h) $(x + 6)^2 = 50$
*(i) $(x - 2)^2 - 3 = 0$ (j) $(x + 5)^2 - 7 = 0$
*(k) $(x + 1)^2 + 1 = 8$ (l) $(x - 2)^2 + 5 = 19$

22.2 COMPLETING THE SQUARE

completing the square

You should now have the appropriate orientation for studying the process called *completing the square*. The equation $x^2 + 4x + 1 = 0$ cannot be solved by factoring. (Try it and you will see.) So another approach is needed. The expression $x^2 + 4x + 1$ is almost the square of $(x + 2)$, since

$$(x + 2)^2 = x^2 + 4x + 4$$

Because we need only add a constant to $x^2 + 4x + 1$ to make it the square of $(x + 2)$, the process of completing the square can be used here. Let's continue. We have

$$x^2 + 4x + 1 = 0$$

and we want

$$x^2 + 4x + 4 = \cdots$$

So add 3 to both sides of the original equation.

$$
\begin{array}{rcr}
x^2 + 4x + 1 = & & 0 \\
+3 & & +3 \\
\hline
x^2 + 4x + 4 = & & 3
\end{array}
$$

Writing $x^2 + 4x + 4$ as $(x + 2)^2$, we obtain

$$(x + 2)^2 = 3$$

which resembles the form of Examples 1 and 2 and should be easy for you to solve. The work is shown next.

$$(x + 2)^2 = 3$$
$$x + 2 = \pm\sqrt{3}$$
$$x = -2 \pm \sqrt{3} \checkmark$$

The roots $-2 \pm \sqrt{3}$ can be written separately as $-2 + \sqrt{3}$ and $-2 - \sqrt{3}$, but there is no advantage in doing so in this instance.

Example 3. *Solve the equation $x^2 + 6x - 2 = 0$ by completing the square.*

The square that includes $x^2 + 6x$ is $(x + 3)^2$, since $(x + 3)(x + 3) = x^2 + 6x + 9$. So add $+11$ to both sides of $x^2 + 6x - 2 = 0$ to get the square $x^2 + 6x + 9$.

$$
\begin{array}{rcr}
x^2 + 6x - 2 = & & 0 \\
+\,11 & & +11 \\
\hline
x^2 + 6x + 9 = & & 11
\end{array}
$$

Since $x^2 + 6x + 9$ is $(x + 3)^2$, our equation can be written as

$$(x + 3)^2 = 11$$

Now
$$x + 3 = \pm\sqrt{11}$$

So
$$x = -3 \pm \sqrt{11} \checkmark$$

You can always complete a square of the form $x^2 + nx + \cdots$ by noting that the square is $\left(x + \dfrac{n}{2}\right)^2$ and working from there. For instance,

$$
\begin{array}{llll}
x^2 + 8x \cdots & \longrightarrow & (x + 4)^2 & +4 \text{ is half of } +8 \\
m^2 - 2m \cdots & \longrightarrow & (m - 1)^2 & -1 \text{ is half of } -2 \\
y^2 + 3y \cdots & \longrightarrow & \left(y + \dfrac{3}{2}\right)^2 & +\dfrac{3}{2} \text{ is half of } +3 \\
x^2 - 50x \cdots & \longrightarrow & (x - 25)^2 & -25 \text{ is half of } -50
\end{array}
$$

Example 4. *Solve the equation $y^2 - 12y + 12 = 0$ by completing the square.*

You can see that $y^2 - 12y + 12$ differs from the square $(y - 6)^2$ by a constant, since $(y - 6)^2 = y^2 - 12y + 36$. Add the constant 24 to both sides to make the expression on the left side equal to the desired square.

$$y^2 - 12y + 12 = 0$$
$$+ 24 = +24$$
$$\overline{y^2 - 12y + 36 = 24}$$
$$(y - 6)^2 = 24$$
$$y - 6 = \pm\sqrt{24}$$
$$y = 6 \pm \sqrt{24}$$

And since $\sqrt{24} = \sqrt{4}\sqrt{6} = 2\sqrt{6}$, replace $\sqrt{24}$ with the simpler form $2\sqrt{6}$ to get

$$y = 6 \pm 2\sqrt{6} \quad \checkmark$$

The next two examples involve fractions and are considerably more involved. They should serve as an incentive to pursue another approach (which we do in the next section). On the other hand, there are situations in which you must be able to complete the square even when fractions arise.

Example 5. *Solve the equation* $x^2 + 7x - 5 = 0$ *by completing the square.*

The $x^2 + 7x$ portion suggests, since $\frac{7}{2}$ is half of the 7, that we use the square

$$\left(x + \frac{7}{2}\right)^2$$

or

$$x^2 + 7x + \frac{49}{4}$$

Since we are beginning with the equation $x^2 + 7x - 5 = 0$, we need to add $+5$ and $+\frac{49}{4}$ to it (both sides) in order to obtain the desired square $x^2 + 7x + \frac{49}{4}$.

$$x^2 + 7x - 5 = 0$$
$$+5 \qquad +5$$
$$+\frac{49}{4} \qquad +\frac{49}{4}$$
$$\overline{x^2 + 7x + \frac{49}{4} = 5 + \frac{49}{4}}$$

Since $x^2 + 7x + \frac{49}{4}$ is $\left(x + \frac{7}{2}\right)^2$ and since $5 + \frac{49}{4} = \frac{20}{4} + \frac{49}{4} = \frac{69}{4}$, we have

$$\left(x + \frac{7}{2}\right)^2 = \frac{69}{4}$$

It follows that

$$x + \frac{7}{2} = \pm\sqrt{\frac{69}{4}}$$

On solving for x, we have

$$x = -\frac{7}{2} \pm \sqrt{\frac{69}{4}}$$

Also, $\sqrt{\frac{69}{4}} = \frac{\sqrt{69}}{\sqrt{4}} = \frac{\sqrt{69}}{2}$. So

$$x = -\frac{7}{2} \pm \frac{\sqrt{69}}{2}$$

or

$$x = \frac{-7 \pm \sqrt{69}}{2} \quad \checkmark$$

Example 6. *Solve the equation $5n^2 - 8n + 2 = 0$ by completing the square.*

In order to complete the square, we must first get a coefficient of 1 for the n^2 term. So multiply both sides of $5n^2 - 8n + 2 = 0$ by $\frac{1}{5}$ to get a coefficient of 1 for n^2.

$$n^2 - \frac{8}{5}n + \frac{2}{5} = 0$$

Since half of $-\frac{8}{5}$ is $-\frac{4}{5}$, our square is

$$\left(n - \frac{4}{5}\right)^2$$

or

$$n^2 - \frac{8}{5}n + \frac{16}{25}$$

We must change $n^2 - \frac{8}{5}n + \frac{2}{5} = 0$ in order to make the left side equal to $n^2 - \frac{8}{5}n + \frac{16}{25}$. We do so by adding $-\frac{2}{5}$ and $+\frac{16}{25}$ so as to change the $\frac{2}{5}$ term to $\frac{16}{25}$.

$$n^2 - \frac{8}{5}n + \frac{2}{5} \underline{- \frac{2}{5} + \frac{16}{25}} = 0 \underline{- \frac{2}{5} + \frac{16}{25}}$$

or

$$n^2 - \frac{8}{5}n + \frac{16}{25} = -\frac{10}{25} + \frac{16}{25}$$

or

$$n^2 - \frac{8}{5}n + \frac{16}{25} = \frac{6}{25}$$

which can be put in the form

$$\left(n - \frac{4}{5}\right)^2 = \frac{6}{25}$$

and then

$$n - \frac{4}{5} = \pm\sqrt{\frac{6}{25}}$$

or

$$n = \frac{4}{5} \pm \sqrt{\frac{6}{25}}$$

and since $\sqrt{\dfrac{6}{25}} = \dfrac{\sqrt{6}}{5}$, we get

$$n = \frac{4}{5} \pm \frac{\sqrt{6}}{5} = \frac{4 \pm \sqrt{6}}{5} \quad \checkmark$$

EXERCISES 22.2

1. Solve each quadratic equation by completing the square.
   *(a) $x^2 + 8x + 1 = 0$ (b) $x^2 + 4x + 1 = 0$
   *(c) $m^2 + 6m + 7 = 0$ (d) $x^2 + 6x + 1 = 0$
   *(e) $y^2 + 12y + 15 = 0$ (f) $t^2 + 12t + 33 = 0$
   *(g) $x^2 - 8x + 8 = 0$ (h) $x^2 + 8x + 5 = 0$
   *(i) $x^2 + 10x = 4$ (j) $x^2 - 4x = 10$

2. Solve each quadratic equation by completing the square.
   *(a) $x^2 - 3x + 1 = 0$ (b) $n^2 + 7n - 4 = 0$
   *(c) $x^2 + 5x + 1 = 0$ (d) $y^2 - 5y - 2 = 0$
   *(e) $x^2 + 3x - 8 = 0$ (f) $m^2 + 9m + 2 = 0$
   *(g) $2x^2 + 6x - 1 = 0$ (h) $2x^2 - 8x - 9 = 0$
   *(i) $5x^2 + 3x = 1$ (j) $3x^2 - 5x = 6$

22.3 THE QUADRATIC FORMULA

quadratic formula

A useful formula, called the *quadratic formula*, has been derived for solving any quadratic equation—that is, any equation of the form

$$ax^2 + bx + c = 0 \qquad a \neq 0$$

The formula was obtained by completing the square for the general equation $ax^2 + bx + c = 0$. The derivation is given as Section 22.4. The result is

$$\boxed{x = \frac{-b \pm \sqrt{b^2 - 4ac}}{2a}}$$

To use the formula, obtain a, b, and c from the equation and substitute them into the formula.

a is the coefficient of x^2.
b is the coefficient of x.
c is the *constant* term.

Here are some examples specifying the a, b, and c.

> In $x^2 + 7x - 10 = 0$, $a = 1$, $b = 7$, $c = -10$.
> In $2x^2 - 3x + 2 = 0$, $a = 2$, $b = -3$, $c = 2$.
> In $5x^2 - 3 = 0$, $a = 5$, $b = 0$, $c = -3$.
> In $x^2 + 2x = 0$, $a = 1$, $b = 2$, $c = 0$.
> In $5x^2 = 6x + 1$,   we must transform the equation into $5x^2 - 6x - 1 = 0$ to see that $a = 5$, $b = -6$, $c = -1$.

Note: *a, the coefficient of x^2, must be positive.* So if the coefficient of x^2, as given, is negative, you can multiply the entire equation (all terms, both sides) by -1 in order to make it positive. For example,

$$-x^2 + 4x - 2 = 0 \xrightarrow{\times -1} x^2 - 4x + 2 = 0$$

Now a is positive. In fact,

$$a = 1$$
$$b = -4$$
$$c = 2$$

Example 7. *Use the quadratic formula to solve the equation $x^2 + 7x + 4 = 0$.*

$$a = 1, \quad b = 7, \quad c = 4$$

$$x = \frac{-b \pm \sqrt{b^2 - 4ac}}{2a}$$

$$= \frac{-7 \pm \sqrt{(7)^2 - 4(1)(4)}}{2(1)}$$

$$= \frac{-7 \pm \sqrt{49 - 16}}{2}$$

$$= \frac{-7 \pm \sqrt{33}}{2} \checkmark$$

Note: You *cannot* split $\sqrt{49 - 16}$ into $\sqrt{49} - \sqrt{16}$. After all, $\sqrt{49 - 16}$ is $\sqrt{33}$, and $\sqrt{49} - \sqrt{16}$ is $7 - 4$, or 3. Surely $\sqrt{33} \neq 3$. Similarly, an expression such as $\sqrt{9 + 16}$ *cannot* be split into $\sqrt{9} + \sqrt{16}$. Observe that $\sqrt{9 + 16} = \sqrt{25} = 5$, yet $\sqrt{9} + \sqrt{16} = 3 + 4 = 7$. And $5 \neq 7$.

Example 8. *Use the quadratic formula to solve the equation*
  $2x^2 + 4x - 3 = 0$.

$$a = 2, \quad b = 4, \quad c = -3$$

$$x = \frac{-b \pm \sqrt{b^2 - 4ac}}{2a}$$

$$= \frac{-4 \pm \sqrt{(4)^2 - 4(2)(-3)}}{2(2)}$$

$$= \frac{-4 \pm \sqrt{16 + 24}}{4}$$

$$= \frac{-4 \pm \sqrt{40}}{4}$$

$$= \frac{-4 \pm 2\sqrt{10}}{4} \qquad \text{since } \sqrt{40} = \sqrt{4 \cdot 10} = \sqrt{4}\sqrt{10} = 2\sqrt{10}$$

$$= \frac{2(-2 \pm \sqrt{10})}{2 \cdot 2} \qquad \text{after factoring out 2's}$$

$$= \frac{-2 \pm \sqrt{10}}{2} \ \checkmark \qquad \text{since the 2's divide out}$$

Example 9. *Solve the equation* $x^2 = 5 + 2x$ *by using the quadratic formula.*

Change $x^2 = 5 + 2x$ to $ax^2 + bx + c = 0$ form in order to select a, b, and c for use in the quadratic formula. First, add $-2x$ to each side.

$$\begin{array}{rcl} x^2 & = & 5 + 2x \\ -2x & & -2x \\ \hline x^2 - 2x & = & 5 \end{array}$$

Next, add -5 to each side.

$$\begin{array}{rcl} x^2 - 2x & = & 5 \\ -5 & & -5 \\ \hline x^2 - 2x - 5 & = & 0 \end{array}$$

From $x^2 - 2x - 5 = 0$ we can see that $a = 1$, $b = -2$, and $c = -5$. These values can be substituted in the quadratic formula to find the roots of the equation.

$$x = \frac{-b \pm \sqrt{b^2 - 4ac}}{2a}$$

$$= \frac{-(-2) \pm \sqrt{(-2)^2 - 4(1)(-5)}}{2(1)}$$

$$= \frac{2 \pm \sqrt{4 + 20}}{2}$$

$$= \frac{2 \pm \sqrt{24}}{2}$$

$$= \frac{2 \pm 2\sqrt{6}}{2} \qquad \text{since } \sqrt{24} = \sqrt{4 \cdot 6} = 2\sqrt{6}$$

$$= \frac{2(1 \pm \sqrt{6})}{2}$$

$$x = 1 \pm \sqrt{6} \checkmark \qquad \text{since } \frac{2}{2} = 1$$

Example 10. *Solve the equation $x^2 + 7x + 12 = 0$ by using the quadratic formula.*

This equation can be solved by factoring and actually should be solved that way. Suppose, however, that you are careless at some time and use the quadratic formula because you do not realize that the quadratic expression could be factored. Let's see what sort of result you can expect. Keep in mind, of course, that *you should, in fact, use the formula only when the quadratic expression cannot be factored by ordinary means.*

$$a = 1, \quad b = 7, \quad c = 12$$

$$x = \frac{-b \pm \sqrt{b^2 - 4ac}}{2a} = \frac{-7 \pm \sqrt{(7)^2 - 4(1)(12)}}{2(1)}$$

$$= \frac{-7 \pm \sqrt{49 - 48}}{2}$$

$$= \frac{-7 \pm \sqrt{1}}{2}$$

$$= \frac{-7 \pm 1}{2}$$

Since no radical is left at this point, we consider both cases ($+$ and $-$) and continue the computations.

$$x = \frac{-7 + 1}{2} \quad \text{and} \quad \frac{-7 - 1}{2}$$

$$= \frac{-6}{2} \quad \text{and} \quad \frac{-8}{2}$$

$$= -3 \quad \text{and} \quad -4 \checkmark$$

So the solutions are $x = -3$ and $x = -4$.

EXERCISES 22.3

*1. Solve each equation by using the quadratic formula.

(a) $x^2 + 3x + 1 = 0$ (b) $x^2 + 5x + 2 = 0$

(c) $x^2 - 5x - 3 = 0$ (d) $y^2 - 3y + 1 = 0$

(e) $3t^2 + 5t - 6 = 0$ (f) $2x^2 + 3x - 3 = 0$

2. Solve each equation by using the quadratic formula.

*(a) $x^2 - 4x + 1 = 0$ (b) $x^2 + 4x - 1 = 0$

*(c) $x^2 + 6x + 1 = 0$ (d) $x^2 - 4x + 1 = 0$

*(e) $3x^2 - 2x - 7 = 0$ (f) $m^2 - 8m + 2 = 0$

*(g) $4x^2 - 8x + 1 = 0$ (h) $4y^2 - 6y - 3 = 0$

3. Solve each equation by using the quadratic formula.

*(a) $x^2 = 5 - 7x$ (b) $x = 2x^2 - 1$

*(c) $x^2 - 8 = 3x$ (d) $m = m^2 - 10$

*(e) $-t^2 - 9t + 2 = 0$ (f) $-2x^2 + 9x + 3 = 0$

*(g) $5y^2 - 7y - 6 = 0$ (h) $2x^2 + 5x + 2 = 0$

*(i) $x^2 + 6x = 0$ (j) $x^2 = 3x$

*(k) $x^2 - 3 = 0$ (l) $x^2 - 10 = 0$

22.4 DERIVATION OF THE QUADRATIC FORMULA (Optional)

In this section we begin with the general form of the quadratic equation, $ax^2 + bx + c = 0$, and find the general solution for x—namely,

$$x = \frac{-b \pm \sqrt{b^2 - 4ac}}{2a}$$

First, begin with

$$ax^2 + bx + c = 0 \qquad a \neq 0$$

and divide both sides by a. (This is why a cannot be zero.)

$$\frac{ax^2}{a} + \frac{bx}{a} + \frac{c}{a} = \frac{0}{a}$$

The result is

$$x^2 + \frac{b}{a}x + \frac{c}{a} = 0$$

Next, add $-\dfrac{c}{a}$ to both sides.

$$x^2 + \frac{b}{a}x = -\frac{c}{a}$$

The left-hand side is almost the square $\left(x + \dfrac{b}{2a}\right)^2$. But $\left(x + \dfrac{b}{2a}\right)^2 = x^2 + \dfrac{b}{a}x + \dfrac{b^2}{4a^2}$; so add $\dfrac{b^2}{4a^2}$ to the left side (and to the right side), in order to complete the square.

$$x^2 + \frac{b}{a}x + \frac{b^2}{4a^2} = \frac{b^2}{4a^2} - \frac{c}{a}$$

or

$$\left(x + \frac{b}{2a}\right)^2 = \frac{b^2}{4a^2} - \frac{c}{a}$$

Now let's combine the fractions that are on the right. Their common denominator is $4a^2$. Thus

$$\frac{b^2}{4a^2} - \frac{c}{a}$$

can be written

$$\frac{b^2}{4a^2} - \frac{c}{a} \cdot \frac{4a}{4a}$$

or

$$\frac{b^2}{4a^2} - \frac{4ac}{4a^2}$$

or

$$\frac{b^2 - 4ac}{4a^2}$$

So far we have

$$\left(x + \frac{b}{2a}\right)^2 = \frac{b^2 - 4ac}{4a^2}$$

It follows that

$$x + \frac{b}{2a} = \pm\sqrt{\frac{b^2 - 4ac}{4a^2}}$$

This equation can easily be solved for x by adding $-\dfrac{b}{2a}$ to both sides.

$$x = -\frac{b}{2a} \pm \sqrt{\frac{b^2 - 4ac}{4a^2}}$$

The radical can be simplified.

$$\sqrt{\frac{b^2 - 4ac}{4a^2}} = \frac{\sqrt{b^2 - 4ac}}{\sqrt{4a^2}} = \frac{\sqrt{b^2 - 4ac}}{2a}$$

Now x can be written

$$x = -\frac{b}{2a} \pm \frac{\sqrt{b^2 - 4ac}}{2a}$$

Finally, we can combine the two fractions to obtain

$$x = \frac{-b \pm \sqrt{b^2 - 4ac}}{2a}$$

And this is the quadratic formula, the formula used to solve equations of the form $ax^2 + bx + c = 0$.

22.5 IMAGINARY NUMBERS (Optional)

All the numbers used so far in this book, and probably all the numbers you have ever used, are classified as *real numbers*. Real numbers possess an important property: The square of any real number is either

real number

positive or zero. If a real number is squared, the result is never a negative number. Are there any numbers whose squares are negative numbers? Solving the quadratic equation $x^2 + 1 = 0$ leads to such a number, for here x must be a number whose square is -1, since x^2 and 1 must add up to zero. The number whose square is -1 is an example **imaginary number**   of a so-called *imaginary number*. If we solve $x^2 + 1 = 0$ mechanically, we first get $x^2 = -1$ and then $x = \pm\sqrt{-1}$. The number $\sqrt{-1}$ is called *i*, for imaginary. The roots of the equation $x^2 + 1 = 0$ are $x = \pm i$.

The number $\sqrt{-1}$ is written as i. In general, $\sqrt{-a}$ for positive numbers a is written $i\sqrt{a}$. Consider some other examples of imaginary numbers: $\sqrt{-4}, \sqrt{-5}, \sqrt{-75}$, and observe their simplification.

$$\sqrt{-4} = i\sqrt{4}$$
$$= i \cdot 2$$
$$= 2i \ \checkmark$$

$$\sqrt{-5} = i\sqrt{5} \ \checkmark$$

$$\sqrt{-75} = i\sqrt{75}$$
$$= i\sqrt{25}\sqrt{3}$$
$$= i \cdot 5\sqrt{3}$$
$$= 5i\sqrt{3} \ \checkmark$$

Example 11. *Solve the equation $x^2 + 2x + 5 = 0$.*

Using the quadratic formula with $a = 1$, $b = 2$, and $c = 5$, we obtain

$$x = \frac{-b \pm \sqrt{b^2 - 4ac}}{2a}$$
$$= \frac{-2 \pm \sqrt{(2)^2 - 4(1)(5)}}{2(1)}$$
$$\frac{-2 \pm \sqrt{4 - 20}}{2}$$
$$= \frac{-2 \pm \sqrt{-16}}{2}$$
$$= \frac{-2 \pm 4i}{2} \qquad \text{since } \sqrt{-16} = i\sqrt{16} = i \cdot 4 = 4i$$
$$= \frac{2(-1 \pm 2i)}{2}$$
$$= -1 \pm 2i \ \checkmark$$

Example 12. *Solve the equation* $x^2 - 4x + 7 = 0$.

Using the quadratic formula with $a = 1$, $b = -4$, and $c = 7$, we have

$$x = \frac{-b \pm \sqrt{b^2 - 4ac}}{2a}$$

$$= \frac{--4 \pm \sqrt{(-4)^2 - 4(1)(7)}}{2(1)}$$

$$= \frac{4 \pm \sqrt{16 - 28}}{2}$$

$$= \frac{4 \pm \sqrt{-12}}{2}$$

$$= \frac{4 \pm 2i\sqrt{3}}{2} \qquad \left\{ \begin{array}{l} \text{since } \sqrt{-12} = i\sqrt{12} \\ = i\sqrt{4}\sqrt{3} = 2i\sqrt{3} \end{array} \right.$$

$$= \frac{2(2 \pm i\sqrt{3})}{2}$$

$$= 2 \pm i\sqrt{3} \ \checkmark$$

EXERCISES 22.5

1. Express each number in terms of i. For example, $\sqrt{-9} = 3i$.
   - *(a) $\sqrt{-16}$
   - (b) $\sqrt{-25}$
   - *(c) $\sqrt{-3}$
   - (d) $\sqrt{-7}$
   - *(e) $\sqrt{-24}$
   - (f) $\sqrt{-20}$
   - *(g) $\sqrt{-27}$
   - (h) $\sqrt{-45}$
   - *(i) $\sqrt{-80}$
   - (j) $\sqrt{-63}$

2. Solve each quadratic equation.
   - *(a) $x^2 + 4 = 0$
   - (b) $x^2 + 9 = 0$
   - *(c) $x^2 - 2x + 5 = 0$
   - *(d) $x^2 + x + 1 = 0$
   - *(e) $m^2 + m + 2 = 0$
   - (f) $x^2 + 4x + 7 = 0$
   - *(g) $2x^2 - 3x + 4 = 0$
   - (h) $2x^2 + 6x + 8 = 0$
   - *(i) $x^2 = 2x - 9$
   - (j) $-x^2 + 2x - 10 = 0$

22.6 NATURE OF THE ROOTS (Optional)

We can determine the types of roots (that is, the nature of the roots) of a quadratic equation by examining $b^2 - 4ac$. The expression $b^2 - 4ac$ is called the *discriminant*.   **discriminant**

$$x = \frac{-b \pm \sqrt{\boxed{b^2 - 4ac}}}{2a} \quad\text{———— discriminant}$$

Case I.   Discriminant $= 0$.

If $b^2 - 4ac = 0$, then $\sqrt{} = 0$ and the roots are $x = \frac{-b}{2a}$; that is, both roots are the same. The roots are therefore *real and equal*. As an example, in $x^2 + 6x + 9 = 0$, $a = 1$, $b = 6$, and $c = 9$. So $b^2 - 4ac$ is $6^2 - 4 \cdot 1 \cdot 9 = 36 - 36 = 0$. The roots of $x^2 + 6x + 9 = 0$ are real and equal. Note

$$x = \frac{-b \pm \sqrt{\text{discriminant}}}{2a}$$

$$= \frac{-6 \pm \sqrt{0}}{2(1)} = \frac{-6 \pm 0}{2} = -\frac{6}{2} = -3 \checkmark$$

Case II.   Discriminant > 0.

If $b^2 - 4ac > 0$, then the number under the $\sqrt{}$ is positive. This means that the roots are real. Because of the $\pm$, the roots are different. Thus the roots are *real and unequal*. To illustrate, the roots of $x^2 + 7x + 12 = 0$ are real and unequal, since $b^2 - 4ac$ is $7^2 - 4 \cdot 1 \cdot 12 = 1$.

$$x = \frac{-b \pm \sqrt{\text{discriminant}}}{2a}$$

$$= \frac{-7 \pm \sqrt{1}}{2(1)} = \frac{-7 \pm 1}{2} = -3 \text{ and } -4 \checkmark$$

Case III.   Discriminant < 0.

If $b^2 - 4ac < 0$, then the number under the $\sqrt{}$ is negative. This means that the roots are imaginary. Because of the $\pm$, the roots are different. Thus the roots are *imaginary and unequal*. As an example, the roots of $x^2 + 3x + 7 = 0$ are imaginary and unequal, because $b^2 - 4ac = 3^2 - 4 \cdot 1 \cdot 7 = -19$.

$$x = \frac{-b \pm \sqrt{\text{discriminant}}}{2a} = \frac{-3 \pm \sqrt{-19}}{2(1)} = \frac{-3 \pm i\sqrt{19}}{2} \checkmark$$

EXERCISES 22.6

1. Check the discriminant to determine the nature of the roots of each quadratic equation below. Do not actually solve the equations.

*(a) $x^2 + 5x + 2 = 0$ (b) $m^2 - 3m + 1 = 0$
*(c) $y^2 + 2y + 6 = 0$ (d) $x^2 - 3x + 2 = 0$
*(e) $x^2 - 6x + 9 = 0$ (f) $x^2 - 6x - 9 = 0$
*(g) $x^2 - 5 = 0$ *(h) $n^2 + 5 = 0$
*(i) $x^2 + 2x = 0$ (j) $3x^2 + 7x + 1 = 0$
(k) $5x^2 - 10x + 6 = 0$ *(l) $x^2 = 3x + 10$
*(m) $x^2 - 1 = x$ *(n) $-x^2 + 7x + 2 = 0$
(o) $-x^2 + 6x - 9 = 0$ (p) $x^2 = 12x$

22.7 PARABOLAS

In Chapter 14 we saw that graphs of straight lines are associated with linear equations. Here we'll see an example of a graph that is associated with a quadratic equation.

Example 13. *Sketch the graph of* $y = x^2 + 3x + 2$.

Obtain points by substituting different values for x into $x^2 + 3x + 2$ in order to get corresponding values for y. Since it is not a straight line, several points should be obtained to see how the curve is supposed to look. The work below shows the substitution of 0, 1, 2, -1, -2, -3, -4, and -5 for x to obtain eight (x, y) points.

$$x = \underline{0}: \qquad y = \underline{0}^2 + 3 \cdot \underline{0} + 2 = 2 \qquad \longrightarrow (0, 2)$$
$$x = \underline{1}: \qquad y = \underline{1}^2 + 3 \cdot \underline{1} + 2 = 6 \qquad \longrightarrow (1, 6)$$
$$x = \underline{2}: \qquad y = \underline{2}^2 + 3 \cdot \underline{2} + 2 = 12 \qquad \longrightarrow (2, 12)$$
$$x = \underline{-1}: \qquad y = (\underline{-1})^2 + 3 \cdot (\underline{-1}) + 2 = 0 \longrightarrow (-1, 0)$$
$$x = \underline{-2}: \qquad y = (\underline{-2})^2 + 3(\underline{-2}) + 2 = 0 \qquad \longrightarrow (-2, 0)$$
$$x = \underline{-3}: \qquad y = (\underline{-3})^2 + 3(\underline{-3}) + 2 = 2 \qquad \longrightarrow (-3, 2)$$
$$x = \underline{-4}: \qquad y = (\underline{-4})^2 + 3(\underline{-4}) + 2 = 6 \qquad \longrightarrow (-4, 6)$$
$$x = \underline{-5}: \qquad y = (\underline{-5})^2 + 3(\underline{-5}) + 2 = 12 \longrightarrow (-5, 12)$$

The graph (shown next) is obtained by passing a smooth curve through the points.

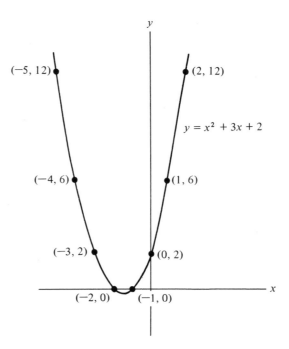

parabola The curve is an example of a *parabola*. Equations of the form

$$y = ax^2 + bx + c$$

are parabolas when graphed.

The parabola $y = x^2 + 3x + 2$ crosses the x axis at $x = -2$ and at $x = -1$. The y value on the x axis is 0; so when $x = -2$ or $x = -1$, then $y = 0$. In other words, the roots of $x^2 + 3x + 2 = 0$ (that is, when y is 0) are -2 and -1 *by graphing*. The roots can also be determined by factoring, as shown next.

$$x^2 + 3x + 2 = 0$$
$$(x + 2)(x + 1) = 0$$

$$x + 2 = 0 \qquad x + 1 = 0$$
$$x = -2 \qquad x = -1$$

If the coefficient of x^2 is negative, then the graph will appear "upside down," as shown in the next example.

Example 14. *Sketch the graph of $y = -x^2 - 2x + 3$.*

We shall use the following values for x: $-4, -3, -2, -1, 0, 1, 2, 3$. Note that in evaluating $-x^2 - 2x + 3$, the $-x^2$ means $-(x^2)$, *not* $(-x)^2$.

x	$y = -x^2 - 2x + 3$	(x, y)
-4	$y = -(16) - 2(-4) + 3 = -5$	$(-4, -5)$
-3	$y = -(9) - 2(-3) + 3 = 0$	$(-3, 0)$
-2	$y = -(4) - 2(-2) + 3 = 3$	$(-2, 3)$
-1	$y = -(1) - 2(-1) + 3 = 4$	$(-1, 4)$
0	$y = -(0) - 2(0) + 3 = 3$	$(0, 3)$
1	$y = -(1) - 2(1) + 3 = 0$	$(1, 0)$
2	$y = -(4) - 2(2) + 3 = -5$	$(2, -5)$
3	$y = -(9) - 2(3) + 3 = -12$	$(3, -12)$

The graph is shown below. Note that the roots of $-x^2 - 2x + 3 = 0$ are -3 and 1.

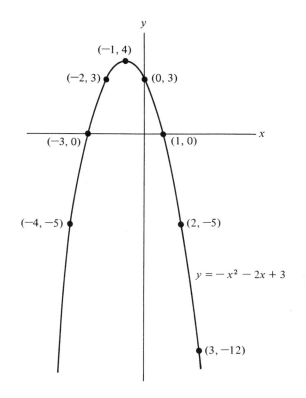

EXERCISES 22.7

*1. Sketch the graph of each parabola.

(a) $y = x^2$

(b) $y = x^2 - 1$

(c) $y = -x^2$

(d) $y = x^2 - 4x + 3$

(e) $y = 2x^2$

(f) $y = -x^2 + 3x + 2$

2. Sketch the graph of each parabola.

*(a) $y = x^2 - 5x + 6$

(b) $y = x^2 + 4x + 3$

*(c) $y = x^2 + 3x - 4$

(d) $y = x^2 + 1$

*(e) $y = 1 - x^2$

(f) $y = x^2 - 9$

*(g) $y = 3x^2 + 1$

(h) $y = 3x^2$

*(i) $y = x^2 + 3x$

(j) $y = x^2 - 2x$

*(k) $y = -x^2 + 2x - 3$

(l) $y = -x^2 + 1$

Chapter 22.   REVIEW EXERCISES

*1. Solve each equation. Do not use the quadratic formula *unless* the expression will not factor.

(a) $3x^2 - 2x - 1 = 0$

(b) $3x^2 - 6x + 2 = 0$

(c) $2t^2 + 9t + 1 = 0$

(d) $4y^2 + 16y + 7 = 0$

2. Solve by completing the square.

*(a) $m^2 + 10m + 1 = 0$

(b) $x^2 + 6x + 2 = 0$

*3. Determine the nature of the roots of each quadratic equation.

(a) $x^2 - 2x + 4 = 0$

(b) $2x^2 + x - 3 = 0$

(c) $x^2 - 8x + 16 = 0$

*4. Sketch the graph of the parabola $y = x^2 - 3x - 4$.

Slopes
and Intercepts
(optional)

23.1 INTRODUCTION

The slope of a line is an indication of the steepness or inclination of the line. Look at the graphs of the following lines.

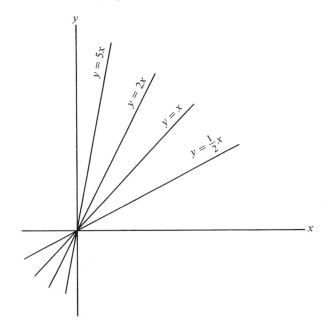

Notice that the steeper the line, the larger the coefficient of x. The coefficient of x represents the slope of the line. To find the slope of any line, write its equation in the form

$$y = mx + b$$

slope   and m (the coefficient of x) is the *slope*.

Example 1. *Determine the slope of $y = 3x - 2$.*

This line is already in $y = mx + b$ form, with $m = 3$. So the slope is 3.

Example 2. *Determine the slope of $2y = 6x - 7$.*

This line is not in $y = mx + b$ form, since the coefficient of y is 2. We want $y = \ldots$, not $2y = \ldots$. So divide both sides of the equation by 2.

$$2y = 6x - 7$$

becomes

$$\frac{2y}{2} = \frac{6x - 7}{2}$$

which simplifies to

$$y = 3x - \frac{7}{2}$$

Clearly $m = 3$; the slope is 3.

Example 3. *Determine the slope of $2x + 3y = 6$.*

Change $2x + 3y = 6$ to $y = mx + b$ form. First, get the y term by itself.

$$
\begin{array}{r}
2x + 3y = 6 \\
-2x \qquad\quad -2x \\
\hline
3y = -2x + 6
\end{array}
$$

Then obtain a coefficient of 1 for y by dividing both sides by 3.

$$y = \frac{-2x + 6}{3}$$

or

$$y = -\frac{2}{3}x + 2$$

Now you can see that $m = -\frac{2}{3}$; the slope is $-\frac{2}{3}$.

The idea of negative slopes will be discussed later in the chapter.

EXERCISES 23.1

Answers to starred exercises are given in the back of the book.

1. Determine the slope of each line.
   *(a) $y = 5x + 1$ (b) $y = 4x - 2$
   *(c) $y = -3x + 6$ (d) $y = -8x + 7$

*(e) $y = -x + 3$
*(g) $3y = 12x - 6$
*(i) $y - x = 7$
*(k) $x = 3y + 2$
*(m) $6y - 3x = 1$
*(o) $5x + 4y = 30$
*(q) $2x + 3y + 7 = 0$

*(f) $y = \frac{1}{3}x$
(h) $2y = 10x + 2$
(j) $y - 6x = -3$
(l) $x = 6 - 2y$
(n) $5y - 10x = 20$
(p) $4x + 5y = 3$
(r) $6x - 5y + 4 = 0$

23.2 *THE PRECISE MEANING OF SLOPE*

So far we have noted that the slope of a line is a measure of its steepness, and we know how to determine the slope of a line from its equation. Now, let's become more precise as to the meaning of slope. To begin, consider the line $y = 2x + 3$. We obtain four points and graph it.

x	y	
0	3	$\longrightarrow$ (0, 3)
1	5	$\longrightarrow$ (1, 5)
2	7	$\longrightarrow$ (2, 7)
4	11	$\longrightarrow$ (4, 11)

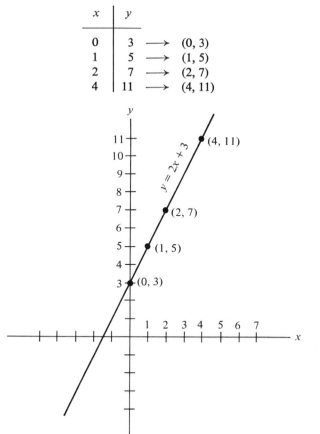

Consider two points on the line, say (1, 5) and (4, 11), as in the diagram below. We can get from (1, 5) to (4, 11) if we change x by 3 (that is, $4 - 1 = 3$) and then change y by 6 (that is, $11 - 5 = 6$).

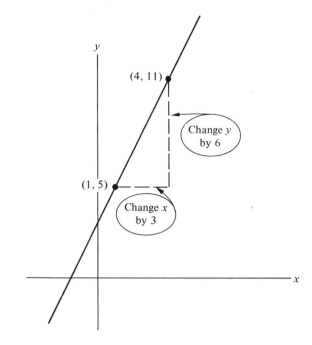

Now look at the following fraction:

$$\frac{\text{change in } y}{\text{change in } x} = \frac{6}{3} = \frac{2}{1} = 2$$

Consider two other points on the line—for example, (0, 3) and (2, 7).

$$\frac{\text{change in } y}{\text{change in } x} = \frac{7 - 3}{2 - 0} = \frac{4}{2} = \frac{2}{1} = 2$$

Consider the changes in going from (0, 3) to (1, 5).

$$\frac{\text{change in } y}{\text{change in } x} = \frac{5 - 3}{1 - 0} = \frac{2}{1} = 2$$

In all three cases, the fraction is the same, 2. This fraction, $\dfrac{\text{change in } y}{\text{change in } x}$, is called the *slope* of the line.† And, as the example suggests, it is constant for a given line.

---

†Note that such usage is consistent with reading the slope as the coefficient of x; for example, in $y = 2x + 3$ both methods yield 2 as the slope.

$$\boxed{\text{slope} = \frac{\text{change in } y}{\text{change in } x}}$$

Since m is often used to represent the slope,

$$m = \frac{\text{change in } y}{\text{change in } x}$$

Also, abbreviate "change in" by using the Greek letter delta, Δ. Thus

$$\boxed{m = \frac{\Delta y}{\Delta x}}$$

In words, slope equals delta y over delta x. Note that the symbol Δy represents one number and should be treated accordingly. It *does not* mean Δ times y. The same is true of Δx.

Example 4. *Compute the slope of the line passing through the points* (2, 3) *and* (7, 5)

$$m = \frac{\Delta y}{\Delta x} = \frac{5 - 3}{7 - 2} = \frac{2}{5} \checkmark$$

The slope is $\frac{2}{5}$. (Not all slopes are whole numbers.)

In this example we were careful to subtract in a consistent manner. Since we subtracted the 3 of (2, 3) from the 5 of (7, 5), we performed a similar subtraction with the x values: we subtracted the 2 of (2, 3) from the 7 of (7, 5) to get

$$\frac{5 - 3}{7 - 2} = \frac{2}{5}$$

If we had subtracted the 7 of (7, 5) from the 2 of (2, 3) instead, there would be an error in sign.

$$\frac{5 - 3}{2 - 7} = \frac{2}{-5} = -\frac{2}{5} \qquad \text{(wrong)}$$

In taking the difference of the y's and the difference of the x's, we must take them in the same direction, as in

$$\frac{5 - 3}{7 - 2} = \frac{2}{5}$$

or

$$\frac{3 - 5}{2 - 7} = \frac{-2}{-5} = \frac{2}{5}$$

but we must be consistent. To avoid confusion, refer to one point as (x_1, y_1) and the other point as (x_2, y_2), and compute m as

$$m = \frac{y_2 - y_1}{x_2 - x_1}$$

Example 5. *Compute the slope of the line passing through the points* $(3, -2)$ *and* $(5, 4)$.

Let $(3, -2)$ be (x_1, y_1) and let $(5, 4)$ be (x_2, y_2). Then

$$x_1 = 3$$
$$y_1 = -2$$
$$x_2 = 5$$
$$y_2 = 4$$

So

$$m = \frac{y_2 - y_1}{x_2 - x_1} = \frac{4 - (-2)}{5 - (3)} = \frac{6}{2} = 3 \checkmark$$

If instead we let $(5, 4)$ be (x_1, y_1) and $(3, -2)$ be (x_2, y_2), then

$$x_1 = 5$$
$$y_1 = 4$$
$$x_2 = 3$$
$$y_2 = -2$$

So

$$m = \frac{y_2 - y_1}{x_2 - x_1} = \frac{-2 - (4)}{3 - (5)} = \frac{-6}{-2} = 3 \checkmark$$

The slope is 3, and we have shown that it does not matter which point is called (x_1, y_1) and which is (x_2, y_2).

Example 6. *Compute the slope of the line through* $(0, -5)$ *and* $(-2, -1)$.

Let $(0, -5)$ be (x_1, y_1) and let $(-2, -1)$ be (x_2, y_2). Then

$$x_1 = 0$$
$$y_1 = -5$$
$$x_2 = -2$$
$$y_2 = -1$$

So

$$m = \frac{y_2 - y_1}{x_2 - x_1} = \frac{-1 - (-5)}{-2 - (0)} = \frac{4}{-2} = -2 \checkmark$$

The slope of the line is -2.

Example 7. *Graph the line that passes through the point* (1, 4) *and has a slope of* 3.

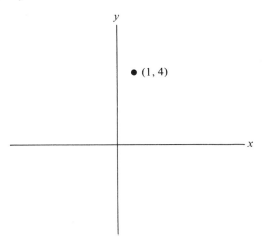

There are an infinite number of lines through (1, 4), but only one such line also has a slope of 3. If the line has a slope of 3, then

$$\frac{\Delta y}{\Delta x} = 3 = \frac{3}{1}$$

This means that if x changes by 1, then y must change by 3 in order to produce a point on the line.

We have the point (1, 4) already. Taking this point and

1. Changing x by 1—making it 2
2. Changing y by 3—making it 7

gives us another point on the line—namely, (2, 7).

Now we can pass a line through the two points.

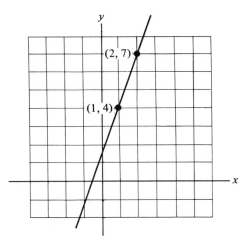

If you want another point, take the point (2, 7) and change x by 1 and y by 3 to get (3, 10).

There are other ways to get more points. Since the slope is 3,

$$\frac{\Delta y}{\Delta x} = \frac{3}{1}$$

You need not change x by 1 and y by 3. For instance, you can

1. Change x by 2 and y by 6, since $\frac{6}{2}$ reduces to 3.
2. Change x by 7 and y by 21, since $\frac{21}{7}$ reduces to 3.
3. Change x by $\frac{1}{2}$ and y by $\frac{3}{2}$, since $\frac{3/2}{1/2}$ reduces to 3.
4. Change x by -1 and y by -3, since $\frac{-3}{-1}$ reduces to 3.

As long as $\frac{\Delta y}{\Delta x}$ reduces to 3, the changes are all right.

Example 8. *Sketch a graph of the line through (0, 3) with slope of -4.*

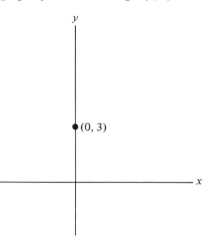

Slope $= -4$ means $\frac{\Delta y}{\Delta x} = -4$. This can be considered as $\frac{-4}{+1}$ or $\frac{+4}{-1}$.

From (0, 3) we can get another point by changing x by $+1$ and y by -4. Doing so gives

$$(0 + 1, 3 - 4) = (1, -1)$$

Although we don't actually need a third point, the fraction $\frac{+4}{-1}$ could be used to find one. If we take the point (0, 3) and change x by -1 and y by $+4$, we obtain

$$(0 - 1, 3 + 4) = (-1, 7)$$

Here is a graph of the line.

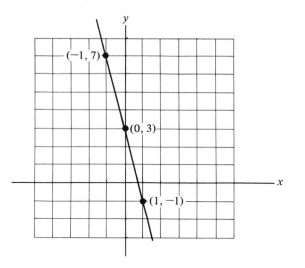

You can see from this example that negative-sloped lines fall as they go from left to right, whereas positive-sloped lines rise as they go from left to right.

EXERCISES 23.2

1. Compute the slope of the line through each pair of points.
   *(a) (5, 7) and (3, 6)   (b) (10, 8) and (6, 4)
   *(c) (4, 2) and (6, 8)   *(d) (5, 1) and (6, 0)
   *(e) (0, 0) and (1, 5)   (f) (7, 0) and (4, 3)
   *(g) (5, 2) and (4, 6)   *(h) (5, 2) and (1, −3)
   *(i) (1, −1) and (−1, 1)   (j) (5, 0) and (0, −6)
   *(k) (2, −3) and (−7, 4)   (l) (−5, 3) and (−3, −1)

2. Sketch a graph of the line through the given point and having the given slope.
   *(a) (0, 1), $m = 2$ (b) (2, 0), $m = 3$
   *(c) (4, 1), $m = \dfrac{5}{3}$ (d) (1, 2), $m = \dfrac{2}{3}$
   *(e) (−2, 1), $m = 4$ (f) (−3, 4), $m = 1$
   *(g) (1, 3), $m = -2$ (h) (4, 5), $m = -3$
   *(i) (5, 3), $m = -\dfrac{1}{2}$ (j) (4, 6), $m = -\dfrac{3}{2}$

23.3 INTERCEPTS

Consider the line $y = 3x - 2$ sketched below. The only points marked are those where the line crosses an axis.

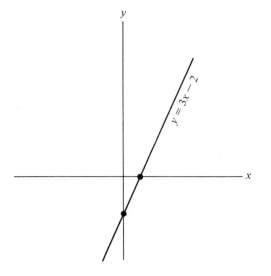

A line crosses the y axis where $x = 0$; that is, $x = 0$ where you are zero units from the y axis, when you are on the y axis. The value of y when $x = 0$ (that is, where the line crosses the y axis) is called the **y intercept**   *y intercept*. The y intercept of the line $y = 3x - 2$ is -2 because when $x = 0$, $y = 3 \cdot 0 - 2 = 0 - 2 = -2$.

The y intercept of the line $y = 7x + 9$ is 9, since $y = 9$ when $x = 0$.

When equations of lines are in the form $y = mx + b$, then mx disappears (that is, is equal to zero) when $x = 0$, and so b is always the y intercept.

$$
\begin{array}{l}
y = mx + b \\
m = \text{slope} \\
b = y \text{ intercept}
\end{array}
$$

The form $y = mx + b$ is often called the slope-intercept form.

Example 9. *Find the slope and y intercept of* $y = 2x + 5$.

$y = 2x + 5$ is already in $y = mx + b$ form.

$$m = 2 \quad \text{(slope)} \checkmark$$
$$b = 5 \quad (y \text{ intercept}) \checkmark$$

Example 10. *Find the slope and y intercept of* $2y - 3x = -5$.

The line $2y - 3x = -5$ is not in $y = mx + b$ form, which means you cannot read off the slope and y intercept. In order to change it to $y = mx + b$ form, begin by adding $3x$ to both sides to isolate the y term.

$$
\begin{array}{rcl}
2y - 3x & = & -5 \\
3x & & 3x \\
\hline
2y & & = 3x - 5
\end{array}
$$

Next, divide by 2.

$$\frac{2y}{2} = \frac{3x - 5}{2}$$

$$y = \frac{3}{2}x - \frac{5}{2}$$

Thus

$$m = \frac{3}{2} \quad \text{(slope)} \checkmark$$

$$b = -\frac{5}{2} \quad \text{(y intercept)} \checkmark$$

Example 11. *Sketch the graph of the line with slope* 2 *and y intercept* 1.

We are given $m = 2$ and $b = 1$. Because $b = 1$, the point $(0, 1)$ is on the line, since the y intercept is the value of y when $x = 0$.

Also, $m = 2 = \frac{2}{1} = \frac{\Delta y}{\Delta x}$.

If we use $(0, 1)$ as a first point and move from it by changing x by 1 and y by 2 $\left(\text{because } \frac{\Delta y}{\Delta x} = \frac{2}{1}\right)$, then we obtain $(1, 3)$, another point on the line. The line can now be sketched as shown below.

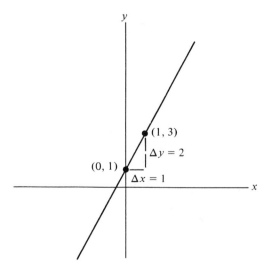

x intercept   Also of some interest is the x intercept. The *x intercept* is the value of x when y is 0. Thus the x intercept is the x value where the line crosses the x axis, as the diagram shows.

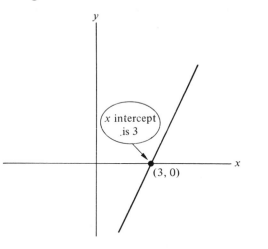

Example 12. *Find the x intercept of* $y = 5x - 1$.

To obtain the x intercept, let $y = 0$ in $y = 5x - 1$. Then solve for x.

$$0 = 5x - 1$$

$$5x = 1$$

$$x = \frac{1}{5} \checkmark$$

The x intercept is $\frac{1}{5}$; that is, x is $\frac{1}{5}$ when y is 0.

EXERCISES 23.3

1. Change each line to $y = mx + b$ form and determine its slope (m) and y intercept (b).

*(a) $y = 2x - 1$ (b) $y = -x + 4$

*(c) $y - 2 = x$ (d) $y = -3x + 2$

*(e) $y = 3(x - 1)$ (f) $x + y = 0$

*(g) $2x - y = 6$ (h) $2x + 5y = 6$

*(i) $x = 3y - 4$ *(j) $x = \dfrac{y}{2}$

*(k) $y - 2x + 3 = 0$ (l) $4x = 2y + 8$

*(m) $2x = 3y + 18$ (n) $3x + y = 5$

*(o) $3(x + y) - 1 = x$ *(p) $y = 4x + 5y - 8$

2. Sketch the graph of the line having the given slope and passing through the given point.

*(a) $m = 2$, $(2, 5)$
(b) $m = 1$, $(1, 4)$
*(c) $m = -3$, $(1, 0)$
(d) $m = \frac{2}{3}$, $(3, 1)$
*(e) $m = -\frac{1}{3}$, $(-3, 0)$
(f) $m = -2$, $(0, 6)$
*(g) $m = -\frac{5}{2}$, $(-1, 6)$
(h) $m = -\frac{7}{3}$, $(1, -2)$

3. Sketch the graph of the line having the given slope and intercept.

*(a) $m = 3$, $b = 2$
(b) $m = 2$, $b = 1$
*(c) $m = 5$, $b = -1$
(d) $m = \frac{2}{3}$, $b = 0$
*(e) $m = -3$, $b = 0$
(f) $m = \frac{3}{5}$, $b = -4$
*(g) $m = 3$, x intercept $= -2$
(h) $m = \frac{1}{2}$, x intercept $= 4$
*(i) $m = -\frac{4}{3}$, $b = 5$
(j) $m = \frac{7}{2}$, x intercept $= 2$

Chapter 23.   REVIEW EXERCISES

*1. Determine the slope of each line.
(a) $y = 3x - 4$
(b) $y = -x + 7$
(c) $5x - 2y = 20$
(d) $x = 3y + 1$

*2. Compute the slope of the line through each pair of points.
(a) $(1, 7)$ and $(3, 19)$
(b) $(-1, 4)$ and $(3, -8)$

*3. Determine the slope and y intercept of each line.
(a) $y = -6x + 3$
(b) $y + 9 = 2x$
(c) $2x - 9y = 16$
(d) $3x = 7y - 21$

Sets
and Functions

24.1 INTRODUCTION

Sets were introduced about a hundred years ago by the German mathematician Georg Cantor. The original theory has been expanded over the years and now serves as a basic tool in many areas of advanced mathematics.

set   To begin, a *set* is a collection of things. In mathematics, the things are often numbers. However, a set can contain anything. Here are some examples of sets.

The set of all students in the classroom
The set of all people who have walked on the moon
The set of integers between one and ten inclusive
The set of manufacturers of automobiles
The set of geometric figures of eight sides or less
The set of all books written by Edgar Allan Poe

element   The things that make up the set are called *elements* or *members* of the set. The elements of the set of integers between one and ten inclusive are 1, 2, 3, 4, 5, 6, 7, 8, 9, 10.

This last set is much easier to discuss and work with if written {1, 2, 3, 4, 5, 6, 7, 8, 9, 10}. Braces { } are used to indicate a set. The elements are placed inside the braces and separated by commas. It does not matter in which order the elements are written. That is, the sets {1, 2, 3, 4, 5} and {3, 2, 1, 5, 4} are the same.

{ }

The symbol ∈ is used to mean "is an element of" or "is in." For instance, 5 ∈ {2, 3, 5, 9, 17} is read as 5 *is in* the set containing 2, 3, 5, 9, 17 or as 5 *is an element of* that set.

∈

Indicate whether the following are true or false.

$$3 \in \{1, 2, 3, 4, 5\}$$
$$3 \in \{1, 3\}$$
$$7 \in \{1, 2, 3, 6\}$$
$$x \in \{a, b, c, x, y, z\}$$
$$\$ \in \{*, /, +, -, \&\}$$

The answers are: true, true, false, true, false.

The statement 7 *is not an element of* {1, 2, 3, 6} can be written 7 ∉ {1, 2, 3, 6}. The symbol ∉ means *is not an element of* or *is not in.*

∉

Indicate whether the following are true or false.

$$5 \in \{1, 2, 3, 4, 6, 7, 8\}$$
$$5 \notin \{1, 2, 3, 4, 6, 7, 8\}$$
$$3 \notin \{1, 5, 9, 17\}$$
$$a \notin \{a, b, c, d, q\}$$
$$x \notin \{u, v, w, x, y, z\}$$
$$e \in \{a, x, y, e, z\}$$

The answers are: false, true, true, false, false, true.

We sometimes use three dots (. . .) to indicate a "continuation in the same manner." This notation is used to avoid writing all the members of a set explicitly. Thus the set of whole numbers

. . .

$$\{1, 2, 3, 4, 5, 6, 7, 8, 9, 10, 11, 12, 13\}$$

can be written

$$\{1, 2, 3, 4, \ldots, 13\}$$

The set {14, 16, 18, 20, 22, 24, 26, 28, 30, 32, 34, 36} of even whole numbers between 14 and 36 can be written {14, 16, 18, . . . , 36}.

The set of all whole numbers greater than 13 can be written {14, 15, 16, 17, . . .}.

EXERCISES 24.1

Answers to starred exercises are given in the back of the book.

1. Write each using set notation.
   *(a) The set of all even numbers from 2 through 24.
   (b) The set of all vowels in the alphabet.
   *(c) The set of integers from 7 through 15.
   (d) The set of all positive integers.
   *(e) The set of all months beginning with the letter M.

2. Write each using set notation.
   *(a) 5 is in the set of all odd numbers between 1 and 15 inclusive.
   *(b) 7 is not in the set of even numbers less than 20.
   *(c) x is not in the set of vowels of the alphabet.
   (d) June is in the set of months of the year.
   (e) Sunday is not in the set of months of the year.

3. Indicate whether the following statements are true or false.
   *(a) $x \in \{a, b, c, x, y, z\}$ *(b) $x \notin \{a, b, c, x, y, z\}$
   *(c) $5 \in \{2, 4, 6, 8, 10\}$ *(d) $5 \in \{1, 3, 15, 17, 19\}$
   *(e) $d \in \{a, e, q, r\}$ *(f) $* \notin \{\square, \triangle, *, \$\}$
   *(g) $*** \in \{+, -, *, /, **\}$ (h) $-19 \in \{\text{integers}\}$
   (i) $t \notin \{R, S, T, U, V\}$ (j) $341 \notin \{\text{odd numbers}\}$

24.2 SET-BUILDER NOTATION

The set $\{2, 4, 6, 8, 10, \ldots\}$ is the set of all positive, even integers. It can also be written as $\{x \mid x$ positive, even integer$\}$, which is read "the set of all numbers x such that x is a positive, even integer." The vertical line $\mid$ is read as "such that." This is particularly useful notation for some statements about sets. For example, how could we represent the set of all numbers that are greater than zero (including integers, decimals, fractions, etc.)? Using the new notation, we simply write this as

| such that

$$\{x \mid x > 0\}$$

the set of all numbers x such that x is greater than zero.† Set notation that uses the "such that" vertical line is called *set-builder notation.*

set-builder notation

---

infinite set
finite set

†The set $\{x \mid x > 0\}$ is an example of an *infinite set*, since it has an unlimited number of elements. Most sets that we discuss will be *finite sets*; they will have a limited (1 or 5 or 1000, etc.) number of elements.

Example 1. *Denote the set of all numbers less than* 17 *except* 0. *Call the set Q.*

$$Q = \{m \mid m < 17, m \neq 0\}$$

Example 2. *Denote the set of all odd numbers greater than or equal to* 7. *Call the set R.*

$$R = \{x \mid x \text{ odd}, x \geq 7\}$$

Example 3. *Denote the set of all numbers that are between* 3 *and* 5—*that is, all that are both greater than* 3 *and less than* 5. *Call the set P.*

$$P = \{n \mid 3 < n < 5\}$$

The notation $3 < n$ indicates *n greater than* 3. Also, $n < 5$ indicates *n less than* 5. They are combined as $3 < n < 5$ to mean all numbers n that are both greater than 3 *and* less than 5.

Example 4. *Explain what is in the set* $\{t \mid 1 < t \leq 7\}$.

The set consists of all numbers that are between 1 and 7, excluding 1 and including 7. In other words, it is the set of all numbers that are greater than 1 and less than or equal to 7.

EXERCISES 24.2

1. Use the notation of this section to denote the following sets. The set of
   *(a) All numbers greater than 13
   *(b) All numbers less than $\frac{1}{2}$
   (c) All numbers greater than or equal to -27
   *(d) All numbers less than or equal to 2
   *(e) All numbers not less than 0
   (f) All odd numbers less than 10
   *(g) All numbers between (and including) -5 and 16
   (h) All numbers between (but not including) 5 and 16
   *(i) All even numbers between 3 and 95
   (j) All even numbers between (and including) 2 and 500
   *(k) All nonzero numbers between -25 and 25 inclusive
   (l) All integers between (but not including) -100 and $+200$

2. Explain what is in each set.
   *(a) $\{x \mid x > 0\}$ *(b) $\{t \mid t \neq 0\}$
   (c) $\{m \mid m < -9\}$ *(d) $\{x \mid x \text{ even}, x \geq 6\}$
   (e) $\{x \mid x \text{ odd}, x \neq 1\}$ *(f) $\{y \mid y \text{ integer}, y \geq 0\}$
   (g) $\{y \mid y \text{ integer}, y > 0\}$ (h) $\{y \mid y \text{ integer}, y < 0\}$
   *(i) $\{x \mid 3 < x < 9\}$ (j) $\{x \mid 3 \leq x \leq 9\}$
   *(k) $\{x \mid 3 \leq x < 9\}$ (l) $\{x \mid 3 < x \leq 9\}$

24.3 UNION AND INTERSECTION

union   The *union* of two sets is computed in several examples below. See if you can determine just what we mean by the union of two sets. The

∪   operation of union is denoted by the U-shaped symbol ∪.

$$\{1, 2, 3\} \cup \{4, 5\} = \{1, 2, 3, 4, 5\}$$
$$\{a, b\} \cup \{x, y\} = \{a, b, x, y\}$$
$$\{5, 6, 17\} \cup \{5, 9\} = \{5, 6, 9, 17\}$$
$$\{1, 2\} \cup \{1, 2\} = \{1, 2\}$$
$$\{2, 4, 6, 8\} \cup \{1, 2, 3\} = \{1, 2, 3, 4, 6, 8\}$$

The examples show that the union of two sets is formed by taking all the elements involved (from both sets) and placing them in one new set. The new set is called their union. More technically, *the union of two sets is a set that contains all elements that are in either of the two original sets.*

intersection   The *intersection* of two sets is computed in several examples below. Read them and determine what we mean by the intersection of two sets.

∩   The symbol ∩ is used to indicate intersection.

$$\{1, 2, 3, 4, 5\} \cap \{1, 2, 3\} = \{1, 2, 3\}$$
$$\{b, g, x, z\} \cap \{b, c, q\} = \{b\}$$
$$\{2, 4, 6, 8\} \cap \{1, 2, 3, 4\} = \{2, 4\}$$
$$\{1, 2, 3\} \cap \{4, 5, 6\} = \{\ \ \}$$

It should be clear from the examples that *the intersection of two sets is a set that contains the elements common to both sets.*

If you are wondering about the example

$$\{1, 2, 3\} \cap \{4, 5, 6\} = \{\ \ \}$$

you should note that the two sets have no elements in common, so their intersection is a set with no elements. The set with no elements can be

empty set ∅   denoted { } or ∅. It is called the *empty set* or *null set*. Thus $\{a, b, c\} \cap \{5, 9\} = \emptyset$. When two sets have no elements in common, they are called

disjoint   *disjoint* sets.

EXERCISES 24.3

1. Determine each union.
   *(a) $\{5, 6, 19, 35\} \cup \{8, 12, 19\}$
    (b) $\{1, 5, 6, 9, 2\} \cup \{3, 4, 8, 7, 10\}$
   *(c) $\{4, 8, 7, 3\} \cup \{14, 3, 7, 12\}$

(d) $\{a, b, c, d, e\} \cup \{e, c, b, a, d\}$

*(e) $\{1, 2, 3, 4, 5, 6\} \cup \{5, 2, 4\}$

2. Determine each intersection.

*(a) $\{6, 4, 19, 35\} \cap \{19, 12, 8\}$

(b) $\{a, b, c, d, e\} \cap \{b, e, a, c, d\}$

*(c) $\{3, 4, 8, 7, 10\} \cap \{1, 5, 6, 9, 2\}$

(d) $\{7, 3, 12, 14\} \cap \{4, 8, 7, 3\}$

*(e) $\{4, 3, 1\} \cap \{3, 5, 1, 2, 4\}$

3. Determine each union or intersection.

*(a) $\{x, y, z\} \cup \}x, y, z\}$

(b) $\{x, y, z\} \cap \{x, y, z\}$

*(c) $\{a, b, c, d\} \cup \{a, b, c, d, e, f\}$

(d) $\{a, b, c\} \cap \{d, e, f\}$

24.4 SUBSETS

$\{1, 2\}$ is a *subset* of $\{1, 2, 19, 50\}$. This means that every element of   **subset**
$\{1, 2\}$ is also in the set $\{1, 2, 19, 50\}$. Similarly,

$\{a, b, c\}$ is a subset of $\{a, b, c, d, e, f\}$.

$\{1, 4\}$ is a subset of $\{1, 2, 3, 4, \ldots, 100\}$. Any set is a

$\{50\}$ is a subset of $\{2, 4, 6, \ldots, 200\}$.   subset of itself.

$\{3, 4, 5\}$ is a subset of $\{3, 4, 5\}$.

$\{\ \ \}$ is a subset of $\{1, 2, 3\}$.   The empty set is a
subset of every set.

The words "is a subset of" are often replaced by the symbol $\subseteq$, $\subseteq$
which means the same thing. For instance,

$$\{3, 5\} \subseteq \{1, 3, 5, 7, 9, 11\}$$

For convenience, we often name sets by using capital letters. If we
name the set $\{1, 2, 3\}$ A and call B the set $\{1, 2, 3, 4, 5, 6\}$, then

$$A = \{1, 2, 3\}$$

$$B = \{1, 2, 3, 4, 5, 6\}$$

and

$$A \subseteq B, \qquad A \text{ is a subset of } B.$$

Example 5. *Let* $A = \{2, 5\}$, $B = \{1, 2, 3, 4, 5\}$, $C = \{1, 3, 6\}$.

Then

$A \subseteq B$ (*A* is a subset of *B*, since both its elements, 2 and 5, are in *B*.)

and

$\not\subseteq$ $C \not\subseteq B$ (*C* is *not* a subset of *B*, since 6 is in *C* but not in *B*.)

Also,

$$A \cap B = \{2, 5\}$$
$$A \cap C = \varnothing$$
$$A \cup C = \{1, 2, 3, 5, 6\}$$
$$A \cup B = \{1, 2, 3, 4, 5\}$$
$$B \cap C = \{1, 3\}$$
$$C \cup B = \{1, 2, 3, 4, 5, 6\}$$

EXERCISES 24.4

1. If $A = \{7, 9, 3, 8, 2\}$, $B = \{3, 8, 10\}$, $C = \{2, 7\}$, which of the following are true statements?

   *(a) $A \subseteq B$ *(b) $A \subseteq C$ (c) $B \subseteq C$

   *(d) $B \not\subseteq A$ (e) $C \subseteq A$ *(f) $C \cap A \subseteq A$

   (g) $C \not\subseteq B$ (h) $A \cap B \subseteq C$ *(i) $B \cap C \subseteq A$

   *(j) $A \cup B \not\subseteq C$

2. List all the subsets of each set.

   *(a) $\{a, b\}$ *(b) $\{x, y, z\}$ (c) $\{p, q, r, s\}$

24.5 COMPLEMENTS

universe   The elements that we select for sets are taken from another set—a *universe*. The universe can change from example to example and from discussion to discussion, and it is determined by the nature of the problem. In each example below a universe is selected, and in each case

$\mathfrak{U}$ it is denoted by $\mathfrak{U}$.

Example 6. *Let* $\mathfrak{U} = \{1, 2, 3, 4, 5, 6, 7, 8, 9, 10, 11, 12\}$.

We will use sets containing numbers selected from 1, 2, 3, 4, 5, 6, 7, 8,

9, 10, 11, 12; for instance, $X = \{1, 2, 3\}$, $Y = \{2, 8, 11, 12\}$, and $Z = \{1, 2, 3, 4, 5, 6, 7, 8, 9, 10, 11, 12\}$. Sets discussed must be subsets of the universe.

$$X \subseteq \mathfrak{U}, \qquad Y \subseteq \mathfrak{U}, \qquad Z \subseteq \mathfrak{U}$$

It was necessary to introduce the concept of "universe" in order to discuss complements. The *complement* of a set X will be denoted by $\overline{X}$ **complement** (read as X-bar). *The complement of a set X is the set $\overline{X}$ containing all elements of the universe that are not in X.*

Example 7. *Let* $\mathfrak{U} = \{a, b, c, d, e, f, g\}$.

   If $X = \{a, b, c, d\}$, then $\overline{X} = \{e, f, g\}$.
   If $Y = \{b, f\}$, then $\overline{Y} = \{a, c, d, e, g\}$.

The sets X and $\overline{X}$ have no elements in common; that is, $X \cap \overline{X} = \varnothing$. Also, X and $\overline{X}$ together contain all elements in the universe; that is, $X \cup \overline{X} = \mathfrak{U}$. The same is true for Y and $\overline{Y}$ as well as for any other set and its complement.

EXERCISES 24.5

*1. If $\mathfrak{U} = \{1, 3, 5, 7, 9, 11, 13, 15, 17\}$, find the complement of each set below.

   (a) $A = \{1, 3, 5\}$ (b) $B = \{7, 15\}$
   (c) $C = \{\ \ \}$ (d) $D = \{1, 3, 5, 7, 9, 11, 13, 15, 17\}$
   (e) $E = \{13\}$

2. If $\mathfrak{U} = \{m, a, t, h, i, s, o, k\}$, find the complement of each set below.

   (a) $F = \{m, a, t, h\}$ (b) $G = \{i, s\}$
   (c) $H = \{o, k\}$ (d) $I = \{\ \ \}$
   (e) $J = \{m, a, t, h, i, s, o, k\}$ (f) $K = \{m, i, o\}$
   (g) $L = \{h, o, t\}$

24.6 VENN DIAGRAMS

Venn diagrams are useful for visualizing relationships between sets.   **Venn diagram** The diagrams are named for John Venn, who used them in 1881 to illustrate logic. They were first used by Leonhard Euler and hence are also called *Euler's circles*. In Venn diagrams, a rectangle represents the universe.

Circles (and their interiors) represent sets.

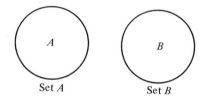

And since the sets are made up of elements from the universe, the circles are placed within the rectangular universe.

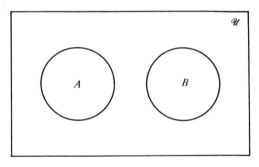

The circles in the rectangle shown are separated, indicating that the sets have no elements in common. If two sets have elements in common, then the circles that represent them overlap, as shown below.

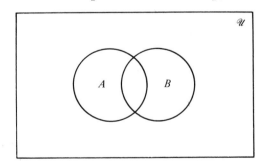

Example 8. *Represent $A \cap B$ in the case that A and B overlap. Shade the intersection.*

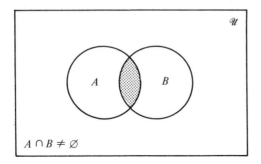

The shaded area represents the intersection of A and B and gives a visual picture of $A \cap B$. Note that $A \cap B \neq \varnothing$ means that the intersection is *not* the empty set; that is, the sets have at least one element in common.

Example 9. *Represent $A \cap B$ for the case $A \cap B = \varnothing$ (that is, the sets have no elements in common).*

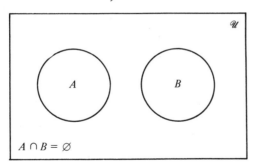

There is nothing to shade, because there are no elements in the intersection.

Example 10. *Represent $A \cup B$ in the case where A and B have elements in common (that is, when $A \cap B \neq \varnothing$).*

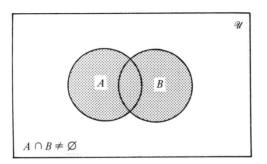

The shaded area is $A \cup B$.

Example 11. *Represent* $A \cup B$ *in the case* $A \cap B = \varnothing$.

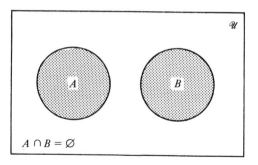

The shaded areas represent $A \cup B$.

Example 12. *Represent* $\bar{A}$ *for some set A in the universe* $\mathcal{U}$.

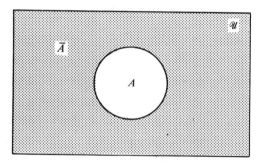

The shaded area is $\bar{A}$, since $\bar{A}$ is all of $\mathcal{U}$ that is not in A.

Example 13. *Represent* $A \cap B$ *for sets A and B such that B is a subset of A.*

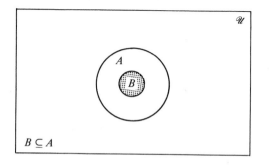

The shaded area is $A \cap B$.

Example 14. *Represent A ∪ B for A and B such that B ⊆ A.*

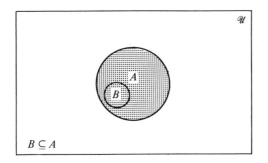

The shaded area is *A ∪ B.*

Example 15. *Represent A ∪ B for A = B.*

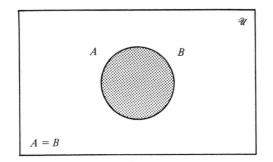

The shaded area is *A ∪ B.*

Example 16. *Represent A ∩ B ∩ C for three sets A, B, and C that overlap each other.*

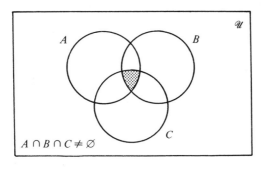

The shaded area is *A ∩ B ∩ C.*

EXERCISES 24.6

*1. Consider the Venn diagram shown below. The numbers 1, 2, 3, and 4 are elements. List the elements that are in the region requested.

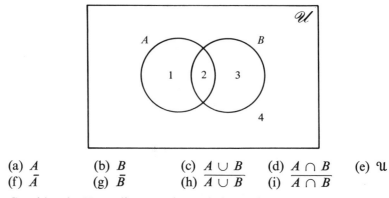

(a) A (b) B (c) $A \cup B$ (d) $A \cap B$ (e) $\mathcal{U}$
(f) $\bar{A}$ (g) $\bar{B}$ (h) $\overline{A \cup B}$ (i) $\overline{A \cap B}$

2. Consider the Venn diagram shown below. The numbers 1, 2, 3, 4, 5, and 6 are elements. List the elements that are in the region requested.

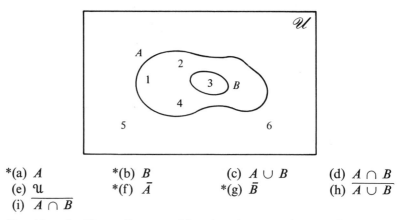

*(a) A *(b) B (c) $A \cup B$ (d) $A \cap B$
(e) $\mathcal{U}$ *(f) $\bar{A}$ *(g) $\bar{B}$ (h) $\overline{A \cup B}$
(i) $\overline{A \cap B}$

3. Consider the Venn diagram. List the elements that are in the region requested.

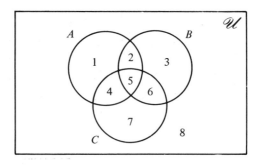

*(a) A	(b) B	(c) C
*(d) $\mathfrak{U}$	*(e) $A \cap B$	(f) $B \cap C$
(g) $A \cap C$	*(h) $A \cap B \cap C$	*(i) $A \cup B$
(j) $B \cup C$	(k) $A \cup C$	*(l) $A \cup B \cup C$

4. Draw Venn diagrams for the indicated sets.

  *(a) $X \cap Y$ for two sets X and Y that do not overlap
  (b) $X \cup Y$ for two sets X and Y that do not overlap
  *(c) $A \cup B \cup C$ for three overlapping sets
  (d) $A \cap B \cap C$ for three overlapping sets
  *(e) $A \cap B$ for two identical sets A and B
  (f) $A \cup B$ for two identical sets A and B
  *(g) $A \cap B$ for $A \subseteq B$
  (h) $A \cap B$ for $B \subseteq A$
  (i) $A \cup B$ for $A \subseteq B$
  *(j) $A \cup B$ for $B \subseteq A$

5. Assuming that A and B are two overlapping sets, represent

  (a) $\overline{A \cap B}$ (b) $\bar{A} \cap \bar{B}$ (c) $\overline{A \cup B}$ (d) $\bar{A} \cup \bar{B}$

  Discuss any apparent relationships.

6. Use the Venn diagram of Exercise 3 to verify that the following statements are true.

  (a) $A \cap B = B \cap A$
  (b) $A \cup B = B \cup A$
  (c) $A \cup (B \cap C) = (A \cup B) \cap (A \cup C)$
  (d) $A \cap (B \cup C) = (A \cap B) \cup (A \cap C)$

24.7 RELATIONS AND FUNCTIONS
(Optional)

  The concept of set can be used to define special sets called relations and functions. Since relations and functions are important in mathematics, we now proceed to have a brief look at them.

  Consider the equation $y = 2x + 1$. For every value of x that you supply, you obtain a corresponding y value. Informally, the equation $y = 2x + 1$ is an example of a *relation*, a correspondence involving two variables. Here are some other relations.

$$x - y = 6$$
$$y = x^2 + 2x + 5$$
$$x^2 + y^2 = 9$$

Each relation includes all x's and y's that satisfy the equation. This means that each relation includes all points that could be graphed by using the equation to determine points. Set notation can be used to formally express the four relations that we have mentioned above.

$$\{(x, y) \mid y = 2x + 1\}$$

$$\{(x, y) \mid x - y = 6\}$$

$$\{(x, y) \mid y = x^2 + 2x + 5\}$$

$$\{(x, y) \mid x^2 + y^2 = 9\}$$

Recall the notation. $\{(x, y) \mid y = 2x + 1\}$ is read "the set of all points (x, y) such that $y = 2x + 1$."

relation   We can now formally define a *relation* as a set of ordered pairs, a set of points. Here are a few simple sets of ordered pairs; each is a relation, just like the more complicated sets above.

$$\{(1, 3), (2, 6), (5, 2)\}$$

$$\{(0, 5), (5, -1), (5, 2)\}$$

$$\{(2, 4)\}$$

All relations can be expressed as sets of ordered pairs, or "points."
function Some relations are also called *functions*. To be a function, a relation cannot have two points in which x (the first coordinate) is the same and y (the second coordinate) is different. For instance, the relation

$$\{(1, 3), (2, 5), (2, 4), (3, -7)\}$$

is not a function because it contains the pairs (2, 5) and (2, 4) in which there are two different y values (5 and 4) for the same x value (2).

The relation

$$\{(1, 3), (2, -2), (4, 3), (7, 5)\}$$

is a function because there are not two y values for a single x value. Note that there are two x values (1 and 4) for the same y value (3), but it does not matter; that is not the test.

You can also examine the graph of a relation to determine whether or not it is a function. An easy test is to pass a vertical line through various parts of the graph. If the line never catches two (or more) points, then the graph is of a function. However, if two (or more) points can be caught, then the relation is not a function. Here are two examples.

Not a function:

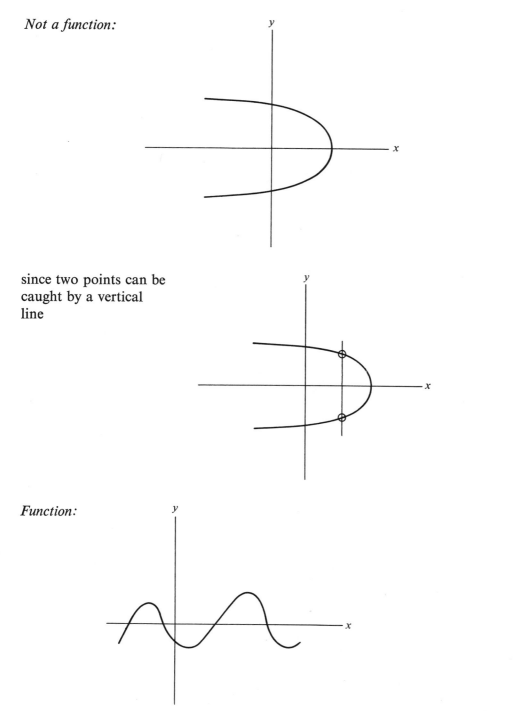

since two points can be
caught by a vertical
line

Function:

EXERCISES 24.7

*1. Which sketches are graphs of functions and which are not?

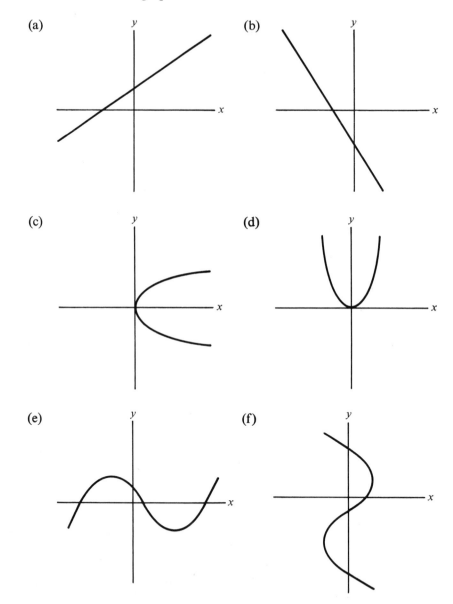

(a)

(b)

(c)

(d)

(e)

(f)

*2. Determine which of the following are functions and which are not.
(a) {(1, 2), (3, 4)}   (b) {(0, 1), (1, 0), (3, 4)}
(c) {(1, 0), (1, 1)}   (d) {(1, 4), (2, 4), (3, 4)}

(e) $y = x + 3$

(f) $y = x^2$

(g) $y = \pm\sqrt{x}$

(h) $y = \pm x$

3. Consider the following statement: "All functions are relations, but not all relations are functions." Use examples in order to explain it.

24.8 FUNCTION NOTATION (Optional)

The famous Swiss mathematician Leonhard Euler (1707–1783) suggested that functions could be written using a special notation—function notation—for some applications. Considering a function such as $y = 2x + 1$, we can say that it represents

$$y \text{ as a function of } x$$

and we can write

$$y = f(x)$$

and

$$f(x) = 2x + 1$$

The difference is that we now use $f(x)$ instead of y. It is only a change in notation. The meaning is the same.

Values of the function $f(x) = 2x + 1$ can be obtained for any x supplied. Since

$$f(\underline{x}) = 2 \cdot \underline{x} + 1$$

we see that

$$f(\underline{3}) = 2 \cdot \underline{3} + 1, \text{ which is } 7$$

Notice that $f(3)$ means the same as *the y value in y = 2x + 1 when x = 3*, but it is an abbreviated form and therefore very simple and useful. Similarly,

$$f(\underline{6}) = 2 \cdot \underline{6} + 1 = 13 \checkmark$$
$$f(\underline{0}) = 2 \cdot \underline{0} + 1 = 1 \checkmark$$
$$f(\underline{m}) = 2 \cdot \underline{m} + 1 = 2m + 1 \checkmark$$
$$f(\underline{x + 5}) = 2 \cdot (\underline{x + 5}) + 1 \quad \text{or} \quad 2x + 11 \checkmark$$
$$f\left(\frac{1}{2}\right) = 2 \cdot \frac{1}{2} + 1 = 2 \checkmark$$

The symbol $f(x)$ is read "f at x" or "f of x." For example, $f(6)$ is f at 6, that is, the value of f when x is 6, or the value of y when x is 6.

Example 17. *If* $f(x) = 3x^2 + 5x + 2$, *determine* $f(0), f(1), f(-1), f(x+1)$
and $f(r-2)$.

$$f(\underline{x}) = 3(\underline{x})^2 + 5(\underline{x}) + 2$$

$$f(\underline{0}) = 3(\underline{0})^2 + 5(\underline{0}) + 2 = 2 \checkmark$$

$$f(\underline{1}) = 3(\underline{1})^2 + 5(\underline{1}) + 2 = 10 \checkmark$$

$$f(\underline{-1}) = 3(\underline{-1})^2 + 5(\underline{-1}) + 2 = 0 \checkmark$$

$$f(\underline{x+1}) = 3(\underline{x+1})^2 + 5(\underline{x+1}) + 2$$
$$= 3(x^2 + 2x + 1) + 5x + 5 + 2$$
$$= 3x^2 + 11x + 10 \checkmark$$

$$f(\underline{r-2}) = 3(\underline{r-2})^2 + 5(\underline{r-2}) + 2$$
$$= 3(r^2 - 4r + 4) + 5r - 10 + 2$$
$$= 3r^2 - 7r + 4 \checkmark$$

EXERCISES 24.8

*1. If $f(x) = 3x - 5$, find the values listed below.
   (a) $f(1)$ (b) $f(2)$ (c) $f(0)$ (d) $f(-5)$
   (e) $f(a)$ (f) $f(\frac{1}{3})$ (g) $f(-\frac{1}{3})$ (h) $f(3m)$
   (i) $f(x+1)$ (j) $f(x-5)$

*2. If $f(x) = x^2 + 8x - 1$, find the values listed below.
   (a) $f(2)$ (b) $f(3)$ (c) $f(10)$ (d) $f(0)$
   (e) $f(-1)$ (f) $f(-3)$ (g) $f(m)$ (h) $f(\frac{1}{2})$
   (i) $f(-\frac{1}{4})$ (j) $f(x+2)$

Chapter 24.   REVIEW EXERCISES

*1. Explain what is in the set $\{x \mid 4 \le x < 9\}$.

2. Let $\mathfrak{U} = \{1, 2, 3, 4, 5, 6, 7\}$, $A = \{1, 2, 3\}$, $B = \{4, 5, 6\}$, $C = \{1, 3, 5, 7\}$.
  *(a) $A \cup B =$ *(b) $A \cap B =$
  *(c) $A \cup C =$ *(d) $B \cap C =$
  *(e) $\bar{B} =$ (f) $\bar{C} =$
  *(g) $\bar{\mathfrak{U}} =$ *(h) Is $A \subseteq \mathfrak{U}$? Explain.
   (i) Is $A \subseteq C$? Explain.

3. Draw a Venn diagram for the indicated sets.
    (a) $A \cup B$ for two sets A and B that overlap
    (b) $A \cap B$ for two sets A and B such that $A \subseteq B$

*4. Is the given relation a function?
    (a) $\{(5, 1), (5, 2), (5, 3)\}$ (b) $y = 7x + 4$

*5. If $f(x) = x^2 + 3x + 5$, determine
    (a) $f(4)$ (b) $f(0)$

REVIEW PROBLEMS FOR PART 3

1. Perform the indicated operation. Reduce if possible.

   *(a) $\dfrac{12xy}{7y} \cdot \dfrac{5}{3x}$

   (b) $\dfrac{5x}{6x^2} \div \dfrac{x^2 - 12x + 36}{2x - 12}$

   *(c) $\dfrac{5}{6xy} + \dfrac{3y}{8x}$

   (d) $\dfrac{5}{x - 3} - \dfrac{3}{x + 2}$

   (e) $\dfrac{x + \dfrac{1}{y}}{y - \dfrac{1}{x}}$

2. Solve each equation.

   (a) $\dfrac{4}{x} + 9 = \dfrac{x + 1}{x}$

   (b) $x^2 + 7x + 1 = 0$

3. Simplify each expression. Write with positive exponents only.

   *(a) $(2x^5 y^3)^6$ (b) $(2m^8)(-3m^6)^3$ *(c) $-5x^{-19} y^6$

   (d) $\sqrt{225x^4 y^{100} z^{14}}$

4. Evaluate each expression.

   *(a) $2^{-1} + 2^0$ (b) $(7 \cdot 6)^0$ *(c) $81^{3/4}$

   (d) $32^{3/5}$

5. Solve each formula for the letter indicated.

   *(a) $y = mx + b$ for m

   (b) $5x + 9y + 7 = 0$ for y

   *(c) $x = \dfrac{ab}{c + d}$ for a

   (d) $x = \dfrac{ab}{c + d}$ for d

   *(e) $a = b + c(d - e)$ for d

6. Simplify.

   *(a) $\sqrt{52}$ (b) $\sqrt{112}$ (c) $\sqrt{\dfrac{13}{25}}$

7. Rationalize the denominator of $\dfrac{1}{6 + \sqrt{11}}$.

*8. Solve $\sqrt{1 - 2x} + 5 = 13$.

9. Solve each equation.

   *(a) $2x^2 - 17x + 21 = 0$ *(b) $x^2 + 4x + 5 = 0$

   (c) $x^2 = 7x - 6$ (d) $2x^2 - 9x - 3 = 0$

10. Solve $x^2 - 6x - 1 = 0$ by completing the square.

11. Determine the slope of each line.

    *(a) $y = 7x - 2$ (b) $3x + 5y = 9$

    (c) The line through $(0, 3)$ and $(1, 5)$

12. Let $\mathcal{U} = \{3, 5, 7, 9, 11, 13, 15\}$, $A = \{3, 7, 11\}$, $B = \{7, 11, 13, 15\}$.
    *(a) $A \cap \mathcal{U} =$ (b) $A \cap B =$ *(c) $\bar{B} =$ (d) $\overline{A \cap B} =$

13. Is $y = 3x + 1$ a function? Explain.

14. Let $f(x) = x^2 - 6x + 2$. Compute the functional values requested.
    *(a) $f(6)$ *(b) $f(0)$ (c) $f(-6)$ (d) $f(-1)$

APPENDICES

Pretests ─────────

Chapter 1. REVIEW OF BASIC OPERATIONS

1. Add: 17
       25
      38

2. Add: 195.63
      218.99

3. Add: 2306
      1524
      1153
      5649

4. Subtract: 9867
      6958

5. Multiply: 83
      21

6. Multiply: 90.8
      62.4

7. Divide: $24 \overline{)\,5928}$

8. Divide: $46.2 \overline{)\,128.629}$ (continue to two decimal places)

9. Round to the nearest whole number.
(a) 23.7
(b) 14.4

10. Divide: $0 \overline{)\,17}$

11. Divide: $29 \overline{)\,0}$

12. Compute the value of 3^4.

Chapter 2.   ORDER OF OPERATIONS

Simplify each expression below to obtain a one-number result.

1. $5 + 8 \cdot 6$

2. $4 \cdot 3 + 7 \cdot 9$

3. $7 + 5(2 + 8) + 9$

4. $5 + 4^3 + \dfrac{2 \cdot 5}{2}$

5. $6 + \dfrac{40}{2} + \dfrac{12}{4} + 2$

6. $2^2 + 4^2 \div 4$

7. $\dfrac{4 \cdot 8}{2 \cdot 4}$

8. $(4 + 3)^2$

Chapter 3.   EVALUATION OF GEOMETRIC FORMULAS

Evaluate each formula by using the values given.

1. $P = 2l + 2w$; $l = 15$, $w = 7.2$
2. $A = \pi r^2$; $\pi = 3.14$ (approximately), $r = 10$
3. $m = \dfrac{a + b + c}{3}$; $a = 10$, $b = 6$, $c = 5$
4. $A = \dfrac{1}{2}(b_1 + b_2)h$; $b_1 = 12$, $b_2 = 14$, $h = 5$
5. $D = b^2 - 4ac$; $b = 9$, $a = 2$, $c = 3$

Chapter 4.   SQUARE ROOTS

1. Fill in the blanks.
   (a) $\sqrt{16} = \underline{\quad}$
   (b) $\sqrt{81} = \underline{\quad}$
   (c) $\sqrt{169} = \underline{\quad}$

2. Determine the length of the hypotenuse of a right triangle if the legs are 3 and 4 inches.

3. If the hypotenuse of a right triangle is 13 and one of the legs is 5, what is the length of the other leg?

Chapter 5.   PROPERTIES OF ARITHMETIC

There are properties that whole numbers satisfy under the operations of addition and multiplication. Among them are the closure, associative, commutative, and distributive properties. In each problem below, indicate which of these properties has been used.

1. $7 + 4 = 4 + 7$
2. $3(5 + 9) = 3 \cdot 5 + 3 \cdot 9$
3. $27 + 12 = 39$
4. $8 \cdot 6 + 8 \cdot 9 = 8(6 + 9)$
5. $6 \cdot (3 \cdot 7) = (6 \cdot 3) \cdot 7$
6. $5(2 + 3) = (2 + 3)5$
7. $120 = 12 \cdot 10$
8. $6 \cdot (2 \cdot 4) = 6 \cdot (4 \cdot 2)$
9. $(9 + 2) \cdot 6 = 11 \cdot 6$

Chapter 6. ARITHMETIC OF FRACTIONS

1. Add: $\dfrac{1}{2} + \dfrac{1}{3}$
2. Add: $\dfrac{7}{8} + \dfrac{3}{4} + \dfrac{2}{5}$
3. Add: $\dfrac{4}{21} + \dfrac{3}{28} + \dfrac{6}{35}$
4. Multiply: $\dfrac{7}{8} \cdot \dfrac{5}{6}$
5. Multiply: $\dfrac{24}{40} \cdot \dfrac{20}{12}$
6. Divide: $\dfrac{6}{17} \div \dfrac{50}{34}$
7. Reduce: $\dfrac{78}{630}$
8. Reduce: $\dfrac{84}{164}$
9. Change $\dfrac{23}{4}$ to a mixed number.
10. Change $7\dfrac{3}{8}$ to an improper fraction.

Chapter 7. FRACTIONS, DECIMALS, AND PERCENT

1. Change .63 to a fraction.
2. Change $\frac{5}{8}$ to a decimal.
3. Change 29% to a fraction.
4. Change 7% to a decimal.
5. Change .43 to a percent.
6. Change $\frac{7}{20}$ to a percent.
7. Change 290% to a fraction.
8. Change 8.3% to a fraction.
9. Change $7\frac{1}{2}$ to a percent.
10. Change 6.9 to a percent.

Chapter 8. UNITS OF MEASURE

Note: There are 5280 feet in a mile.

1. Change 4 hours to seconds.
2. Change 3520 yards to miles.

3. Add: 7 inches $+$ 9 inches $+$ 23 inches

4. Multiply: 5 inches $\times$ 6 inches $\times$ 2 inches

5. Change 60 miles per hour to feet per second.

Chapter 9.   METRIC SYSTEM

The following information should be helpful for the conversions requested. 1 meter $=$ 39.37 inches; 1 kilometer $=$ .6 mile; 1 inch $=$ 2.54 centimeters; 1 liter $=$ 1.06 quarts; 1 ounce $=$ 28.35 grams; 1 kilogram $=$ 2.2 pounds.

Two decimal places is sufficient for answers when decimal calculations are used.

1. Change 73 inches to meters.

2. Change 65 kilometers to miles.

3. Change 7 meters to millimeters.

4. Change 19 centimeters to inches.

5. Change 13 quarts to liters.

6. Change 623 milliliters to liters.

7. Change 58 centimeters to meters.

8. Change 23 kilograms to pounds.

9. Change 42 kilograms to milligrams.

10. Change 5.6 ounces to grams.

Chapter 10.   SIGNED NUMBERS

1. Add: $(-9) + (+6)$

2. Subtract: $(+6) - (-2)$

3. Combine: $-3 - 2 - 5 + 10$

4. Combine: $+7 - 2 + 4 - 9 + 1 - 3$

5. Multiply: $(-8)(-4)$

6. Multiply: $(-2)(+1)(-3)(+2)(-1)$

7. Divide: $(-8) \div (+1)$

8. Simplify: $(-2)^3 + 3$

9. Simplify: $(+2)(-1)^2 + (-3)(-4)(-1) + (-3)(+2)$

10. Simplify: $\dfrac{+6}{-2} - \dfrac{-8}{+4} + \dfrac{+2}{-1}$

11. Determine the value of $b^2 - 4ac$ when $a = -5$, $b = 3$, and $c = -1$.

12. Compute: $|+7 - 9| =$

13. True or false: $-7 < -3$

Chapter 11. ALGEBRAIC EXPRESSIONS

1. Combine like terms: $4x + 5x$

2. Combine like terms: $3x - 7x + x$

3. Combine like terms: $4x - 3 + 7x - y + 2$

4. Simplify: $2y + 5(3 - 2y)$

5. Simplify: $-4(5 + 2x - 2) + 2(6 - x) - 5(x + 4)$

6. Simplify: $6 - 2[5 + 3(2x - 1)] + 7x$

7. What is the coefficient of x in $-4x$?

8. Multiply: $(-3x^2)(7x^4)$

9. Multiply: $(2x + 5)(3x - 1)$

10. Divide: $x + 2 \overline{)\, x^3 + 7x^2 + 7x - 6}$

Chapter 12. SOLVING LINEAR EQUATIONS

1. What is the reciprocal of $\frac{2}{3}$?   **2.** What is the reciprocal of -2?

3. Solve: $3x = 12$ **4.** Solve: $\frac{2}{3}x = -7$

5. Solve: $x + 8 = 19$ **6.** Solve: $x - 7 = -5$

7. Solve: $3y + 2 = 1$ **8.** Solve: $-\frac{1}{2}x + 6 = 9$

9. Solve: $5m - 6 = 2m + 9$ **10.** Solve: $2(x + 3) - 4 = x + 9$

11. Solve: $5x - 2x + x - 4 = 7 - 2x + 3x$

12. Solve: $x - 4(2 - 3x) + 5x = 1 - 6x$

Chapter 13. WORD PROBLEMS

1. How many cents are there in n nickels?

2. How many inches are there in x feet?

3. Represent the number that is seven less than four times x.

4. What is the annual interest on m dollars at 7% interest per year?

5. A wire 31 feet long is to be cut into two pieces, one being 4 feet longer than the other. How long should each piece be?

6. June buys 13¢ stamps, 17¢ stamps, and 20¢ stamps. She buys the same number of 13¢ and 17¢ stamps, and she buys five more 20¢ stamps than 13¢ stamps. All together, she spends \$8.50. How many of each does she buy?

Chapter 14. GRAPHING STRAIGHT LINES

1. Graph the line $y = 3x$. 2. Graph the line $y = -2x + 4$.

3. Graph the lines $3x - y = 3$ and $y = 2x + 1$ and determine approximately where they meet.

Chapter 15. SYSTEMS OF LINEAR EQUATIONS

1. Solve: $\begin{cases} 4x - 6y = 0 \\ 7x + 5y = 31 \end{cases}$ by addition or elimination

2. Solve: $\begin{cases} x - 7y = 9 \\ y = x - 3 \end{cases}$ by substitution

3. Is $x = 2$, $y = 3$ the solution of $\begin{cases} 5x + 3y = 19 \\ 2x + 7y = 24 \end{cases}$?

4. Solve: $\begin{cases} 7x + 3y + 2z = 9 \\ 3x - y + 3z = 17 \\ 5x + 4y + 4z = 13 \end{cases}$

Chapter 16. INEQUALITIES

1. Solve: $2x + 5 < 9$ 2. Solve: $5x - 7 - 7x \le 4$

3. Solve: $4(x + 5) - 7 > 2x - 5$ 4. Sketch the graph of $y < 4x - 1$.

5. Find graphically the solution to the system below.

$$\begin{cases} y \ge 3x + 2 \\ x + y \le 6 \end{cases}$$

Chapter 17. FACTORING

1. Factor: $6mn^2 - 3m^2n + 12m^2n^2$

2. Factor: $x^2 + 7x + 10$ 3. Factor: $x^2 - 4x - 12$

4. Factor: $2y^2 + 10y - 28$ 5. Factor: $x^2 + 9x$

6. Solve: $x^2 - 9x + 18 = 0$

7. Solve: $y^2 - 16 = 0$

8. Solve: $5x^2 - 17x + 6 = 0$

Chapter 18.   FRACTIONS CONTAINING VARIABLES

1. Combine: $\dfrac{3}{x} + \dfrac{2}{y}$

2. Combine: $\dfrac{4}{3x^2y} + \dfrac{2}{15xy^2}$

3. Add: $\dfrac{4}{x^2 + 3x + 2} + \dfrac{10}{x + 1}$

4. Reduce: $\dfrac{x + 1}{x^2 + 6x + 5}$

5. Reduce: $\dfrac{2x - 2y}{y - x}$

6. Multiply: $\dfrac{3x^2}{2} \cdot \dfrac{2x^2 - 8}{x^2 - 7x + 10}$

7. Divide: $\dfrac{5}{x} \div \dfrac{10x + 20}{x^2 + 2x}$

8. Simplify: $\dfrac{5 + \dfrac{x}{y}}{2 - \dfrac{y}{x}}$

9. Solve: $x + \dfrac{x + 3}{5} = \dfrac{1}{3}$

Chapter 19.   PROPERTIES OF EXPONENTS

Simplify. Leave answers with positive exponents.

1. $x^8 \cdot x^5$

2. $(x^3)^4$

3. $3m \cdot 4m^5 \cdot m^3$

4. $(3x^5y^9)^4$

5. a^{-7}

6. $5x^{-2}$

7. $\dfrac{m^{12}}{m^5}$

8. $\left(\dfrac{x^3}{y^4}\right)^6$

9. $\dfrac{a^8b^4}{a^6b^6}$

10. $5a^{-5}bc^3d^{-2}$

11. $9^{1/2} + 5^0$

12. $(2 \cdot 3)^0 + 8^{2/3}$

13. $\sqrt{64x^4y^{16}}$

Chapter 20.   FORMULA AND EQUATION MANIPULATION

Solve each formula for the letter indicated.

1. $y = mx$ for m

2. $P = 2l + 2w$ for l

3. $ax + by + c = 0$ for x

4. $C = \dfrac{nE}{R + nr}$ for E

5. $\dfrac{1}{f} = \dfrac{1}{p} + \dfrac{1}{q}$ for p

Chapter 21. RADICALS AND ROOTS

1. Simplify $\sqrt{48}$.   **2.** Combine $7\sqrt{2} + \sqrt{8}$.

3. Remove the radical from the denominator of $\dfrac{1}{\sqrt{5}}$.

4. Simplify $\sqrt{\dfrac{1}{3}}$ and leave it with no radical in the denominator.

5. Rationalize the denominator of $\dfrac{5}{\sqrt{5}}$.

6. Multiply $(1 + \sqrt{3})(1 - \sqrt{3})$.

7. Solve $\sqrt{10x - 2} = 2\sqrt{x}$.

8. Reduce the fraction $\dfrac{3 + \sqrt{45}}{3}$.

9. Remove the radical from the denominator of $\dfrac{1}{5 + \sqrt{2}}$.

Chapter 22. MORE QUADRATIC EQUATIONS

1. Solve $x^2 + 8x + 5 = 0$ by completing the square.

2. Solve $3x^2 + 5x - 6 = 0$ by using the quadratic formula.

3. Solve $m^2 + 6m - 4 = 0$ by using the quadratic formula.

4. Solve $x^2 - 3x + 5 = 0$ by using the quadratic formula.

5. Check the discriminant to determine the nature of the roots of
$$x^2 - 3x + 1 = 0$$

6. Sketch the graph of $y = x^2 - 5x + 4$.

Chapter 23. SLOPES AND INTERCEPTS

Find the slope and the y intercept of each line described below.

1. $y = 5x + 2$ **2.** $y = -x$

3. $2x + y = 6$ **4.** $4x - 5y - 20 = 0$

5. The line passing through $(10, 9)$ and $(0, 4)$

Chapter 24. SETS AND FUNCTIONS

1. True or false: $x \notin \{a, b, c\}$ **2.** True or false: $4 \in \{x \mid x \text{ odd}\}$

3. True or false: $6.3 \in \{x \mid 2 < x < 7\}$

4. $\{1, 2, 3\} \cup \{3, 4, 5\} =$ **5.** $\{1, 2, 3\} \cap \{3, 4, 5\} =$

6. True or false: $\{1, 2, 3\} \subseteq \{3, 4, 5\}$

7. If $A = \{1\}$, $\mathcal{U} = \{1, 2, 3, 4\}$, $\bar{A} =$

8. Draw a Venn diagram to represent $A \cap B$ in the case that $B \subseteq A$.

9. Is $y = 4x - 7$ a function?

10. Is $\{(2, 3), (3, 4), (5, 4)\}$ a function?

11. If $f(x) = 2x + 7$, what is $f(-3)$?

12. If $f(x) = x^2 - 9x + 3$, what is $f(-1)$?

APPENDIX **B**

Answers
to Pretests

Chapter 1. REVIEW OF BASIC OPERATIONS

1. 80   **2.** 414.62   **3.** 10,632   **4.** 2909   **5.** 1743   **6.** 5665.92
7. 247   **8.** 2.78   **9.** (a) 24   (b) 14   **10.** impossible   **11.** 0   **12.** 81

Chapter 2. ORDER OF OPERATIONS

1. 53   **2.** 75   **3.** 66   **4.** 74   **5.** 31   **6.** 8   **7.** 4   **8.** 49

Chapter 3. EVALUATION OF GEOMETRIC FORMULAS

1. 44.4   **2.** 314   **3.** 7   **4.** 65   **5.** 57

Chapter 4. SQUARE ROOTS

1. (a) 4   (b) 9   (c) 13   **2.** 5 inches   **3.** 12

396

Chapter 5. PROPERTIES OF ARITHMETIC

1. commutative   **2.** distributive   **3.** closure   **4.** distributive
5. associative   **6.** commutative   **7.** closure   **8.** commutative
9. closure

Chapter 6. ARITHMETIC OF FRACTIONS

1. $\frac{5}{6}$ **2.** $\frac{81}{40}$ **3.** $\frac{197}{420}$ **4.** $\frac{35}{48}$ **5.** 1   **6.** $\frac{6}{25}$ **7.** $\frac{13}{105}$ **8.** $\frac{21}{41}$
9. $5\frac{3}{4}$ **10.** $\frac{59}{8}$

Chapter 7. FRACTIONS, DECIMALS, AND PERCENT

1. $\frac{63}{100}$ **2.** .625   **3.** $\frac{29}{100}$ **4.** .07   **5.** 43%   **6.** 35%   **7.** $\frac{290}{100}$ or $\frac{29}{10}$
8. $\frac{83}{1000}$ **9.** 750%   **10.** 690%

Chapter 8. UNITS OF MEASURE

1. 14,400 seconds   **2.** 2 miles   **3.** 39 inches   **4.** 60 cubic inches
5. 88 feet per second

Chapter 9. METRIC SYSTEM

1. 1.85 meters   **2.** 39 miles   **3.** 7000 millimeters   **4.** 7.48 inches
5. 12.26 liters   **6.** .623 liter   **7.** .58 meter   **8.** 50.6 pounds
9. 42,000,000 milligrams   **10.** 158.76 grams

Chapter 10. SIGNED NUMBERS

1. -3 **2.** $+8$ **3.** 0   **4.** -2 **5.** 32   **6.** -12 **7.** -8 **8.** -5
9. -16 **10.** -3 **11.** -11 **12.** 2   **13.** true

Chapter 11. ALGEBRAIC EXPRESSIONS

1. $9x$ **2.** $-3x$ **3.** $11x - y - 1$ **4.** $15 - 8y$ **5.** $-15x - 20$
6. $-5x + 2$ **7.** -4 **8.** $-21x^6$ **9.** $6x^2 + 13x - 5$
10. $x^2 + 5x - 3$

Chapter 12. SOLVING LINEAR EQUATIONS

1. $\dfrac{3}{2}$ **2.** $-\dfrac{1}{2}$ **3.** 4   **4.** $-\dfrac{35}{2}$ **5.** 11   **6.** 2   **7.** $-\dfrac{1}{3}$ **8.** -6

9. 5   **10.** 7   **11.** $\dfrac{11}{3}$ **12.** $\dfrac{9}{24}$ or $\dfrac{3}{8}$

Chapter 13. WORD PROBLEMS

1. $5n$ **2.** $12x$ **3.** $4x - 7$ **4.** $\dfrac{7m}{100}$ or $.07m$ **5.** $13\frac{1}{2}$ feet, $17\frac{1}{2}$ feet

6. 15 13¢, 15 17¢, 20 20¢

Chapter 14. GRAPHING STRAIGHT LINES

1. 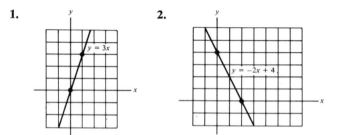   **2.**

3. Meet at $(4, 9)$

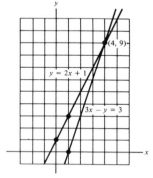

Chapter 15. SYSTEMS OF LINEAR EQUATIONS

1. $x = 3, y = 2$ **2.** $x = 2, y = -1$ **3.** no   **4.** $x = 1, y = -2, z = 4$

Chapter 16. INEQUALITIES

1. $x < 2$ **2.** $x \geq -\frac{11}{2}$ **3.** $x > -9$

4.   **5.**

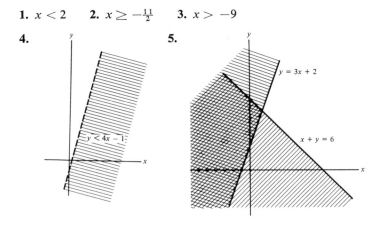

Chapter 17. FACTORING

1. $3mn(2n - m + 4mn)$ **2.** $(x + 2)(x + 5)$ **3.** $(x - 6)(x + 2)$
4. $2(y + 7)(y - 2)$ **5.** $x(x + 9)$ **6.** $x = 3, 6$ **7.** $y = \pm 4$
8. $x = \frac{2}{5}, 3$

Chapter 18. FRACTIONS CONTAINING VARIABLES

1. $\dfrac{2x + 3y}{xy}$ **2.** $\dfrac{2x + 20y}{15x^2 y^2}$ **3.** $\dfrac{10x + 24}{(x + 2)(x + 1)}$ **4.** $\dfrac{1}{x + 5}$ **5.** -2

6. $\dfrac{3x^2(x + 2)}{x - 5}$ **7.** $\dfrac{1}{2}$ **8.** $\dfrac{5xy + x^2}{2xy - y^2}$ **9.** $-\dfrac{2}{9}$

Chapter 19. PROPERTIES OF EXPONENTS

1. x^{13} **2.** x^{12} **3.** $12m^9$ **4.** $81x^{20}y^{36}$ **5.** $\dfrac{1}{a^7}$ **6.** $\dfrac{5}{x^2}$ **7.** m^7

8. $\dfrac{x^{18}}{y^{24}}$ **9.** $\dfrac{a^2}{b^2}$ **10.** $\dfrac{5bc^3}{a^5d^2}$ **11.** 4 **12.** 5 **13.** $8x^2y^8$

Chapter 20. FORMULA AND EQUATION MANIPULATION

1. $m = \dfrac{y}{x}$ **2.** $l = \dfrac{P - 2w}{2}$ **3.** $x = \dfrac{-by - c}{a}$ **4.** $E = \dfrac{C(R + nr)}{n}$

5. $p = \dfrac{fq}{q - f}$

Chapter 21. RADICALS AND ROOTS

1. $4\sqrt{3}$ **2.** $9\sqrt{2}$ **3.** $\dfrac{\sqrt{5}}{5}$ **4.** $\dfrac{\sqrt{3}}{3}$ **5.** $\sqrt{5}$ **6.** -2

7. $x = \dfrac{1}{3}$ **8.** $1 + \sqrt{5}$ **9.** $\dfrac{5 - \sqrt{2}}{23}$

Chapter 22. MORE QUADRATIC EQUATIONS

1. $x = -4 \pm \sqrt{11}$ **2.** $x = \dfrac{-5 \pm \sqrt{97}}{6}$ **3.** $m = -3 \pm \sqrt{13}$

4. $x = \dfrac{3 \pm i\sqrt{11}}{2}$ **5.** real, unequal   **6.**

$$y = x^2 - 5x + 4$$

Chapter 23. SLOPES AND INTERCEPTS

1. $m = 5, b = 2$ **2.** $m = -1, b = 0$ **3.** $m = -2, b = 6$
4. $m = \frac{4}{5}, b = -4$ **5.** $m = \frac{1}{2}, b = 4$

Chapter 24. SETS AND FUNCTIONS

1. true   **2.** false   **3.** true   **4.** $\{1, 2, 3, 4, 5\}$ **5.** $\{3\}$ **6.** false
7. $\{2, 3, 4\}$ **8.**   **9.** yes   **10.** yes

11. 1   **12.** 13

Ratio
and Proportion

C.1 RATIO AND PROPORTION

A fraction can be regarded as a comparison of two numbers—that is, a *ratio*. For instance, $\frac{5}{6}$ compares 5 with 6. It can be called the ratio of 5 to 6. If Bob reads 5 books in the time that Don reads 3 books, then **ratio**

$\frac{5}{3}$ is the ratio of Bob's reading to Don's reading

$\frac{3}{5}$ is the ratio of Don's reading to Bob's reading

In other words, Bob reads $\frac{5}{3}$ as much as Don; Don reads $\frac{3}{5}$ as much as Bob.†

In this example, if Don reads 12 books, then Bob reads $\frac{5}{3}$ as many as 12.

$$\text{Bob's reading} = \frac{5}{3} \times 12 = \frac{5}{3} \cdot \frac{12}{1} = 20 \text{ books}$$

Similarly, if *Bob* reads 35 books, then Don reads $\frac{3}{5}$ as many as 35.

$$\text{Don's reading} = \frac{3}{5} \times 35 = \frac{3}{5} \cdot \frac{35}{1} = 21 \text{ books}$$

---

†The ratio of 5 to 3 can be written $\frac{5}{3}$ or 5 : 3.

proportion   An equation which states that two fractions (or ratios) are equal is called a *proportion*. When you change a fraction to one with a different denominator, you are working with a proportion.

Example 1. *Change* $\frac{2}{5}$ *to tenths.*

$$\frac{2}{5} = \frac{?}{10}$$

Using elementary arithmetic,

$$\frac{2}{5} = \frac{4}{10}$$

This simple equation is an example of a proportion.

Any time you form an equation in which one fraction equals another, you have a proportion. The following are examples of proportions.

$$\frac{3}{4} = \frac{75}{100} \qquad \frac{20}{25} = \frac{4}{5} \qquad \frac{3}{10} = \frac{x}{12} \qquad \frac{n}{17} = \frac{2}{5}$$

In the proportion

$$\frac{a}{b} = \frac{c}{d}$$

extreme a and d are called the *extremes* of the proportion, b and c are called the **mean**   *means*.

$$\frac{\text{extreme}}{\text{mean}} = \frac{\text{mean}}{\text{extreme}}$$

We will show that

product of extremes = product of means

In symbols, the statement is: $ad = bc$.

If you multiply both sides of

$$\frac{a}{b} = \frac{c}{d}$$

by bd, the result is

$$ad = bc$$

Example 2. *Determine n:* $\dfrac{3}{7} = \dfrac{18}{n}$.

Use the fact that the product of the extremes equals the product of the means to obtain

$$3 \cdot n = 7 \cdot 18$$

or
$$3n = 126$$

Finally,
$$n = \frac{126}{3} = 42 \ \checkmark$$

Example 3. *Determine y:* $\dfrac{12}{y} = \dfrac{15}{17}.$

The product of the extremes equals the product of the means. Therefore
$$12 \cdot 17 = y \cdot 15$$

Next, multiply $12 \cdot 17$ to get 204 and write $y \cdot 15$ as $15y$.
$$204 = 15y$$

Interchange 204 and $15y$.
$$15y = 204$$

Finally, divide both sides by 15 to get $y =$
$$y = \frac{204}{15} = \frac{68}{5} \ \checkmark$$

Example 4. *Solve for m:* $\dfrac{m}{6} = \dfrac{10}{500}.$

$$m \cdot 500 = 6 \cdot 10$$
$$500m = 60$$
$$m = \frac{60}{500} = \frac{3}{25} \ \checkmark$$

Example 5. *Solve for m:* $\dfrac{124}{2} = \dfrac{m}{8}.$

$$124 \cdot 8 = 2 \cdot m$$
$$992 = 2m$$
$$2m = 992$$
$$m = 496 \ \checkmark$$

Example 6. *In order to cut a 20-foot board into two parts in the ratio of 2 to 3 (2:3), how long must each piece be?*

One approach:

If x represents the size of the larger piece, then $\frac{2}{3}x$ represents the size of the smaller, and

$$x + \frac{2}{3}x = 20$$

Then, after multiplying both sides by the denominator 3,

$$3x + 2x = 60$$

Next,

$$5x = 60$$

$$x = 12 \qquad \text{The larger piece is 12 feet. } \checkmark$$

$$\frac{2}{3}x = 8 \qquad \text{The smaller piece is 8 feet. } \checkmark$$

Another approach:

To avoid fractions, let $3x$ be the length of the larger piece and $2x$ the length of the smaller. As you can see, the ratio of the smaller to the larger is still $\frac{2}{3}$, since

$$\frac{2x}{3x} = \frac{2}{3} \cdot \frac{x}{x} = \frac{2}{3} \cdot 1 = \frac{2}{3}$$

Now

$$2x + 3x = 20$$

$$5x = 20$$

$$x = 4$$

$$3x = 12 \qquad \text{larger } \checkmark$$

$$2x = 8 \qquad \text{smaller } \checkmark$$

Example 7. *If* 12 *pounds of a fertilizer will cover a lawn of* 1500 *square feet, how much fertilizer is needed for a* 2000-*square foot lawn?*

Begin by letting x represent the amount of fertilizer needed for a 2000-square foot lawn. Next, establish a proportion, such as

$$\frac{12}{1500} = \frac{x}{2000}$$

This proportion states that 12 pounds of fertilizer compares with a 1500-square foot lawn as x pounds of fertilizer compares with a 2000-square foot lawn.

Since the product of the means equals the product of the extremes, we have

$$(1500)(x) = (12)(2000)$$

or

$$1500x = 24{,}000$$

This means that

$$x = \frac{24{,}000}{1500} = 16 \checkmark$$

Thus, 16 pounds of fertilizer is needed to cover a 2000-square foot lawn.

Example 8. *Two brothers decide to go into business and share expenses in the ratio of 3 to 5 with the older brother paying the larger share. If the older brother pays $4500, how much does the younger brother pay?*

The *only* thing we don't know is how much the younger brother must pay. So let x represent that amount. Since they share expenses in a $\frac{3}{5}$ ratio, and they share expenses in the ratio of $\frac{x}{4500}$, we get the proportion

$$\frac{3}{5} = \frac{x}{4500}$$

We solve by first multiplying the means and then the extremes. We set these products equal to each other to obtain

$$5 \cdot x = 3 \cdot 4500$$

or

$$5x = 13{,}500$$

Thus

$$x = 2700 \checkmark$$

The younger brother pays $2700.

EXERCISES C.1

Answers to starred exercises are given in the back of the book.

1. Solve each proportion for the unknown.

   *(a) $\dfrac{4}{5} = \dfrac{x}{20}$

   (b) $\dfrac{5}{7} = \dfrac{x}{21}$

   *(c) $\dfrac{x}{15} = \dfrac{2}{3}$

   (d) $\dfrac{x}{25} = \dfrac{80}{40}$

   *(e) $\dfrac{12}{30} = \dfrac{5}{m}$

   (f) $\dfrac{15}{20} = \dfrac{4}{n}$

   *(g) $\dfrac{3}{x} = \dfrac{7}{4}$

   (h) $\dfrac{4}{y} = \dfrac{17}{12}$

   *(i) $\dfrac{14}{15} = \dfrac{5}{x}$

   (j) $\dfrac{m}{6} = \dfrac{12}{5}$

*2. John, Bill, Tom, and Sam all travel. John walks at 4 miles per hour, Bill runs at 10 miles per hour, Tom bicycles at 20 miles per hour, and Sam drives his car at 40 miles per hour.

   (a) How many miles can Sam drive in the time that Bill runs 20 miles?

   (b) How many miles can Sam drive in the time that John walks 20 miles?

   (c) How many miles can Sam drive in the time that Tom bicycles 60 miles?

   (d) How many miles can Tom bicycle in the time that John walks 12 miles?

   (e) How many miles can Bill run in the time that Tom bicycles 35 miles?

3. Solve each problem.

*(a) A 100-foot wire is cut into two parts in the ratio of 7 to 3. How long is each piece?

(b) Two numbers are in the ratio of $3:5$ and their sum is 872. What are the two numbers?

*(c) A factory turns out defective products in the ratio of $1:1000$ (defective: total). If 25 defective products were made in one week, how many total products were probably made?

(d) A factory turns out defective products in the ratio of $13:500$ (defective: total). If 9500 products were produced, how many would you expect to be defective?

*(e) If 3 pounds of bananas cost 87¢, how much will 5 pounds of bananas cost?

*(f) If 3 gallons of paint will cover 450 square feet, how much will 7 gallons cover?

*(g) If 3 pounds of bananas cost 87¢, how many pounds of bananas will you get for $1.16?

*(h) If 3 gallons of paint will cover 420 square feet, how many gallons are needed to cover 1120 square feet?

*(i) A 9 by 12 inch picture is enlarged so that the shorter side is 15 inches. What is the length of the other side of the enlarged photo?

(j) A 16 by 10 inch photo is reduced so that the larger side is 12 inches. How long is the shorter side of the reduced photo?

*(k) A recipe calls for 5 teaspoons of soy sauce and 2 teaspoons of lemon juice. To make a larger portion, Ron uses 12.5 teaspoons of soy sauce. How much lemon juice should he use?

(l) Doris has invested $3000 in stocks and $7000 in bonds. She then receives a check for $1400 and decides to invest in stocks and bonds, using the same ratio as before. How much should she invest in each?

C.2 DIRECT AND INVERSE VARIATION

direct variation   A quantity y *varies directly* with respect to another quantity x if the ratio of y to x is constant. In other words, y varies directly with respect to x if $\frac{y}{x} = c$ for some constant c. This can also be seen as $y = cx$.

$$y = cx \qquad \text{(direct variation)}$$

When the variation $\frac{y}{x} = 2$ is considered as $y = 2x$, you readily see that y is always twice x. If x becomes larger, then y becomes correspondingly larger. If x becomes smaller, so does y. Also, if y becomes larger or smaller, x changes accordingly. Thus you can visualize that y and x vary *directly*.

Example 9. *Suppose that y varies directly as x. If y is 20 when x is 5, what is the value of y when x is 12?*

Because y varies directly as x, we know that

$$y = cx$$

Since $y = 20$ when $x = 5$, c can be determined by substituting 20 for y and 5 for x. Thus,

$$20 = c \cdot 5$$

or $$5c = 20$$

or $$c = 4 \checkmark$$

Thus $y = cx$ becomes

$$y = 4x$$

Now if x is 12, y is determined from $y = 4x$.

$$y = 4(12)$$
$$y = 48 \checkmark$$

Example 10. *A worker's wages (y) vary directly as the number of hours worked (x). If he works 10 hours he earns $45. Determine the constant (in y = cx) and determine how much he earns for working 17 hours.*

Since y varies directly as x, we know that

$$y = cx$$

Furthermore, we know that y is 45 when x is 10, which means

$$45 = c \cdot 10$$

or $$10c = 45$$

or $$c = \frac{45}{10} = 4.5 \checkmark$$

Thus $$y = 4.5x$$

To determine the y value when x is 17, substitute 17 for x in $y = 4.5x$.

$$y = 4.5(17) = 76.5 \checkmark$$

He earns $76.50 for working 17 hours.

inverse variation   A quantity (y) *varies inversely* with respect to another quantity (x) if their product (xy) is constant. In other words, x varies inversely with respect to y if $xy = c$.

$$\boxed{xy = c \quad \text{(inverse variation)}}$$

Suppose that $xy = 12$:

If $x = 6$, y must be 2, since $6 \cdot 2 = 12$.

If $x = 1$, y must be 12, since $1 \cdot 12 = 12$.

If $x = \frac{1}{2}$, y must be 24, since $\frac{1}{2} \cdot 24 = 12$.

You can see that if y increases, x must decrease; if y decreases, x must increase. This is what you should understand as an *inverse variation*.

Example 11. *Suppose that y varies inversely as x. If y is 5 when x is 4, what is x when y is 2?*

Because y varies inversely as x, we know that

$$xy = c$$

Since $y = 5$ when $x = 4$, c can be determined by substituting 5 for y and 4 for x. Thus

$$4 \cdot 5 = c$$

or $$c = 20 \checkmark$$

Thus $xy = c$ becomes $xy = 20$. Now if y is 2, x can be determined from

$$xy = 20$$
$$x(2) = 20$$
$$x = \frac{20}{2} = 10 \checkmark$$

Example 12. *When a gas is held at constant temperature, pressure (P) varies inversely with respect to volume (V). If P is 50 when V is 12, determine the constant (in $PV = c$) and determine the pressure when the volume is 30.*

Since P varies inversely with respect to V, we know that

$$PV = c$$

Since $P = 50$ when $V = 12$, we have

$$(50)(12) = c$$

or
$$c = 600 \checkmark$$

Thus, $PV = c$ becomes $PV = 600$. Now if V is 30, P can be determined from

$$PV = 600$$

$$P(30) = 600$$

$$P = \frac{600}{30} = 20 \checkmark$$

The pressure is 20 when the volume is 30.

EXERCISES C.2

*1. Suppose that y varies directly as x. If y is 21 when x is 3, what is
   (a) The value of y when x is 2?
   (b) The value of y when x is 1?
   (c) The value of y when x is 19?
   (d) The value of x when y is 14?
   (e) The value of x when y is 10?

*2. Suppose that y varies inversely as x. If y is 6 when x is 5, what is
   (a) The value of y when x is 3?   (b) The value of y when x is 10?
   (c) The value of y when x is 1?   (d) The value of x when y is 60?
   (e) The value of x when y is $\frac{1}{2}$?

3. Determine the constant in each *direct* variation between x and y.
   *(a) y is 88 when x is 4.   (b) y is 13 when x is 156.
   *(c) y is 63 when x is 1/2.   (d) y is 42 when x is 42.

4. Determine the constant in each *inverse* variation between x and y.
   *(a) x is 16 when y is 7.   (b) x is 23 when y is 40.
   *(c) y is 2/3 when x is 12.   (d) y is 9 when x is 9.

5. Solve each problem.
   *(a) A worker's wages (y) vary directly as the number of hours worked
       (x). If he works 12 hours he earns $72. Determine the constant (in
       $y = cx$) and determine how much he earns for working 32 hours.

*(b) Same as part (a), except determine how many hours he must work in order to earn $102.

(c) Assume the worker in part (a) earns $69 for working 15 hours. Determine c and the number of hours he must work in order to earn $124.20.

*(d) Assume $PV = c$, as in Example 12. If P is 40 when V is 70, determine c and determine the pressure when the volume is 16. Also determine the volume when the pressure is 8.

(e) Assume $PV = c$, as in Example 12. If the constant is known to be 1200, determine the volume when the pressure is 30. Determine the pressure when the volume is 15.

Right Triangle Trigonometry

D.1 RIGHT TRIANGLE TRIGONOMETRY

Right triangle trigonometry was developed in order to provide an indirect means of measuring distances that could not be measured directly—for instance, the distance from the earth to the moon. Hipparchus (about 140 B.C.) and Ptolemy (about A.D. 150) are considered the founders of trigonometry. They devised techniques to catalog the positions of stars relative to the earth. Much of the early interest in trigonometry was due to its natural application to astronomy, map making, and navigation. Before we can consider applications, however, some definitions are needed.

In a right triangle the side opposite the right angle is called the *hypotenuse*. The side opposite a specified nonright angle (call the angle θ) is called the *opposite side*. The remaining side next to angle θ is called the *adjacent side*. Which side is opposite and which side is adjacent depends on the angle that you choose to refer to, as illustrated by the diagrams at the top of page 412.

hypotenuse

opposite

adjacent

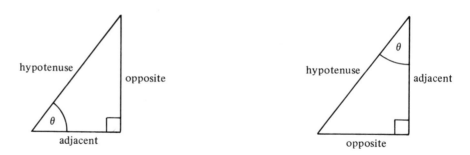

Three trigonometric functions are defined in terms of the lengths of the hypotenuse, opposite, and adjacent sides.

sine

$$\text{sine of } \theta = \sin \theta = \frac{\text{opposite}}{\text{hypotenuse}}$$

cosine

$$\text{cosine of } \theta = \cos \theta = \frac{\text{adjacent}}{\text{hypotenuse}}$$

tangent

$$\text{tangent of } \theta = \tan \theta = \frac{\text{opposite}}{\text{adjacent}}$$

Example 1. *Find* $\sin \theta$, $\cos \theta$, *and* $\tan \theta$ *corresponding to the triangle shown.*

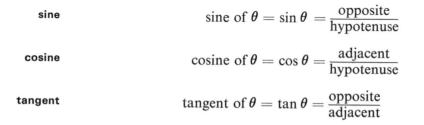

$$\sin \theta = \frac{\text{opposite}}{\text{hypotenuse}} = \frac{4}{5} = .8000$$

$$\cos \theta = \frac{\text{adjacent}}{\text{hypotenuse}} = \frac{3}{5} = .6000$$

$$\tan \theta = \frac{\text{opposite}}{\text{adjacent}} = \frac{4}{3} = 1.3333 \qquad \text{approximately}$$

Example 2. *Find* $\sin x$, $\cos x$, *and* $\tan x$ *corresponding to the triangle shown.*

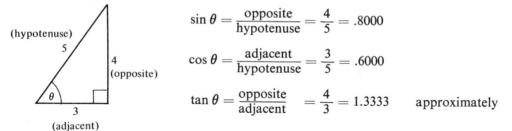

$$\sin x = \frac{\text{opposite}}{\text{hypotenuse}} = \frac{5}{13} = .3846 \qquad \text{approximately}$$

$$\cos x = \frac{\text{adjacent}}{\text{hypotenuse}} = \frac{12}{13} = .9231 \qquad \text{approximately}$$

$$\tan x = \frac{\text{opposite}}{\text{adjacent}} = \frac{5}{12} = .4167 \qquad \text{approximately}$$

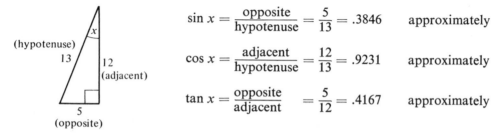

Example 3. *Find* tan θ *in the diagram shown.*

$$\tan \theta = \frac{\text{opposite}}{\text{adjacent}} = \frac{?}{4}$$

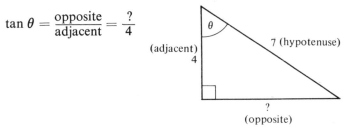

The opposite side is not known, but if we apply the Pythagorean theorem (from Chapter 4), we can determine the length of this unknown side.

Since

$$c^2 = a^2 + b^2 \qquad \text{the Pythagorean theorem}$$

We have

$$7^2 = 4^2 + b^2 \qquad \text{Call the unknown side } b.$$

or $\qquad 49 = 16 + b^2$

So

$$b^2 = 33$$

which means

$$b = \sqrt{33}$$

Thus

$$\tan \theta = \frac{\text{opposite}}{\text{adjacent}} = \frac{\sqrt{33}}{4} \quad \checkmark$$

Approximate values for the sine, cosine, and tangent of an angle can be looked up in Table 1 (page 414). In order to be sure that you can read the table, look up sin 24°, cos 81°, and tan 53°. Then check your results with those given below. The symbol $\doteq$ means approximately equal to.

$$\sin 24° \doteq .4067$$

$$\cos 81° \doteq .1564$$

$$\tan 53° \doteq 1.3270$$

Table 1

Values of Trigonometric Functions

Angle	Sin	Cos	Tan	Angle	Sin	Cos	Tan
0°	.0000	1.0000	.0000	46°	.7193	.6947	1.0355
1°	.0175	.9998	.0175	47°	.7314	.6820	1.0724
2°	.0349	.9994	.0349	48°	.7431	.6691	1.1106
3°	.0523	.9986	.0524	49°	.7547	.6561	1.1504
4°	.0698	.9976	.0699	50°	.7660	.6428	1.1918
5°	.0872	.9962	.0875				
				51°	.7771	.6293	1.2349
6°	.1045	.9945	.1051	52°	.7880	.6157	1.2799
7°	.1219	.9925	.1228	53°	.7986	.6018	1.3270
8°	.1392	.9903	.1405	54°	.8090	.5878	1.3764
9°	.1564	.9877	.1584	55°	.8192	.5736	1.4281
10°	.1736	.9848	.1763				
				56°	.8290	.5592	1.4826
11°	.1908	.9816	.1944	57°	.8387	.5446	1.5399
12°	.2079	.9781	.2126	58°	.8480	.5299	1.6003
13°	.2250	.9744	.2309	59°	.8572	.5150	1.6643
14°	.2419	.9703	.2493	60°	.8660	.5000	1.7321
15°	.2588	.9659	.2679				
				61°	.8746	.4848	1.8040
16°	.2756	.9613	.2867	62°	.8829	.4695	1.8807
17°	.2924	.9563	.3057	63°	.8910	.4540	1.9626
18°	.3090	.9511	.3249	64°	.8988	.4384	2.0503
19°	.3256	.9455	.3443	65°	.9063	.4226	2.1445
20°	.3420	.9397	.3640				
				66°	.9135	.4067	2.2460
21°	.3584	.9336	.3839	67°	.9205	.3907	2.3559
22°	.3746	.9272	.4040	68°	.9272	.3746	2.4751
23°	.3907	.9205	.4245	69°	.9336	.3584	2.6051
24°	.4067	.9135	.4452	70°	.9397	.3420	2.7475
25°	.4226	.9063	.4663				
				71°	.9455	.3256	2.9042
26°	.4384	.8988	.4377	72°	.9511	.3090	3.0777
27°	.4540	.8910	.5095	73°	.9563	.2924	3.2709
28°	.4695	.8829	.5317	74°	.9613	.2756	3.4874
29°	.4848	.8746	.5543	75°	.9659	.2588	3.7321
30°	.5000	.8660	.5774				
				76°	.9703	.2419	4.0108
31°	.5150	.8572	.6009	77°	.9744	.2250	4.3315
32°	.5299	.8480	.6249	78°	.9781	.2079	4.7046
33°	.5446	.8387	.6494	79°	.9816	.1908	5.1446
34°	.5592	.8290	.6745	80°	.9848	.1736	5.6713
35°	.5736	.8192	.7002				
				81°	.9877	.1564	6.3138
36°	.5878	.8090	.7265	82°	.9903	.1392	7.1154
37°	.6018	.7986	.7536	83°	.9925	.1219	8.1443
38°	.6157	.7880	.7813	84°	.9945	.1045	9.5144
39°	.6293	.7771	.8098	85°	.9962	.0872	11.4301
40°	.6428	.7660	.8391				
				86°	.9976	.0698	14.3007
41°	.6561	.7547	.8693	87°	.9986	.0523	19.0811
42°	.6691	.7431	.9004	88°	.9994	.0349	28.6363
43°	.6820	.7314	.9325	89°	.9998	.0175	57.2900
44°	.6947	.7193	.9657	90°	1.0000	.0000	
45°	.7071	.7071	1.0000				

The same table can be used to determine the angle when either the sine, or the cosine, or the tangent is known. Just reverse the table look-up process used above. If sin θ is .9744, how many degrees are in angle θ? If cos θ = .5878, what is the value of θ? If tan θ = .2867, what is θ? When you have finished using Table 1 to determine these values of θ, check your answers with the results given below.

$$\text{If } \sin \theta = .9744, \text{ then } \theta \doteq 77°.$$

$$\text{If } \cos \theta = .5878, \text{ then } \theta \doteq 54°.$$

$$\text{If } \tan \theta = .2867, \text{ then } \theta \doteq 16°.$$

Example 4. *Tom walks out 35 feet from the side of a building. He then lies down and looks up at the top of the building. He measures his line of vision to be at an angle of 40° with respect to the ground. Approximately how high is the building?*

Here is the situation.

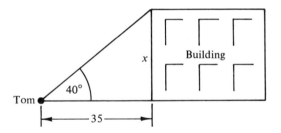

With respect to the angle of 40°, the side of the triangle of length 35 feet is the adjacent side, and the unknown side labeled x is the opposite side. The trigonometric relationship involving these two sides is tangent.

$$\tan 40° = \frac{\text{opposite}}{\text{adjacent}} = \frac{x}{35}$$

or

$$.8391 \doteq \frac{x}{35}$$

so

$$x \doteq 35(.8391) \doteq 29.4$$

The building is about 29.4 feet high.

EXERCISES D.1

Answers to starred exercises are given in the back of the book.

*1. For the triangle at the right, find each of the following.
    (a) $\sin \theta$
    (b) $\cos \theta$
    (c) $\tan \theta$

*2. For the triangle at the right, find each of the following.
    (a) $\sin x$
    (b) $\cos x$
    (c) $\tan x$
    (d) $\sin y$
    (e) $\cos y$
    (f) $\tan y$

*3. For the triangle at the right, find each of the following.
    (a) $\sin y$
    (b) $\cos y$
    (c) $\tan y$
    (d) $\sin x$
    (e) $\cos x$
    (f) $\tan x$

4. For the triangle at the right, find each of the following.
    (a) $\sin m$
    (b) $\cos m$
    (c) $\tan m$
    (d) $\sin n$
    (e) $\cos n$
    (f) $\tan n$

*5. In forestry, right triangle trigonometry is sometimes used to determine the height of a tree. The observer measures his distance (d) from the tree and the angle of inclination (θ) of his line of sight to the top of the tree. To the nearest foot, what is the height of the tree if the distance of the observer from the tree is 65 feet and the angle of inclination is 53°?

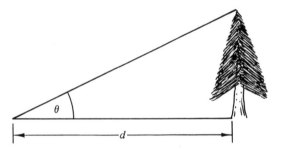

*6. The string of a kite is attached to the ground. The angle formed between the string and the ground is 58°. If the length of the string is 1000 feet (that is, you have let out 1000 feet of string), how high is the kite?

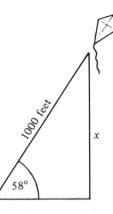

7. A 43-foot ladder is placed against a wall at an angle of 50° with respect to the ground. How high on the wall does the ladder reach?

8. The sun shines on a 25-foot vertical telephone pole, creating a 15-foot shadow on the ground. Determine the angle that the sun's rays make with respect to the ground.

*9. Having climbed to the top of a flagpole, Bob looks downward at a chalk mark that he made 150 feet from the base of the pole. He estimates that the angle between his line of vision to the chalk mark and the flagpole is 65°. About how high is the pole?

Number
Systems

E.1 DIFFERENT NUMBER SYSTEMS

This appendix presents a brief study of number systems. Specifically, we are interested in the decimal, binary, octal, and hexadecimal number systems.

The Decimal Number System (Base 10)

decimal   Several observations can be made about the decimal number system (base 10), the system most familiar to us.

1. All numbers are made up of digits selected from 0, 1, 2, 3, 4, 5, 6, 7, 8, 9.
2. The contribution of a digit to the total value of a number depends not only on what it is (that is, 1, 2, 3, etc.) but also on *where* it is.

This chapter was taken from Daniel D. Benice, *Introduction to Computers and Data Processing*, © 1970, pp. 41–65. Reprinted by permission of Prentice-Hall, Inc., Englewood Cliffs, N.J.

Let us examine the number 372, which can be written

$$3 \times 100 + 7 \times 10 + 2 \times 1$$

The 2 contributes 2 to the total value of the number.
The 7 contributes 70 to the total value of the number.
The 3 contributes 300.

Consider the place values of another whole number, 25,164.

...	10,000	1000	100	10	1
...	2	5	1	6	4

$$\begin{cases} 2 \times 10,000 \\ +5 \times 1000 \\ +1 \times 100 \\ +6 \times 10 \\ +4 \times 1 \end{cases}$$

This can also be written

...	10^4	10^3	10^2	10^1	10^0
...	2	5	1	6	4

Let us list some properties of *base 10* numbers.

base 10

1. *Ten* digits are used. They are 0, 1, 2, 3, 4, 5, 6, 7, 8, 9.
2. Column (place) values begin at 1 and increase by factors of *10* as we go from right to left.
3. The total value of a number is "computed" by multiplying the digits in the columns by their column values and adding all the products.

The Binary Number System (Base 2)

base 2

We establish the *binary* number system by using *2* as the base **binary** instead of 10 and by using *two* digits instead of ten. We list the properties of this different number system.

1. *Two* digits are used. They are 0, 1.
2. Column (place) values begin at 1 and increase by factors of *2* as we go from right to left.

We have charts as indicated below.

...	16	8	4	2	1

or

...	2^4	2^3	2^2	2^1	2^0

The binary number 1101, which we will write as $(1101)_2$ to indicate that it is a base two number, can be placed in a chart

8	4	2	1
1	1	0	1

and considered

$$1101 = 1 \times 8 + 1 \times 4 + 0 \times 2 + 1 \times 1$$
$$= \quad 8 \quad + \quad 4 \quad + \quad 0 \quad + \quad 1$$
$$= 13$$

In other words, $(1101)_2$ is the same as the base ten number 13, or

$$(1101)_2 = (13)_{10} \checkmark$$

Example 1. *Write $(1001010)_2$ as a decimal number.*

The chart

64	32	16	8	4	2	1
1	0	0	1	0	1	0

yields

$$1 \times 64 + 0 \times 32 + 0 \times 16 + 1 \times 8 + 0 \times 4 + 1 \times 2 + 0 \times 1$$

When computed, $(1001010)_2 = (74)_{10}$. $\checkmark$

Example 2. *Convert $(11101)_2$ to base 10.*

The chart

16	8	4	2	1
1	1	1	0	1

yields

$$1 \times 16 + 1 \times 8 + 1 \times 4 + 0 \times 2 + 1 \times 1$$

This is $(29)_{10}$. $\checkmark$

It should be clear that the reverse process will convert a decimal number to binary. The problem is that of fitting the number to a binary chart.

Example 3. *Change $(52)_{10}$ to binary.*

We need only go as far as the 32's column in the binary chart, since 64 is greater than the number, 52, which we are converting.

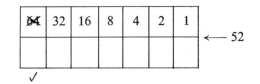

We now indicate, by a 1 in the proper column, that one 32 is contained in 52. Removing this 32 leaves a remainder of 20 to be placed in the other columns.

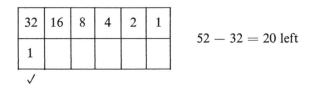

$52 - 32 = 20$ left

There is a 16 contained in the 20 which remains. Placing a 1 in the 16's column leaves a remainder of 4.

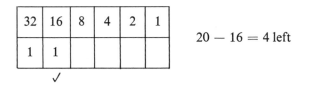

$20 - 16 = 4$ left

There are no 8's in 4, so we have zero 8's in our number.

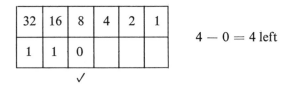

$4 - 0 = 4$ left

There is a 4 in the 4 that remains. Marking this in the proper column leaves us with no remainder.

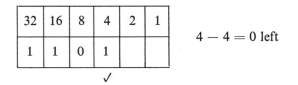

$4 - 4 = 0$ left

Since 0 is left, we have no 2's and no 1's; but 0's must be put in these columns to serve as placeholders.

32	16	8	4	2	1
1	1	0	1	0	0

Thus $(52)_{10} = (110100)_2$. ✓

Example 4. *Change $(67)_{10}$ to binary.*

There is a 64 in 67. When removed and charted, we have 3 left ($67 - 64 = 3$). The 3, of course, is a 2 and a 1. Thus we have a 64, a 2, a 1, and the rest zeros.

64	32	16	8	4	2	1
1	0	0	0	0	1	1

$(67)_{10} = (1000011)_2$ ✓

base 8   ### The Octal Number System (Base 8)

If we use *8* as the base and *eight* digits, we have the foundation of octal   the *octal* number system.

1. The *eight* digits 0, 1, 2, 3, 4, 5, 6, 7 are used.
2. Column (place) values begin at 1 and increase by factors of *8* as we go from right to left.

We have charts as indicated below.

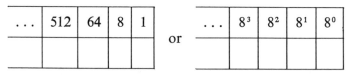

...	512	64	8	1

or

...	8^3	8^2	8^1	8^0

Accordingly, the octal number $(327)_8$ can be placed in a chart as

64	8	1
3	2	7

and considered

$$327 = 3 \times 64 + 2 \times 8 + 7 \times 1 = 215$$

In other words,

$$(327)_8 = (215)_{10}$$

and we have a convenient method for changing octal numbers to their decimal equivalents.

Consider the reverse process—that of converting a base 10 number to base 8. The mechanics are the same as for the conversion from base 10 to base 2 except that we fit the decimal number to an *octal* chart.

Example 5. *Change* $(59)_{10}$ *to octal.*

We use the chart

and note that it is unnecessary to extend the chart to include the 64's column, since 64 is larger than the number, 59, which we are fitting to the chart.

There are seven 8's in 59. Thus we have a 7 in the 8's column and a remainder of 3.

8	1
7	

$59 - 56 = 3$ left

✓

Clearly the 3 that remains is three 1's. We place a 3 in the 1's column, leaving a remainder of zero.

8	1
7	3

$3 - 3 = 0$ left

✓

Thus $(59)_{10} = (73)_8$. ✓

Example 6. *Convert* $(135)_{10}$ *to octal.*

There are two 64's in 135. The remainder is 7.

64	8	1
2		

$135 - 128 = 7$ left

✓

There are no 8's in 7. The remainder is 7.

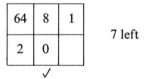

7 left

There are seven 1's in 7. The remainder is 0.

64	8	1
2	0	7

Thus $(135)_{10} = (207)_8$. ✓

base 16   *The Hexadecimal Number System (Base 16)*

hexadecimal   There are *16* digits in the *hexadecimal* number system, and *16* is used as the base.

1. The *16* digits, 0, 1, 2, 3, 4, 5, 6, 7, 8, 9, A, B, C, D, E, F, are used. We use A to represent 10, B for 11, C for 12, D for 13, E for 14, and F for 15. This convention is used because the numbers 10, 11, 12, 13, 14, 15 are not digits but rather combinations of two digits each. Their use as "digits" would be confusing, as is shown in Example 7 below.

2. Column (place) values begin at 1 and increase by factors of *16* as we go from right to left.

We have the following charts.

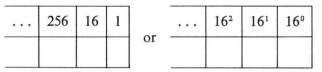

So the hexadecimal number $(2A4)_{16}$ can be charted as

256	16	1
2	A	4

and converted.

$$2 \times 256 + 10 \times 16 + 4 \times 1 = 676$$

Thus $(2A4)_{16} = (676)_{10}$, and we have a way of converting a hexadecimal number to decimal.

Example 7. *Justification of the use of symbols A, B, C, D, E, F.*

Suppose that $(2A4)_{16}$ were written $(2104)_{16}$—that is, without the use of the symbol A. The reader might understand it to mean

$$2 \times 16^3 + 1 \times 16^2 + 0 \times 16^1 + 4 \times 16^0$$

rather than

$$2 \times 16^2 + 10 \times 16^1 + 4 \times 16^0$$

which is intended.

Example 8. *Change $(523)_{10}$ to hexadecimal.*

Using a hexadecimal chart, we begin by recording the two 256's that are in 523. The remainder is 11.

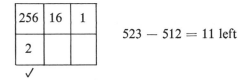

$523 - 512 = 11$ left

There are no 16's in 11. The remainder is 11.

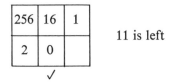

11 is left

There are eleven 1's in 11. We use "B" for 11.

Thus $(523)_{10} = (20B)_{16}$. ✓

EXERCISES E.1

*1. Change each to a number in the indicated base.

(a) $(1100)_2$ to base 10　　　　　　(b) $(10011)_2$ to base 10

(c) $(11010)_2$ to base 10　　　　　　(d) $(15)_{10}$ to base 2

(e) $(33)_{10}$ to base 2　　　　　　　(f) $(87)_{10}$ to base 2

(g) $(17)_8$ to base 10   (h) $(54)_8$ to base 10

(i) $(77)_8$ to base 10   (j) $(29)_{10}$ to base 8

(k) $(50)_{10}$ to base 8   (l) $(99)_{10}$ to base 8

(m) $(AB)_{16}$ to base 10   (n) $(F9)_{16}$ to base 10

(o) $(13C)_{16}$ to base 10   (p) $(38)_{10}$ to base 16

(q) $(96)_{10}$ to base 16   (r) $(208)_{10}$ to base 16

(s) $(139)_{10}$ to base 2   (t) $(139)_{10}$ to base 8

(u) $(101110)_2$ to base 10   (v) $(347)_8$ to base 10

(w) $(5DE)_{16}$ to base 10

*2. The first 12 counting numbers in base 10 are 1, 2, 3, 3, 5, 6, 7, 8, 9, 10, 11, 12. What are the first 12 counting numbers in base 3? In base 4? In base 5?

*3. In what bases is the number 12305 unacceptable?

E.2 ADDITION IN DIFFERENT NUMBER SYSTEMS

A further understanding of number systems can be gained by studying addition of numbers in different bases.

Example 9. *Addition in base* 10.

This example might seem trival, but if you can *understand* the procedure (not just the mechanics), then addition in other number bases will be nearly as simple.

$$
\begin{array}{r}
\text{Add:} \quad 29 \\
48 \\
15 \\
\underline{13} \\
\end{array}
$$

If the addition is performed correctly, the result obtained will be 105. But what can we learn about number systems from this? Again, if you *understand* how to add, you realize that, on adding the digits in the first column, you get $9 + 8 + 5 + 3 = 25$. But 25 is not *one* digit, and so 25 cannot appear in *one* column. So you do something like "put down 5 and carry 2." What this means is that $25 = 2 \times 10 + 5$; there are two tens in 25, with 5 left over. The 5 left over is five *ones*, so this is placed in the first (one's) column. The "two tens" indicates that a 2 should be added to the second (ten's) column. Thus the phrase "carry 2."

Perhaps we have already learned enough to apply these ideas to addition in base 2.

Example 10. *Addition in binary.*

$$
\begin{array}{r}
10 \\
11 \\
\hline
\end{array}
$$

In the first column we have $0 + 1 = 1$. So our result at this point is

$$
\begin{array}{r}
10 \\
11 \\
\hline
1
\end{array}
$$

In the second column we have $1 + 1$. This is "2". But we don't have a digit called 2 in base 2 (only 0 and 1). In fact, since the base is 2, we find that $2 = 1 \times 2 + 0$, one two and no ones, or "put down 0 and carry 1." Our result at this stage is

$$
\begin{array}{r}
^1 10 \\
11 \\
\hline
01
\end{array}
$$
The small boldface **1** is the 1 being carried.

Moving left to the next column and adding, we obtain our final result.

$$
\begin{array}{r}
^1 10 \\
11 \\
\hline
101 \ \checkmark
\end{array}
$$

Example 11. *Addition in base 2.*

$$
\begin{array}{cccccc}
 & ^1 & ^{11} & ^{111} & ^{111} \\
111 & 111 & 111 & 111 & 111 \\
101 & 101 & 101 & 101 & 101 \\
\hline
 & 0 & 00 & 100 & 1100
\end{array}
$$
The result is 1100.

Example 12. *Addition in base 2.*

$$
\begin{array}{cccccc}
 & ^1 & ^{11} & ^{101\,1} & ^{1\,0\,1\,1} \\
101 & 101 & 101 & 101 & 101 \\
111 & 111 & 111 & 111 & 111 \\
101 & 101 & 101 & 101 & 101 \\
\hline
 & 1 & 01 & 001 & 10001 \ \checkmark
\end{array}
$$

Example 13. *Addition in base 8.*

$$
\begin{array}{cc}
 & ^1 \\
37 & 37 \\
24 & 24 \\
\hline
 & 3
\end{array}
$$
$\begin{cases} 7 + 4 \text{ is "11" in base } 10, \text{ but this is } 1 \times 8 + 3, \\ \text{since we are adding in base 8.} \end{cases}$

$$
\begin{array}{r}
^1 \\
37 \\
24 \\
\hline
63 \ \checkmark
\end{array}
$$

Example 14. *Addition in base* 8.

$$
\begin{array}{r}
2 \\
746 \\
157 \\
\underline{567} \\
4
\end{array}
\qquad 6 + 7 + 7 = \text{``20''} = 2 \times 8 + 4
$$

$$
\begin{array}{r}
2\,2 \\
746 \\
157 \\
\underline{567} \\
14
\end{array}
\qquad 2 + 4 + 5 + 6 = \text{``17''} = 2 \times 8 + 1
$$

$$
\begin{array}{r}
1\,2\,2 \\
746 \\
157 \\
\underline{567} \\
714
\end{array}
\qquad 2 + 7 + 1 + 5 = \text{``15''} = 1 \times 8 + 7
$$

$$
\begin{array}{r}
1\,2\,2 \\
746 \\
157 \\
\underline{567} \\
1714 \checkmark
\end{array}
$$

Example 15. *Addition in base* 16.

$$
\begin{array}{r}
A9 \\
\underline{89}
\end{array}
\qquad
\begin{array}{r}
1 \\
A9 \\
\underline{89} \\
2
\end{array}
\qquad 9 + 9 = \text{``18''} = 1 \times 16 + 2
$$

$$
\begin{array}{r}
1\,1 \\
A9 \\
\underline{89} \\
32
\end{array}
\qquad 1 + A + 8 = \text{``19''} = 1 \times 16 + 3
$$

$$
\begin{array}{r}
1\,1 \\
A9 \\
\underline{89} \\
132 \checkmark
\end{array}
$$

Example 16. *Addition in base* 16.

$$
\begin{array}{r}
B3E \\
127 \\
\underline{1F3}
\end{array}
\qquad
\begin{array}{r}
1 \\
B3E \\
127 \\
\underline{1F3} \\
8
\end{array}
\qquad E + 7 + 3 = \text{``24''} = 1 \times 16 + 8
$$

$$
\begin{array}{r}
\text{\scriptsize 1\,1} \\
\text{B3E} \\
127 \\
\underline{1F3} \\
58
\end{array}
\qquad 1 + 3 + 2 + F = \text{``21''} = 1 \times 16 + 5
$$

$$
\begin{array}{r}
\text{\scriptsize 1\,1} \\
\text{B3E} \\
127 \\
\underline{1F3} \\
\text{E58} \quad \checkmark
\end{array}
$$

EXERCISES E.2

Perform the additions in the bases indicated.

*1. Base 2.

(a) 101
  111
	(b) 1000
	  1111
		(c) 1011
		   111

(d) 10011
  1100
  10001
	(e) 101
	  110
	  100
		(f) 1100
		  1011
		    11

*2. Base 8.

(a) 73
   6
	(b) 347
	  450
		(c) 54
		  36
		  21

(d) 103
  235
  777
  111
	(e) 64
	  37
	  12	

*3. Base 16.

(a) 89
  25
	(b) AB
	  CD
		(c) 1DF
		  AB8
		  E14

(d) 5CE
  F72
  123
	(e) A34
	  17E
	  101	

4. Base 4.

(a) 31
  20
	(b) 222
	  101
	  231
		(c) 333
		  111
		  230

APPENDIX F

Answers
to Selected
Exercises

EXERCISES 1.1 (page 5)

1. (a) $(1 \times 100) + (4 \times 10) + (3 \times 1)$ (c) $(5 \times 10) + (0 \times 1)$
(d) $(1 \times 1000) + (6 \times 100) + (0 \times 10) + (3 \times 1)$
(e) $(7 \times 100) + (7 \times 10) + (7 \times 1)$ (g) $(6 \times 100) + (0 \times 10) + (0 \times 1)$
(h) $(1 \times 10,000) + (9 \times 1000) + (7 \times 100) + (3 \times 10) + (4 \times 1)$
(i) $(3 \times 10,000) + (4 \times 1000) + (8 \times 100) + (5 \times 10) + (1 \times 1)$
(k) $(6 \times 100,000) + (4 \times 10,000) + (8 \times 1000) + (7 \times 100) + (1 \times 10)$
$+ (3 \times 1)$
(m) $(4 \times 1,000,000) + (9 \times 100,000) + (8 \times 10,000) + (0 \times 1000) + (6 \times 100)$
$+ (1 \times 10) + (7 \times 1)$

2. (a) seven hundred fifty-four   (c) one thousand three hundred forty-one
(e) fifty-four thousand six hundred seventeen
(g) five hundred sixty-three thousand four hundred thirty-two
(i) one million eight hundred ninety-seven thousand four hundred
(k) seventy-eight million eight hundred thousand one hundred sixty-seven
(m) six hundred fifty-four million four hundred fifty-six thousand five hundred
forty-six
(o) eight billion ninety-eight million seven hundred sixty-four thousand one hundred
twenty-three

3. Only 9, 0, 3, and 2 are digits.

4. Only 6, 16, 160, and 10,000,000 are whole numbers.

EXERCISES 1.2 (page 8)

1. (a) 9   (b) 6   (c) 8   (d) 13   (e) 8   (f) 14   (g) 18   (h) 15
(i) 9   (j) 10   (k) 11   (l) 13   (m) 10   (n) 16   (o) 10   (p) 13
(q) 12   (r) 14
2. (a) 69   (b) 89   (c) 75   (f) 153   (h) 1302   (i) 1590
(k) 7869   (m) 10,802   (n) 146   (o) 1284   (r) 18,923   (t) 13,858
(u) 18,310   (w) 107,603   (x) 233,574
3. (a) 158.1   (b) 9.57   (d) 169.082   (e) 1629.7   (g) 128.5
(j) 10,000

EXERCISES 1.3 (page 11)

1. (a) 5   (b) 1   (c) 6   (d) 2   (e) 7   (f) 10   (g) 14   (h) 17
(i) 27   (j) 9   (k) 19   (l) 8
2. (a) 24   (c) 29   (e) 275   (g) 599   (i) 5992   (j) 6569
(k) 9317   (m) 12,020   (o) 566,667
3. (a) 14.5   (b) 63.9   (c) 46.94   (e) 6.999   (g) 47.68   (i) 79.9

EXERCISES 1.4 (page 14)

1. (a) 72   (b) 81   (c) 35   (d) 8   (e) 0   (f) 28   (g) 18   (h) 0
(i) 12   (j) 12
2. (a) 1737   (b) 1320   (c) 3588   (d) 11,850   (g) 3,645,158
(i) 22,990,110   (k) 26,499,200   (l) 27,016,000   (n) 5,272,978,200
3. (a) 492.4668   (b) 75.23988   (e) 261.5459224   (g) 4,711,312.52312
(i) 805.6450008

EXERCISES 1.5 (page 19)

1. (a) 4   (b) 9   (c) 7   (d) 3   (e) 7   (f) 6   (g) 8   (h) 9
(i) 5   (j) 7   (k) 9   (l) 9
2. (a) 711   (b) 8423   (c) 921   (f) 845   (g) 94
3. (a) 8 R6   (c) 127 R7   (e) 253 R8   (g) 232 R38
4. (a) 311.70   (b) 175.42   (f) 1282.03   (g) 96.89
5. (a) 32.810   (c) 126.85   (e) 23,828   (g) 16,450

EXERCISES 1.6 (page 22)

1. (a) 73.77   (b) 15.22   (d) 98.14   (f) 765.92   (g) 56.90
2. (a) 17.8   (b) 99.7   (c) 187.2   (d) 40.0   (f) 456.8   (h) 46.0
3. (a) 20   (b) 23   (f) 10   (g) 34
4. (a) 58.644   (b) 46.163   (d) 98.002   (g) 10.000   (i) 5.765
5. (a) 90   (b) 150   (c) 540   (e) 5680   (g) 90
6. (a) 1800   (b) 1900   (c) 5400   (e) 2000   (h) 10,000

EXERCISES 1.7 (page 24)

1. (a) 0   (c) indeterminate   (e) impossible   (g) indeterminate
(i) impossible   (k) indeterminate

EXERCISES 1.9 (page 25)

1. (a) 6^2 (c) 2^3 (d) 9^7 (f) 3^{10} (h) m^2 (i) x^5 (j) $(x + y)^3$
2. (a) $8 \cdot 8$ (b) $2 \cdot 2 \cdot 2$ (d) $6 \cdot 6 \cdot 6 \cdot 6 \cdot 6 \cdot 6 \cdot 6 \cdot 6$ (g) $x \cdot x$
(i) $(a + b)(a + b)(a + b)(a + b)$
3. (a) 8   (c) 625   (f) 343   (g) 1   (i) 243   (k) 0

Chapter 1. REVIEW EXERCISES (page 26)

1. (a) 2351   (c) 149.93   (e) 223,715
2. (a) 4413   (c) 18,990   (d) 13.088
3. (a) 345,796   (c) 41,090,778   (d) 613.224
4. (a) 164.2   (b) 0   (c) impossible   (d) 14.71
5. (a) 6.0   (c) 9.9   (e) .9
6. (a) 16   (b) 0   (c) 125   (d) 144   (e) 1.69

EXERCISES 2.1 (page 31)

1. (a) 20   (b) 27   (c) 11   (d) 14   (e) 9   (f) 4   (g) 25   (h) 8
(i) 61   (j) 41   (k) 11   (l) 32   (m) 28   (n) 19   (o) 30   (p) 27
2. (a) 5   (c) 66   (e) 57   (g) 36
3. (a) 24   (b) 25   (c) 13   (d) 1   (e) 2   (f) 6   (g) 36   (h) 1
(i) 16   (j) 24   (k) 9   (l) 52   (m) 9   (n) 476

Chapter 2. REVIEW EXERCISES (page 31)

1. (a) 24   (b) 25   (c) 9   (d) 1
2. (a) 25   (b) 57   (c) 27   (d) 8   (e) 12   (f) 8
3. (a) 6   (b) 1   (c) 20   (d) 13   (e) 10   (f) 2   (g) 25   (h) 16
(i) 33   (j) 27   (k) 6   (l) 124

EXERCISES 3.1 (page 36)

1. (a) 1036   (c) 420   (e) 120   (g) 87.92   (i) 249.316   (j) 59
2. (a) 30   (b) 1800
3. (a) 22.8   (b) 153.86   (c) 1017.36   (d) 523.33   (e) 1004.8
4. (a) area 177, perimeter 72   (b) area 128, perimeter 56

(c) area 450, perimeter 100   (d) area 1104, perimeter 168
(e) area 132, perimeter 74   (f) area 656, perimeter 248

Chapter 3. REVIEW EXERCISES (page 38)

1. (a) 204   (b) 85   (c) 125.6   (d) 180   (e) 314

EXERCISES 4.1 (page 42)

3. (a) 3   (b) $\sqrt{5}$ (c) 12   (e) 100   (f) $\sqrt{61}$ (i) 25   (j) $\sqrt{40}$
(k) 8   (l) 25
4. (a) $t = 2$ (c) $x = 4$ (e) $A = 6$

Chapter 4. REVIEW EXERCISES (page 44)

2. (a) 4   (b) 13   (c) $\sqrt{13}$ (d) $\sqrt{18}$

EXERCISES 5.1 (page 48)

5. (a) commutative property for addition   (b) associative property for addition
(c) commutative property for addition   (d) distributive property
(e) distributive property   (f) closure property for addition
(g) associative property for addition   (h) commutative property for addition
(i) closure property for addition
(j) commutative property for multiplication   (k) distributive property
(l) distributive property
7. (a) associative property for addition   (b) commutative property for addition
(c) commutative property for addition   (d) closure property for addition
(e) closure property for addition
8. (a) distributive property   (b) commutative property for addition
(c) distributive property   (d) commutative property for multiplication
(e) closure property for multiplication

Chapter 5. REVIEW EXERCISES (page 50)

1. (a) associative property for multiplication
(b) commutative property for addition   (c) distributive property
(d) closure property for multiplication   (e) closure property for addition
(f) distributive property
2. (a) distributive property   (b) commutative property for multiplication
(c) closure property for multiplication   (d) commutative property for addition
(e) closure property for addition

EXERCISES 6.2 (page 53)

1. (a) 3/10   (b) 8/27   (c) 6/35   (e) 15/28   (g) 1/81   (h) 12/25
(j) 60/91   (k) 180/299   (m) 7/8   (o) 8/9   (p) 3/8   (r) 50/99
(s) 0
2. (a) 10/21   (c) 8/35

EXERCISES 6.3 (page 56)

1. (a) 2/3   (b) 1   (c) 3/4   (d) 3/4   (g) 2/3   (h) 1/2   (k) 3/4
(l) 5/12   (m) 1/3
2. (a) 1/9   (b) 7/20   (c) 1/12   (d) 18/35   (e) 2/3   (f) 1   (h) 4/11
(j) 3   (k) 1/2   (m) 4/27   (o) 3/10

EXERCISES 6.4 (page 59)

1. (a) 2, 4, 5, 10   (b) 2, 3, 5, 6, 10   (c) 3, 5, 9   (f) 3
(g) none of them   (h) 2   (j) 2, 3, 6   (l) 3, 5   (m) none of them
(o) 3   (p) 2, 3, 4, 6   (q) 2, 3, 4, 6, 9   (s) 2, 4
2. (a) 2/5   (b) 7/11   (c) 1/3   (d) 14/15   (e) 4/5   (g) 7/8
(i) 1/4   (j) 1/24   (m) 5/27   (n) cannot be reduced   (p) 7/8
(r) 13/17
3. (a) 4/7   (b) 1/25   (c) 2/9   (d) 7/8   (f) 7/11   (h) 13/24
(i) 3/16   (k) 8/31
4. (a) 7/31   (b) 6/49   (c) 30/49   (e) 1   (f) 2   (h) 1/40
(j) 5/192   (k) 2/5

EXERCISES 6.5 (page 62)

1. (a) 20/21   (b) 2/3   (d) 5/6   (f) 35/24   (g) 10/7   (h) 24/49
(j) 1   (l) 16/25   (n) 2/21   (p) 17/24   (q) 15

EXERCISES 6.6 (page 66)

1. (a) 6/13   (b) 4/7   (c) 5/9   (d) 4/5
2. (a) 3/7   (b) 2/11   (c) 1/3   (d) 1/9
3. (a) 1   (c) 1/2   (e) 1/2   (g) 23/20   (i) 67/91   (k) 91/132
(m) 23/16   (o) 73/42
4. (a) 1/3   (c) 16/35   (e) 1/12   (g) 1/4   (i) 1/6

EXERCISES 6.7 (page 68)

1. (a) $1\frac{1}{6}$ (c) $4\frac{1}{2}$ (e) $7\frac{1}{4}$ (g) $3\frac{4}{5}$ (i) $25\frac{1}{2}$
2. (a) 5/3   (c) 20/7   (d) 28/9   (e) 27/5   (f) 17/2   (g) 53/12
(h) 65/9
(i) 12 is not a mixed number, but it can be written as an improper fraction, for example, as 12/1.
(j) 66/7
3. (a) 147/16 or 9 3/16   (c) 7/3 or 2 1/3   (e) 18/7 or 2 4/7
(g) 9/8 or 1 1/8
4. (a) 17/4 or 4 1/4   (c) 139/12 or 11 7/12   (e) 9/35   (g) 23/12 or 1 11/12

EXERCISES 6.8 (page 71)

1. (a) $3 \cdot 7$ (b) $2 \cdot 5$ (d) $3 \cdot 11$ (e) $2 \cdot 3 \cdot 7$ (f) $3 \cdot 5 \cdot 7$
(h) $2^3 \cdot 5$ (j) $2 \cdot 5^2$ (l) $2^2 \cdot 3^2$ (m) 2^6 (p) $2 \cdot 3^2 \cdot 5$ (s) $2^4 \cdot 3^2$
(t) $2^3 \cdot 3^3$
2. (a) 30   (b) 210   (c) 100   (d) 480   (e) 105   (g) 60   (h) 18
(i) 120   (l) 180   (m) 600   (o) 180

EXERCISES 6.9 (page 74)

1. (a) 5/18   (b) 77/120   (c) 29/72   (f) 83/120   (h) 131/420
(j) 403/504   (l) 97/162   (m) 73/100   (o) 131/80   (q) 259/180
(s) 49/100   (u) 601/90

Chapter 6. REVIEW EXERCISES (page 76)

1. (a) 3/5   (b) 11/40   (c) 3/5   (e) 2/5   (g) 3/14   (i) 56/75
2. (a) 1   (c) 1/2   (e) 8/7   (g) 391/377   (i) 2/35   (k) 65/44
(m) 187/20   (n) 22/7
3. (a) 1   (c) 6/5   (e) 2/3   (g) 76/33   (i) 16/27   (k) 15/124
(m) 23/10   (n) 77/51
4. (a) $3^3 \cdot 5$ (c) $2^3 \cdot 11$ (e) $2^5 \cdot 3$ (g) $2 \cdot 5 \cdot 11$ (i) 3^5
5. (a) 30   (c) 120   (e) 120   (g) 315
6. (a) 13/20   (c) 25/96   (e) 203/280   (g) 113/252   (i) 589/630
(k) 1312/180   (m) 79/8 or $9\frac{7}{8}$ (n) 26/15 or $1\frac{11}{15}$

EXERCISES 7.1 (page 80)

1. (a) 17/100   (b) 31/100   (c) 1/2   (d) 17/50   (e) 3/4   (g) 1/5
(h) 63/200   (i) 88/125   (k) 19/20   (m) 12/25   (n) 3/10
(o) 1131/2000   (q) 33/40   (s) 19/20   (t) 999,999/1,000,000
2. (a) .1250   (b) .6000   (c) .7777   (e) .5714   (f) .6666   (g) .8461
(i) .6250   (k) .1818   (m) .1111   (n) .2666

EXERCISES 7.2   (page 84)

1. (a) 7/20   (c) 7/25   (e) 1/5   (g) 9/25   (h) 1/20   (i) 3/100
(j) 16/25   (k) 21/50   (m) 17/20   (o) 2/25
2. (a) .4   (c) .25   (e) .15   (f) .06   (g) .19   (i) .54
3. (a) 73%   (c) 65%   (e) 7%   (g) 8%   (h) 80%   (i) 80%
(k) 72.3%   (m) 6.2%   (o) 90.1%   (q) .7%
4. (a) 80%   (c) 33.3% (approximately)   (e) 26.6% (approximately)
(g) 85.7% (approximately)   (h) 36.3% (approximately)   (i) 40%
(k) 58.3% (approximately)   (m) 69.2% (approximately)
(o) 21.4% (approximately)   (p) 37.5%

EXERCISES 7.3   (page 86)

1. (a) 59/1000   (b) 2   (c) 153/1000   (f) 0   (h) 2/125   (i) 4
(k) 9/2000   (l) 3/10,000   (m) 12/5   (o) 269/200   (q) 313/100
(s) 10
2. (a) .035   (c) .0725   (e) .0275   (g) .05375
3. (a) 3   (c) 1.5   (e) 1.73   (g) 13.5   (i) 14.25
4. (a) 230%   (c) 150%   (e) 350%   (g) 1%   (i) 20%   (k) .2%
(m) 7.1%   (o) 999%   (q) 740%   (s) 212.5%

EXERCISES 7.4   (page 88)

1. (a) 32.25   (b) 46.4   (c) 164   (d) 500   (e) 12.5%
(f) 14.28% (approximately)   (g) 263.15 (approximately)
2. (a) 70%   (c) 6 windows   (e) $40,000

Chapter 7.   REVIEW EXERCISES   (page 89)

1. (a) .8, 80%   (b) .1666, 16.66%   (e) 6, 600%   (g) 1.6666, 166.66%
(i) 9.514, 951.4%
2. (a) .55, 11/20   (c) .72, 18/25   (e) 7.25, 29/4   (f) .163, 163/1000
(h) .0009, 9/10,000
3. (a) 11.1% (approximately)   (b) 40.6
4. (a) 30%   (b) 28,000 men   (c) $11,000

EXERCISES 8.1   (page 95)

1. (a) 123 feet   (b) 276 inches   (c) 720 minutes   (d) 94 hours
(g) 80 quarts   (h) 54 feet
2. (a) 10,800 seconds   (b) 316,800 inches   (c) 1680 hours
(d) 256,000 ounces   (e) 7 hours   (h) 512 gallons

3. (a) 3 miles per minute   (b) 15,840 feet per minute
(c) 264 feet per second   (d) 45 miles per hour   (e) 170 feet per minute
(f) 17 inches per second
4. (a) 720 square inches   (b) 16 square feet   (e) 29,376 cubic inches
5. (a) 182 square inches   (b) 272 cubic feet
(c) 91.56 square miles (approximately)   (e) 100 square inches
(g) 4186 cubic inches (approximately)   (i) 90 cubic feet

Chapter 8. REVIEW EXERCISES (page 96)

1. (a) 174 yards   (b) 19 minutes   (c) 5,443,200 seconds   (d) 1.4 weeks
(e) 2430 cubic feet
2. (a) 339.12 cubic feet   (b) 126 inches *or* 10.5 feet

EXERCISES 9.1 (page 102)

1. (a) 1.34 m (approximately)   (c) 8.83 meters (approximately)
(e) 150 km (approximately)   (g) 48.26 cm (approximately)
(i) 76.2 mm (approximately)
2. (a) 275.59 in. (approximately)   (c) 141.07 ft (approximately)
(e) 9.44 in. (approximately)   (g) 84 mi. (approximately)
(i) 9.25 in. (approximately)
3. (a) 21.69 l (approximately)   (c) 360 ml (approximately)
(e) 12.84 l (approximately)   (g) 67.92 l (approximately)
(i) 1.13 kl (approximately)
4. (a) 39.22 qt (approximately)   (c) 3 oz (approximately)
(e) 500 oz (approximately)   (g) 6.09 gal (approximately)
(i) 2385 gal (approximately)
5. (a) 10.45 kg (approximately)   (c) 198.45 g (approximately)
(e) 85,050 mg (approximately)
6. (a) 61.6 lb (approximately)   (c) 3.42 oz (approximately)
(e) .02 oz (approximately)
7. (a) 7000 mm   (c) .158 m   (e) 45,000 ml   (g) .7 liter   (h) 19,000 g
(i) 70,000 mg   (l) 3,000,000 ml   (m) 700,000 cm   (n) 7,000,000 mm
8. (a) 300 grams   (b) 1900 grams   (c) 50 grams   (d) 170 grams
(e) 25 grams   (f) 3.3 grams   (g) 740 liters   (h) 15 decaliters
(i) 20 meters   (j) 3000 decimeters   (k) 4000 decaliters   (l) 300 deciliters
9. (a) 35°C   (c) 5°C   (e) 68°F   (g) 113°F   (i) 24°C   (k) 138°F

Chapter 9. REVIEW EXERCISES (page 104)

1. (a) 12.7 centimeters   (b) 538.65 grams   (c) 12.26 liters
(d) 500 kilometers
2. (a) 88 pounds   (b) 166.67 fluid ounces   (c) 12.6 inches   (d) 55.77 feet
3. (a) 8520 mm   (b) 85.2 cm   (c) .052 km   (d) .005 kg
4. (a) 11°C   (b) 160°F

REVIEW PROBLEMS FOR PART 1   (page 105)

1. (a) 1632   (c) 978   (e) 786,477   (g) 1196.70   (i) 343
2. (a) 44   (b) 9
3. (a) $A = 20$ (b) $S = 452.16$
4. (a) 13   (c) $\sqrt{2}$
5. (a) commutative property for addition
(c) closure property for multiplication
(d) commutative property for multiplication
6. (a) 16/17   (c) 9/5   (e) 393/50   (g) 16/42 or 8/21
7. (a) 105   (c) 360
8. (a) 40%   (b) 87.5%
9. (a) 45/100 or 9/20   (b) 83/100
10. (a) 18,000 seconds   (c) 4 miles per minute
11. (a) 56.60 liters   (b) 1.10 pounds   (c) 5000 millimeters

EXERCISES 10.2   (page 115)

1. (a) +33   (b) +39   (c) +5   (d) −9   (e) +4   (f) −7
(g) −27   (h) −27   (i) −5   (j) +5   (k) −77   (l) −18
2. (a) +11   (b) −11   (c) −4   (d) +6   (e) 0   (g) +11
(i) −23   (k) +7   (m) +45   (o) −9   (q) −31   (s) −31
3. (a) +1/12   (b) −1/10   (c) −19/15   (d) +17/28
4. (a) +16   (b) −3   (c) +6   (d) 0   (e) −6   (g) +11   (i) −5
(k) −27
5. (a) −3.1   (b) +7.7

EXERCISES 10.3   (page 119)

1. (a) +25   (b) −35   (c) −60   (d) +2   (e) +51   (f) +29
(g) −98   (h) +79   (i) −115   (j) −12   (k) −76   (l) +911
(m) −239   (n) −341   (o) +95
2. (a) +117.75   (b) +92.1   (c) +94.72
3. (a) +3   (b) −9   (c) +9   (d) −15   (e) +17   (f) −3
(h) +15   (k) −35
4. (a) −5.12   (b) 113.43   (c) +5/7   (d) −3/4   (e) −4/15
(f) +1/12
5. (a) +4   (c) +9   (e) −13   (g) +4   (i) −8   (k) −40
(m) +4

EXERCISES 10.4   (page 125)

1. (a) −6   (b) −10   (c) +35   (d) −18   (e) +24   (f) +63
(g) +9   (h) −40   (i) +20   (j) +36   (k) −4   (l) −48   (m) +9
(n) −49   (o) −90   (p) −80   (q) +84   (r) +76   (s) −120

(t) $+64$ (u) -57 (v) $+99$ (w) $+240$ (x) -105
2. (a) -5.12 (b) $+33.28$ (c) -74.305 (d) $+663.48$ (e) $-1/21$
(f) $+6/55$ (g) $+1/2$ (h) $-35/48$
3. (a) -24 (b) -6 (c) $+216$ (d) -60 (e) -96 (g) $+16$
(h) -1 (i) $+360$ (k) -24 (m) $+144$ (o) $+24$
4. (a) -1 (b) $+16$ (c) -27 (d) $+32$ (e) $+81$ (f) -243
(g) -128 (h) $+1$ (i) -1 (j) -64 (k) $+1$ (l) $+27$
5. (a) $+15$ (b) -7 (c) -64 (d) -82 (e) -1 (f) -983
(g) -3 (h) $+18$ (i) $+24$ (j) -32

EXERCISES 10.5   (page 129)

1. (a) -5 (b) -18 (c) $+22$ (d) -5 (e) $+10$ (f) $+16$
(g) $+14$ (h) $+34$
2. (a) -6 (c) $+2$
3. (a) -13 (b) $+15$ (c) $+10$ (d) -15 (e) $+17$ (f) $+3$
(g) $+6$ (h) -2 (i) $+12$ (j) -5 (k) -3 (l) $+8$ (m) $+2$
(n) -3 (o) $+5$ (p) 0
4. (a) -17 (b) -2 (c) $+22$ (e) $+78$ (g) -31 (i) -7
(j) -4
5. (a) $+32$ (b) 0 (c) 0 (e) $+6$ (g) -103 (h) $+20$ (i) 0
(j) -2 (k) -22

EXERCISES 10.6   (page 133)

1. (a) $+5$ (b) -3 (c) -3 (d) $+4$ (e) $+4$ (f) -6 (g) -4
(h) -5 (i) $+12$ (j) -1 (k) -6 (l) $+3$ (m) $+27$ (n) -6
(o) 0 (p) 0
2. (a) $-3/5$ (b) $-4/7$ (c) $+7/9$ (d) $-6/7$ (e) $+5/8$ (f) $-1/3$
(g) $-5/6$ (h) $+3/4$ (i) $+6/7$
3. (a) $+1$ (c) $+1$ (e) -11 (g) 0
4. (a) -12 (b) -2 (c) $+3$ (d) $+1$ (g) $+30$ (h) $-11/24$

EXERCISES 10.7   (page 135)

1. (a) 21 (b) -7 (c) 3 (d) 47
3. (a) 41 (c) 16 (e) -44 (g) 29 (i) 9
4. (a) 4 (b) 5 (d) 2

EXERCISES 10.8   (page 137)

1. (a) 30 (b) 0 (c) 2 (d) 17 (e) 7 (f) 0 (g) 12 (h) 3
2. (a) true   (b) true   (c) true   (d) false   (e) true   (f) true
(g) false   (h) false   (i) false   (j) false   (k) true   (l) true   (m) false
(n) true   (o) false   (p) false   (q) true   (r) false

Chapter 10.   REVIEW EXERCISES   (page 138)

1. (a) $+3$ (b) -15 (c) -1 (d) $+10$ (e) -3 (f) -16
(g) $+4$ (h) $+4$ (i) -45 (j) $+42$ (k) $+1$ (l) -1
(m) $+6$ (n) $+49$ (o) $+37$ (p) -71 (q) -20 (r) $+1$
(s) -32 (t) -3 (u) -27 (v) -5
2. (a) $+18$ (b) $+4$ (c) -12 (d) -15 (e) $+16$ (f) -62
(g) -9 (h) $+7$ (i) -37 (j) $+36$ (k) $+9$ (l) -17 (m) $+54$
(n) -22 (o) -24 (p) -2 (q) -10 (r) $+11$
3. (a) -16 (b) $+20$ (c) $+17$ (d) $+10$ (e) -21 (f) $+9$

EXERCISES 11.1   (page 143)

1. (a) $8x$ (b) $17y$ (c) $14m$ (d) $8y$ (e) $5x$ (f) $-2x$ (g) $7x$
(h) $-9t$ (i) $5x$ (j) 0 (k) n (l) $-3x$
2. (a) $12x + 7$ (c) 0 (e) $6mn + 5m$ (g) $6x^2 + 5x$
(i) $2x + 3y + 9z$ (k) 7
3. (a) 2 (b) 1 (c) -5 (d) -1 (e) y (f) $3y$ (g) $-5y$
(h) $-y$ (i) $25my^2$ (j) w
4. (a) $13.2x$ (c) $\dfrac{17}{12}x$ (e) $\dfrac{5}{8}x$

EXERCISES 11.2   (page 148)

1. (a) x^{12} (b) y^{13} (c) m^{14} (d) t^5 (e) cannot simplify   (f) x^{21}
(g) a^{40} (h) cannot simplify   (i) y^{120} (j) x^{21}
2. (a) $24y^2$ (c) $5m^{10}$ (e) $3x^{11}$ (g) $8x^{11}$ (i) $-6a^3b^7$ (k) $10x^{10}y^7$
(m) $-30u^{10}v$
3. (a) $6x + 7$ (b) $114 - 42x$ (c) $-y - 10$ (e) $6x$ (g) -4
(h) $6 + 16t$ (k) $-11x + 11$
4. (a) $7x^2 + 12x - 1$ (c) $6x^2 - 22x + 1$ (d) $-2x^2 - 6x + 20$
(e) $-15x + 11$
5. (a) $-15x + 20$ (b) $-44x + 54$ (c) $53 - 43x$ (e) $5x - 75$
(f) $8x + 4$ (h) $13x + 7$
6. (a) $-x + 14y - 78$ (c) $28x - 26y$

EXERCISES 11.3   (page 151)

1. (a) $15x^6 + 30x^2$ (c) $56y + 24y^6$ (e) $5x^9 + 8x^7$ (g) $48m^{16} - 96m^7$
2. (a) $x^2 + 5x + 6$ (c) $x^2 + 2x - 35$ (e) $y^2 - 10y + 16$
(g) $ce - 6c + 5e - 30$ (i) $ab + 6a + 2b + 12$ (k) $m^2 - 1$
(m) $9 - x^2$ (o) $14 + 5x - x^2$
3. (a) $20x^2 + 17x + 3$ (c) $8x^2 + 2x - 1$ (e) $3x^2 - 11x + 10$
(g) $7m^2 + 30m - 25$ (i) $6mn + 10m + 21n + 35$ (k) $10 + 23x + 9x^2$
(m) $6x^2 - 24$ (o) $2x^2 + 3xy + y^2$

4. (a) $m^2 + 2mn + n^2$ (c) $a^2 - 2ab + b^2$ (e) $x^2 + 14x + 49$
(g) $y^2 - 10y + 25$
5. (a) $x^3 + 5x^2 + 10x + 12$ (c) $x^3 - 9x^2 + 20x - 12$
(e) $x^3 + 6x^2 + 3x - 10$ (g) $2x^4 - 11x^3 + 15x^2 - 17x - 5$
(i) $15x^2 - 44xy + 32y^2 + 41x - 64y + 14$

EXERCISES 11.4   (page 156)

1. (a) $x^2 + 6x + 1$ (c) $3y^2 - 2y - 1$ (e) $4x^3 + x + 2$
(g) $4t^4 - 2t^3 - t + 3$ (i) $3x^2 + 9x - 10$
2. (a) x^{13} (b) x^6 (c) y^2 (d) m^8 (e) a^7 (f) x^3 (g) b^5
(h) u^9 (i) n^8 (j) t^8
3. (a) $x + 8$ (c) $x + 5$ (e) $x^2 + 5x - 3$ (g) $7y^2 - 19$
(i) $4x - 3 - \dfrac{5}{2x + 1}$
4. (a) $x^2 + x + 2$ (c) $3x^3 - 2x^2 + 2x + 5 + \dfrac{2}{x + 1}$ (e) $x^2 - x + 1$
(g) $x^3 + x^2 + x + 1$
5. (a) $x + 5 + \dfrac{14}{x - 2}$ (c) $x^2 + 10x + 43 + \dfrac{239}{x - 5}$ (e) $5x^2 + x$
(g) $x^2 + 5x - 3 - \dfrac{5}{x - 2}$

Chapter 11.   REVIEW EXERCISES   (page 157)

1. (a) $-x$ (b) $-4x - 4y$ (c) $13x - 12$
2. (a) x^{17} (c) x^7 (e) $7x^{13}$ (f) $3u^{14}$
3. (a) $-x - 17$ (b) $7 + 9x$ (c) $6x + 11$ (d) $16 - 47x$
4. (a) $6x^5 + 21x^2$ (b) $4x^6 + 36x^2$ (c) $m^2 + 10m + 21$
(e) $x^2 + 20x + 100$ (g) $9y^2 - 18y + 5$ (i) $x^3 + 5x^2 - 33x + 27$
(j) $10x^3 + 29x^2 - 23x - 7$
5. (a) $2x^2 - 3x - 1$
6. (a) $x - 10$ (b) $3x^2 + 2x + 5 - \dfrac{3}{x - 4}$ (c) $x^2 + 2x + 4$
(e) $x^2 + 2x - 2 + \dfrac{12}{x + 3}$

EXERCISES 12.2   (page 161)

1. (a) 5   (b) 3   (c) 9   (d) 17   (e) -3 (f) -7 (g) -9
(h) 160   (i) 0   (j) 2   (k) -6 (l) -6 (m) 9   (n) -26 (o) 0
(p) 5
2. (a) 3/8   (b) 1/12   (c) 5.6   (d) 1.6   (e) 15/8   (g) 0   (i) -1.2
(k) -14.1

EXERCISES 12.3 (page 163)

1. (a) 3 (b) 6 (c) 48 (d) 21 (e) 17/4 (f) 3/8 (g) 2/5
(h) 19/3 (i) −5 (j) −14 (k) −12 (l) −6 (m) −17/5
(n) −5/8 (o) 10 (p) 92 (q) 9/5 (r) 1/6 (s) −19 (t) 7
(u) 0 (v) 0 (w) 20 (x) 22.6 (y) −500 (z) 6.5
2. (a) 65/4 (c) −16 (e) −63/8 (g) 17/65 (i) 0
3. (a) −20 (c) −1/4 (e) 9/2 (g) 0 (i) 37/6

EXERCISES 12.4 (page 166)

1. (a) 14 (b) 28/3 (c) 15 (d) −24 (e) −9/10 (f) 8/5
(g) −28 (h) 99/10 (i) 4/15 (j) 11/91
2. (a) 45/2 (c) 14 (e) −13/42 (g) −35/12
3. (a) −3/7 (c) −54 (e) −27/44 (g) −6/5 (i) −27/7

EXERCISES 12.5 (page 169)

1. (a) 10/3 (b) 2 (c) −15/7 (d) −1 (e) −3/7 (f) 0 (g) −1
(h) −44/7 (i) 6 (j) −75 (k) 6 (l) 5/7 (m) 4 (n) 5
(o) −35 (p) 75/2
2. (a) 25/3 (b) 20 (c) 12/7 (d) 2/3 (e) 1/15 (f) 0
3. (a) 39 (c) −6 (e) −4 (g) 1/12 (i) 4 (k) 2

EXERCISES 12.6 (page 173)

1. (a) 5/3 (b) −11/3 (c) −3/2 (d) 4 (e) 7/3 (f) −8
(g) 19/18 (h) −2 (i) 5/7 (j) −14/11
2. (a) 1/5 (c) −5/3 (e) −6 (g) 17 (i) 7/2 (k) 1/5 (m) −8
(o) 5/3
3. (a) −1 (b) −32 (c) 72/17 (d) 1

EXERCISES 12.7 (page 176)

1. (a) yes (b) no (c) no (d) yes (e) yes (f) yes (g) no
(h) no (i) yes (j) yes
3. (a) 4 (b) 3 (c) 0 (d) 12/5 (e) 3 (f) 1 (i) −1/7

EXERCISES 12.8 (page 177)

1. (a) none (b) infinite (c) one (e) none (g) infinite

Chapter 12.   REVIEW EXERCISES   (page 177)

1. (a) 43   (b) 17/5   (c) $-19/6$ (d) 4/7   (e) -32 (g) $-11/9$
(i) $-11/4$ (k) 7   (m) 35/4   (n) $-9/5$ (o) no solution   (q) 12/7
(s) $-5/3$ (u) 4

EXERCISES 13.2   (page 182)

1. (a) $y + 4$ (b) $w - 17$ (c) $5x + 10$ (d) $\frac{1}{2}m$ or $\frac{m}{2}$ (e) $2t$
(f) $2x + 3$ (g) $3y - 5$ (h) $n + m$ (i) $y - x$ (j) xy
(k) $7 \cdot 6c$, which simplifies to $42c$ (l) $.23x$ or $\frac{23}{100}x$ (m) $\frac{k}{100}$ or $.01k$
(n) $\frac{k}{100}x$ or $.01kx$ or $\frac{kx}{100}$
2. (a) $5t$ (c) $v/100$ (e) $5x + 50y$ (g) $150x + 300y$
3. (a) $2p$ (c) $60H$ (e) $T/24$ (g) $15y$ (i) $p/12$ (k) mx
(m) $ax + by$ (o) $x - y$ (*not* $y - x$)   (p) $\$5000 - x$
4. (a) $.06d$ or $6d/100$ or $\frac{6}{100}d$ (b) $2dp/100$ or $.02dp$ (e) $.01mnw$ or $mnw/100$
(f) $x + .23x$ or $1.23x$ or $123x/100$
5. (a) $1000x$ (b) $y/1000$ (c) $300x$ (d) $2000y$ (e) $m/1000$
(f) $1000x$ (g) $y/1000$
6. (a) $x + y + z$ yards   (b) $2x + 2y$ or $2(x + y)$ (c) xy (d) $4x$
(f) $(x + 2)7$ or $7x + 14$ (h) $x \cdot 2x$, which simplifies to $2x^2$
(j) $2x + 2(x + 4)$, which simplifies to $4x + 8$ (l) $a + b + c$

EXERCISES 13.3   (page 189)

1. 37, 38, 39   **3.** 114, 228   **5.** 73, 80   **7.** $7\frac{1}{2}$, $9\frac{1}{2}$ meters   **9.** 225
11. 260   **12.** $55   **14.** $43.75   **16.** $15.40   **18.** 16, 32
20. 312, 104 meters   **21.** 91 centimeters

EXERCISES 13.4   (page 193)

1. 78 dimes, 39 nickels   **3.** 33 dimes, 11 quarters   **4.** 23
5. 19 20¢ and 23 17¢   **8.** 5 quarters, 10 dimes, 14 nickels
10. 7 nickels, 8 dimes   **11.** $2200 at 5%, $2300 at 6%
13. $2600 at 8%, $2800 at 5%   **15.** train 90 mph, car 65 mph
16. faster: 115 mph; slower: 85 mph

Chapter 13. REVIEW EXERCISES (page 195)

1. (a) $ab - 15$ (b) $10x + 100y$ (c) $17x/3$
(d) $2x + 2(x - 4)$, which simplifies to $4x - 8$ (e) $213cx/100$ or $2.13cx$
(f) $1.37y$ or $137y/100$
2. (a) 42 and 126   (b) 105
3. (a) 17 nickels, 15 quarters   (b) \$6500 at 9%, \$3500 at 8%

EXERCISES 14.1 (page 199)

2. (a) $(1, 1)$ (b) $(1, -3)$ (c) $(-5, 3)$ (d) $(6, 4)$ (e) $(-3, -1)$
(f) $(-1, 3)$ (g) $(7, -1)$ (h) $(2, 6)$ (i) $(-6, -2)$ (j) $(-2, -3)$
(k) $(-4, 0)$ (l) $(0, 6)$ (m) $(0, 0)$ (n) $(6, 0)$ (o) $(0, -1)$
(p) $(4, -2)$

EXERCISES 14.2 (page 208)

1. (a)

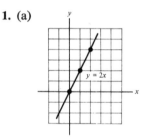

(c)

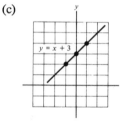

(e)

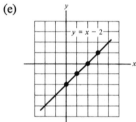

(g)

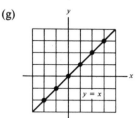

(i)

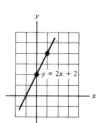

(k)

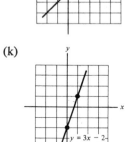

(m)

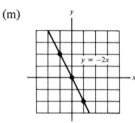

(o)

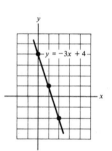

(q)

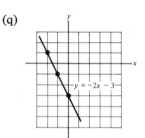

(s)

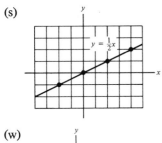

(u)

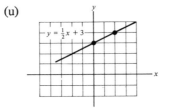

(w)

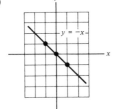

2. (a)

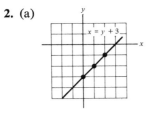

(c)

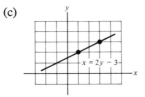

(e)

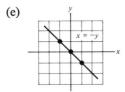

(g)

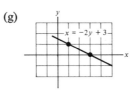

3. (a)

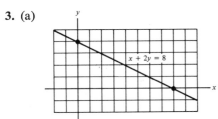

(c)

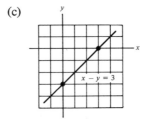

(e)

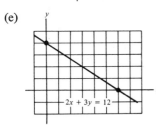

(g)

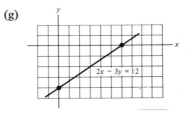

(i)

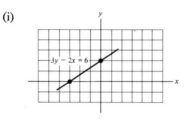

(k)

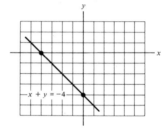

(m)

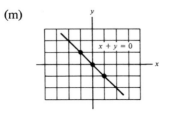

4. (a)

(c)

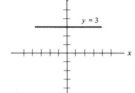

5. (a)

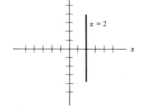

(c)

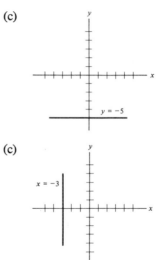

6. (a)

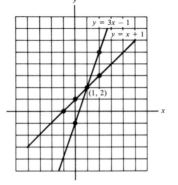

(b)

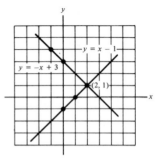

(c)

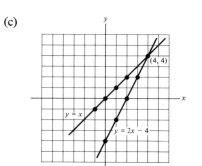

(e)

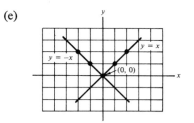

(g)

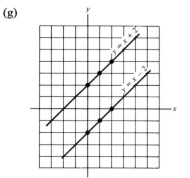

(i)

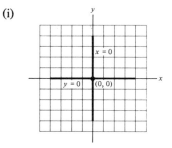

(k)

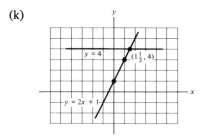

(m)

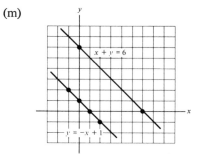

Chapter 14. REVIEW EXERCISES (page 210)

1. (a)

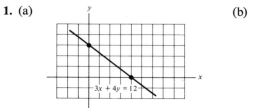

(b)

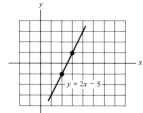

(c)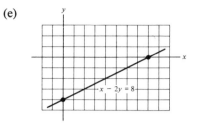

(d)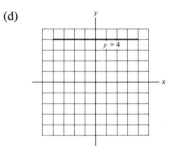

(e)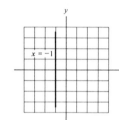

(f)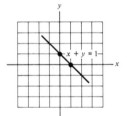

(g)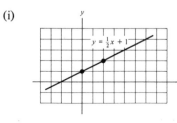

(h)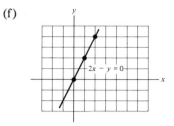

(i)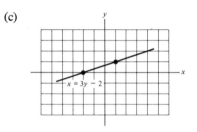

(j)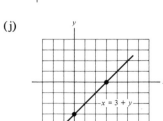

EXERCISES 15.2 (page 214)

1. (a) $x = 6, y = 7$ (c) $x = 4, y = 1$ (d) $x = -2, y = -1$
(e) $x = 2, y = 1$ (f) $x = 1, y = -3$ (g) $x = -2, y = 3$
(i) $m = 5, n = 0$
2. (a) $x = 2, y = 1$ (b) $x = 2, y = 3$ (c) $x = 3, y = 2$
(e) $x = 4, y = -3$ (g) $m = 5, n = 3$ (i) $x = \frac{1}{2}, y = 3$

EXERCISES 15.3 (page 218)

1. (a) $x = 2, y = 3$ (c) $x = 4, y = 2$ (e) $x = -1, y = -2$
(g) $x = 4, y = 4$ (i) $x = 7/19, y = 54/19$ (k) $x = 3, y = 2$
(m) $a = -2, b = 3$
2. (a) $x = 0, y = -1$ (c) $x = -2, y = 1/2$ (e) $x = -7, y = 8$

EXERCISES 15.4 (page 219)

1. (a) $x = -1, y = 4$ (c) $x = 5, y = 3$ (e) $x = 11, y = 7$

EXERCISES 15.5 (page 220)

1. (a) $x = 4, y = 1$ (b) $x = 3, y = 1/2$
(c) Infinite number of solutions; both lines are the same.   (d) $x = 6, y = 2$
(e) No solution; parallel lines.   (f) No solution; parallel lines.
(g) $x = 3, y = 3$ (h) Infinite number of solutions; both lines are the same.

EXERCISES 15.6 (page 222)

1. (a) $x = 1, y = 2, z = 3$ (b) $x = 1, y = -1, z = 1$
(c) $x = 2, y = 0, z = -3$ (d) $x = 5, y = 1, z = 0$
(e) $x = 1, y = -2, z = 4$ (f) $x = 4, y = 1, z = -2$
(g) $r = 2, s = -3, t = 4$ (h) $m = 0, n = -1, p = 2$

Chapter 15. REVIEW EXERCISES (page 222)

1. (a) $x = -4, y = -6$ (b) $a = 4, b = -1$
2. (a) $x = 3, y = -1$ (b) $x = 4, y = 5$
3. (a) $m = -1, n = 1$ (b) $x = 1/4, y = -2$

EXERCISES 16.1 (page 225)

1. (a) $(3, 8)$ (b) $[5, 6]$ (c) $[0, 5)$ (d) $(1, 19]$ (e) $[10, 100]$
(f) $[-6, 0)$ (g) $(-4, -1]$ (h) $(-10, 10)$
2. (a) $2 < x < 5$ (b) $2 \leq x \leq 5$ (c) $1 < x \leq 7$ (d) $3 \leq x < 11$
(e) $-3 \leq x \leq 0$ (f) $-2 < x \leq 4$ (g) $-7 \leq x \leq -3$ (h) $0 < x < 7$

3. (a)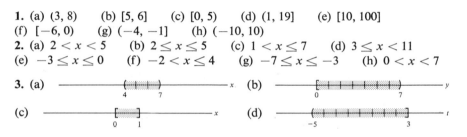

(e) 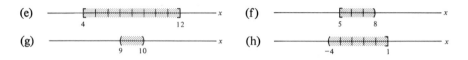 (f)

(g) (h)

EXERCISES 16.2 (page 228)

1. (a) $x < 5$ (b) $t > -24$ (c) $x > 6/5$ (d) $x \le -4/3$
(e) $y < -7/4$ (f) $n > -9/14$ (g) $x < -1$ (h) $x \ge -13/7$
(i) $x > 11/3$ (j) $x \le -1/4$
2. (a) $y > 6$ (c) $x > -3/2$ (e) $x \ge 6$ (g) $t > -16/3$ (i) $x \ge 9/2$
(k) $x < 1/4$

EXERCISES 16.3 (page 232)

1. (a)

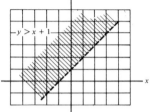

(c)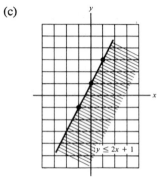

(e)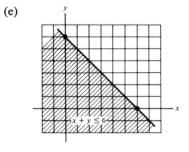

(g)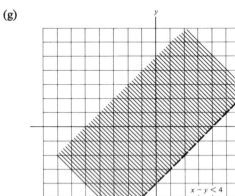

(i)

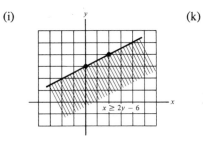

$x \geq 2y - 6$

(k)

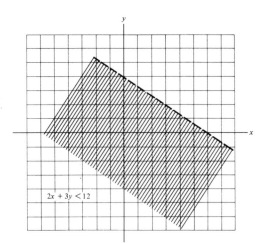

$2x + 3y < 12$

(m)

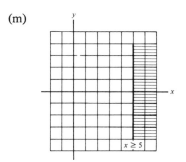

$x \geq 5$

(o)

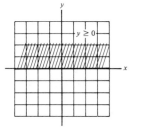

$y \geq 0$

2. (a)

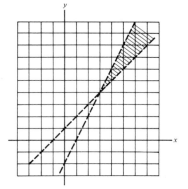

(c)

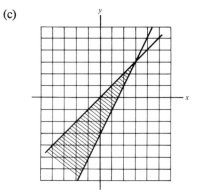

(e)

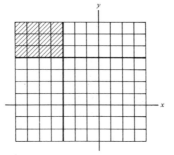

(g)

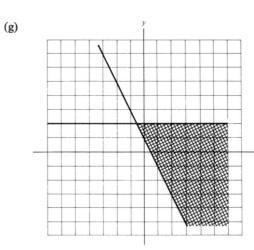

(i)

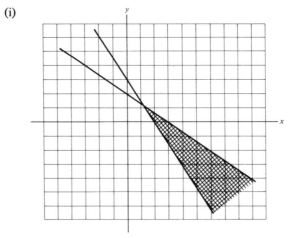

3. (a) $x \geq 40$ (b) $x < 100$ (e) $x > y$
(f) $x \geq 40$, $x \leq 60$ or $40 \leq x \leq 60$ (h) $x \geq 20$, $y \geq 40$, $x + y \leq 75$

Chapter 16. REVIEW EXERCISES (page 232)

1. (a) $(-4, 6)$ (b) $[-5, 0)$ (c) $[78, 100]$ (d) $(-9, 7]$
2. (a) $-3 \leq x \leq 10$ (b) $2 < x \leq 11$
3. (a) $x < 5$ (b) $x < -13/2$ (c) $x \leq 3/5$ (d) $x > 7/8$
(e) $x > -13/28$ (f) $x \geq -2$

4. (a)

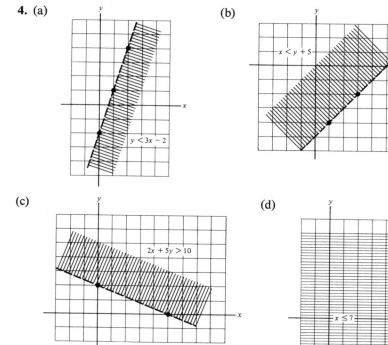

$y < 3x - 2$

(b)

$x < y + 5$

(c)

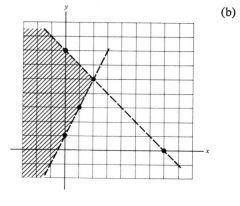

$2x + 5y > 10$

(d)

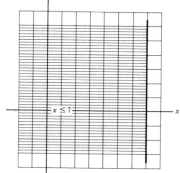

$x \le 7$

5. (a)

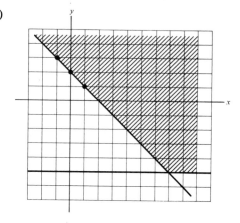

(b)

EXERCISES 17.1   (page 236)

1. (a) $(7 + n)x$ (b) $(m + n)x$ (c) $(p + q)r$ (d) $(3 - t)y$
(e) $5w(x + 2y)$ (f) $3x(y + 4z)$ (g) $6y(2x - 3z)$ (h) $5mn(3 + 4p)$
(i) $x(x + 6)$ (j) $y(y - 8)$ (k) $a(a + 1)$ (l) $y(y - 1)$ (m) $m(5 - m)$
(n) $x(3x - 1)$ (o) $5mn(4m + 3n)$ (p) $6xy^2(3x - 4)$ (q) $x(6 + 7x)$
(r) $t(17t - 6)$ (s) $3(x + 1)$ (t) $10(1 - y)$
2. (a) $5(2x + 1)$ (c) $8(1 + 5x^2)$ (e) $x^3(x^2 + 1)$ (g) $ab(a - 2c^4)$
(i) cannot be factored   (k) $(p + q + r)c$ (m) $m(m^2 + m + 1)$
(o) $(x + 3 + w)yz$ (q) $10m^2(10n + 5n^2 + 2)$ (r) $ab(5b + 10a + ab)$
(s) $3u(5v^2 - 2u + 4v)$ (t) cannot be factored   (u) $mx(m + x + 1)$
(v) cannot be factored

EXERCISES 17.2   (page 242)

1. (a) $x^2 + 5x + 6$ (b) $x^2 + 14x + 45$ (c) $m^2 + 10m + 24$
(d) $m^2 + 2m + 1$ (e) $x^2 - 7x + 10$ (g) $m^2 - 10m + 24$
(i) $y^2 + 2y - 15$ (k) $x^2 - 49$ (l) $x^2 - 25$ (m) $m^2 - 5m - 24$
(o) $y^2 + 5y - 24$ (q) $x^2 - 7x - 18$ (s) $m^2 - 12m + 20$
2. (a) $(x + 3)(x + 2)$ (b) $(x + 5)(x + 1)$ (c) $(x + 2)(x + 2)$
(d) $(x - 5)(x - 1)$ (e) $(t - 7)(t + 1)$ (f) $(x - 10)(x - 2)$
(g) $(m - 6)(m + 1)$ (h) $x(x + 3)$ (i) $(x + 4)(x - 3)$ (j) $a(a - 9)$
(k) $(y - 6)(y + 2)$ (l) $(m + 6)(m - 5)$ (m) $(x + 2)(x - 1)$
(n) $(y + 8)(y - 3)$
3. (a) $(x - 4)(x - 5)$ (c) cannot be factored   (e) $(x - 1)(x - 1)$
(g) $(x + 6)(x - 6)$ (i) $(x + 9y)(x - 9y)$ (k) cannot be factored
(m) $(x + 5)(x + 5)$ (o) cannot be factored   (q) $(5 + 2t)(5 - 2t)$
(s) $x(x - 9)$ (u) $(x - 5)(x - 9)$ (w) $m(m - 8)$
(y) $(10y + 7x)(10y - 7x)$

EXERCISES 17.3   (page 245)

1. (a) $3(m + 3)(m + 2)$ (c) $2(x - 10)(x + 1)$ (e) $3(x + 1)(x - 1)$
(g) $5(x - 3)(x - 3)$ (i) $7x(x - 3)$
2. (a) $(2x + 3)(x + 5)$ (b) $(3x + 2)(x + 1)$ (c) $(3x + 1)(x - 2)$
(d) $(5y - 2)(y - 3)$ (e) $(5n - 2)(n + 3)$ (g) $(3x + 2)(2x - 3)$
3. (a) $(5y + 2)(y + 4)$ (b) $(2m - 3)(m + 4)$ (c) $(6x + 1)(x + 10)$
(d) cannot be factored   (e) $2(y - 3)(y - 3)$ (g) $3(m + 5)(m - 5)$
(i) $2(3x - 5)(x - 1)$ (k) $(4x - 9)(x - 1)$ (m) $2(x^2 + 1)$

EXERCISES 17.4   (page 250)

1. (a) $-1, -6$ (b) $-1, -3$ (c) $1, 5$ (d) $2, 5$ (e) $-2, 3$
(f) $-8, 1$ (g) $-1, -1$ (h) $-1, 3$ (i) $-4, 4$ (j) $-7, 7$

(k) 0, −7   (l) 0, 8
2. (a) −4, 5   (b) 0, 1   (c) −1, 1   (d) cannot be factored
(e) cannot be factored   (g) 1, 8   (i) −5, −1/2
3. (a) −1, 3/5   (b) −2/3, −2   (c) −5/2, 1/2   (e) −3, 3   (g) 4, 6
(i) −4/3, −3/2   (k) cannot be factored   (m) −5, 3   (n) 0, 8
4. (a) cannot be factored   (c) 1, 7   (e) −2, 5/3   (f) −8, −4
(g) −1/2, 7   (i) 3/4, 2   (k) −1, −2/3   (m) −3, 3   (o) −2, 9
5. (a) length 21 meters, width 7 meters   (c) length 12 feet, width 8 feet
6. (a) 3 or −12   (b) −9 or 9   (c) −20 or 0

Chapter 17. REVIEW EXERCISES (page 251)

1. (a) $(m + n + p)x$ (b) $ab^2c(a + c)$ (c) $13b(2a + 3c)$
(d) $x(x − 10y)$ (e) $10(1 − 3y)$ (f) cannot be factored
(g) $(x + 5)(x − 2)$ (h) $(x − 13)(x − 1)$ (i) $x(x − 21)$ (j) $m(m + 30)$
(k) $(n + 10)(n − 10)$ (l) $(y + 1)(y − 1)$ (m) $(2x + 3)(2x + 1)$
(n) $2(x − 7)(x + 2)$ (o) $(4x − 9)(x + 1)$ (p) $3(x + 4)(x − 4)$
2. (a) −12, −1   (b) 1, 1   (c) 5, 5   (d) −2, 7   (e) cannot be factored
(f) −8, −2   (g) −12, −2   (h) −2, −7   (i) −2, 3
(j) cannot be factored   (k) 3, 5   (l) −2, 7/2   (m) −3, 1/5   (n) −1, 5
(o) −7, 7   (p) 0, 12

REVIEW PROBLEMS FOR PART 2 (page 252)

1. (a) +74   (c) −33   (e) +1
2. (a) −x   (c) −7x + 6
3. (a) −11/3   (c) −27/2   (e) 9/4
4. (a) x^{14} (b) y^{38} (c) m^{21} (d) x^4 (e) y^8 (f) m^6
5. (a) $x^2 + x + 7$, R3   (b) $x^2 − 6x + 3$
6. (a) $z + 10$ (b) $5x − 9$ (c) $7x/12$ (d) .07qr or 7qr/100

7. (a)   (c)

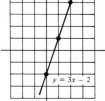

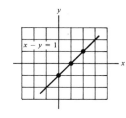

8. (a) $x = 3, y = −1$ (b) $x = −1, y = 1$
9. (b) $x \leq −6/5$

10. (a)

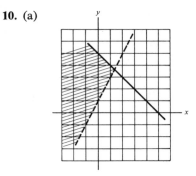

11. (a) $x(x + 1)$ (c) cannot be factored   (e) $(x + y)(x - y)$
(g) $(3x - 4)(x + 7)$
12. (a) 6, 6   (c) $-3, 5$ (e) 2, 9   (g) $-7, 7$

EXERCISES 18.1   (page 258)

1. (a) $\dfrac{10}{21}$ (b) $\dfrac{8}{45}$ (c) $\dfrac{xy}{15}$ (d) $\dfrac{7b}{6t}$ (e) $\dfrac{5x + 5}{14}$ (f) $\dfrac{2x - 8}{5x}$
(g) $\dfrac{4x + 20}{x^2}$ (h) $\dfrac{x^2 + 5x + 4}{x^2}$ (i) $\dfrac{3m^2 n^2}{5t}$ (j) $\dfrac{x^2 y^2}{16w^2}$
2. (a) $\dfrac{6x^2 y}{35}$ (c) $\dfrac{10r^3 w^2}{3x^2}$ (e) $\dfrac{35x}{y^5}$ (g) $-\dfrac{15x^2 y^2}{14z}$
3. (a) $-\dfrac{7x^4 + 7}{12x^2}$ (c) $\dfrac{x^2 + 4x + 3}{(x - 4)^2}$ (e) $\dfrac{5x^2 + 32x - 21}{3x^2}$

EXERCISES 18.2   (page 263)

1. (a) $\dfrac{1}{5}$ (b) $\dfrac{1}{50}$ (c) $\dfrac{3}{4}$ (d) $\dfrac{4}{5}$ (e) $\dfrac{x}{4}$ (f) $\dfrac{5}{2x^2}$ (g) $\dfrac{3}{x - 1}$
(h) $\dfrac{1}{2(1 - m)}$ (i) -1 (j) cannot be reduced   (k) $\dfrac{1}{x + 1}$ (l) $\dfrac{1}{x - 2}$
2. (a) $\dfrac{x + 3}{x + 1}$ (c) cannot be reduced   (e) $\dfrac{x + y}{4}$ (g) -2
(i) cannot be reduced   (k) $\dfrac{10}{3}$ (m) -3 (o) $\dfrac{x + 1}{2}$

EXERCISES 18.3   (page 265)

1. (a) 6   (b) $\dfrac{xy}{4}$ (c) $\dfrac{3}{2}$ (d) $\dfrac{10x}{a^2 b^2 c^2}$ (e) $\dfrac{x - 5}{x + 1}$ (f) $\dfrac{20}{(x + 3)7}$
(g) $\dfrac{1}{2x^2}$ (h) $\dfrac{2n - 8}{n + 4}$
2. (a) $x + 3$ (c) $-\dfrac{y}{2x}$ (e) $\dfrac{x + 4}{x}$ (g) $2x$ (i) $-\dfrac{4x}{3}$

EXERCISES 18.4   (page 268)

1. (a) $\dfrac{3x + 15}{7}$ (b) $\dfrac{1}{4}$ (c) $\dfrac{15xy}{x + 3}$ (d) $\dfrac{x + 1}{3x(x - 1)}$ (e) $-\dfrac{1}{2}$

(f) $\dfrac{m - 4}{3m}$

2. (a) $\dfrac{15}{x}$ (c) $\dfrac{x}{3}$ (e) $\dfrac{2x^2 + 2x + 24}{5(x + 3)(x + 4)}$ (g) $-\dfrac{2x}{x + 4}$

(i) $\dfrac{2(x + y)(x + 4)^2}{xy(x + 8)(x - 8)}$

EXERCISES 18.5   (page 272)

1. (a) $\dfrac{17}{30}$ (b) $\dfrac{27}{60}$ or $\dfrac{9}{20}$ (c) $\dfrac{7y + 5x}{xy}$ (d) $\dfrac{4x + 1}{x}$ (e) $\dfrac{3m + 5}{m}$

(f) $\dfrac{2 - 9c}{c}$ (g) $\dfrac{11}{2x}$ (i) $\dfrac{15w + 2x^3}{5w^2x^2}$ (k) $\dfrac{32cx + 9y}{60abc}$

(l) $\dfrac{3bc + 4ac + 5ab}{abc}$ (n) $\dfrac{6adf + 4bcf - 3bde}{12bdf}$ (o) $\dfrac{10y + 21x}{30x^2y^2}$

(q) $\dfrac{2bx + 3ay}{36a^3b^2}$ (r) $\dfrac{5xy + 4xz + 3yz}{x^2y^2z^2}$ (s) $\dfrac{3 + 4abc - 5ac}{a^2bc^2}$

2. (a) $\dfrac{26x + 14}{(3x + 2)x}$ (c) $\dfrac{2x^2 + 2x + 7}{x + 1}$ (e) $\dfrac{2x - 1}{x - 2}$ (g) $\dfrac{8x + 13}{(x + 1)(x + 2)}$

(i) $\dfrac{38 - 6x}{(x - 5)(x + 3)}$ (k) $\dfrac{9x - 1}{6x(x - 1)}$ (m) $\dfrac{8x + 15}{2x(x + 3)}$

3. (a) $\dfrac{2x + 12}{x^2 - 4}$ (c) $\dfrac{10x + 21}{(x + 1)(x + 2)}$ (d) $\dfrac{5x^2 - 3x + 6}{x(x - 7)(x - 2)}$

(e) $\dfrac{2x - 6}{(x + 4)(x + 2)(x + 2)}$ (g) $\dfrac{-21x^2 + 34x + 64}{2x^2(x + 4)}$

EXERCISES 18.6   (page 276)

1. (a) $\dfrac{33}{20}$ (c) $\dfrac{ad}{bc}$ (e) $\dfrac{3}{7x}$ (g) $\dfrac{xy + 3y}{xy + 2x}$ (i) $x^2 - 4x + 3$

2. (a) $\dfrac{x + 1}{x - 1}$ (c) $\dfrac{b + a}{b - a}$ (e) $\dfrac{3 + xy}{2y + 3x}$ (g) $\dfrac{3xy - 42y}{48x + 2xy}$ (i) $\dfrac{mxy + x^2}{y^2 - mxy}$

(k) $\dfrac{9b}{a + 4b}$

EXERCISES 18.7   (page 279)

1. (a) 4/45   (c) 19/4   (d) $-2/9$ (e) 8   (f) 31/10   (g) $-251/40$
(i) -5
2. (a) 5/2   (c) $-68/9$ (e) -4
3. (a) 76/11   (b) 5/4   (c) 39/8   (d) $-13/5$

Chapter 18. REVIEW EXERCISES (page 280)

1. (a) $\dfrac{x}{2y}$ (b) $-4x$ (c) $\dfrac{3x^2}{2}$ (d) $\dfrac{2(x-2)}{x+3}$ (e) $\dfrac{wz+xy}{xz}$ (f) $\dfrac{29}{35x}$

(g) $\dfrac{9x+xy+y}{(x+1)x}$ (h) $\dfrac{25z+45y-21x}{15xyz}$ (i) $\dfrac{x+5}{6x}$ (j) $-\dfrac{x+1}{2}$

(k) $12bxy^2$ (l) $\dfrac{bce+ade+acf}{ace}$ (m) $\dfrac{2t^2-2tu}{5u}$ (n) 2

(o) $\dfrac{10x+114y+3x^2}{6x^2y}$ (p) $\dfrac{25y+24x}{90xy}$ (q) -1 (r) $\dfrac{15+4x}{5x(x-1)}$

2. (a) $\dfrac{x^2+x}{xy-y}$ (b) $\dfrac{3m-1}{2+m}$

3. (a) 17 (b) 5

4. (a) $2x-2$ (c) $\dfrac{x-3}{x+2}$ (e) $\dfrac{x+2}{2(x-2)}$

EXERCISES 19.1 (page 286)

1. (a) x^5 (c) d^5 (d) 2^{25} (e) x^9 (g) m^{25} (h) x^{589} (i) 5^{615}
(j) cannot simplify   (k) a^{250} (l) cannot simplify   (m) cannot simplify
(o) $40x^{11}$ (p) $12x^9y^{11}$ (q) $-56x^{18}y^{13}$ (r) $-6x^{13}$ (s) $63a^7b^7$
(u) $-90m^9n^{11}$

2. (a) x^{12} (c) r^{28} (e) $r^{14}s^{14}$ (g) $x^{50}y^{50}$ (h) $p^{17}q^{17}r^{17}$
(i) $x^{20}y^{30}$ (k) a^8c^{14} (l) $125x^9y^{24}$ (m) cannot simplify   (o) $64x^2y^{12}$
(p) $-27x^{15}y^{30}$ (q) $16m^{12}n^4p^8$ (s) $-125a^9b^{21}$

3. (a) $75m^9n^{16}$ (b) $648x^{31}y^{42}$ (c) $-8p^{52}q^{13}$ (e) cannot simplify

(g) $-288b^{11}c^{33}d^8$ (h) cannot simplify   (i) $\dfrac{x^9}{y^9}$ (k) $\dfrac{32m^5}{n^5}$ (m) $\dfrac{x^8y^{10}}{z^6}$

(n) $\dfrac{b^6c^{21}}{a^{12}d^{15}}$ (o) $\dfrac{m^{28}n^{32}p^8}{625t^8}$

4. (a) 8 (b) 64 (c) 625 (d) 243 (e) 18 (f) 48 (g) 256
(h) 729 (i) $16/81$ (j) $27/64$ (k) 625 (l) 125

5. (a) 2^{10}

EXERCISES 19.2 (page 290)

1. (a) $1/8$ (c) $1/27$ (e) $1/6$ (g) $3/5$ (i) $7/8$ (k) $1/36$

2. (a) $\dfrac{1}{x^5}$ (c) $\dfrac{4}{m^6}$ (e) $-\dfrac{7}{t^4}$ (g) $\dfrac{1}{9x^2}$ (i) $\dfrac{1}{4n^7}$ (k) $\dfrac{2}{7m^4}$

(m) $\dfrac{5}{x^7y^4}$ (o) $\dfrac{7n^7}{m^6}$ (q) $\dfrac{a^2c^4}{b^3d^5}$ (s) $-\dfrac{8n^2}{m^7p^9}$

3. (a) $\dfrac{3y}{x^4z^5}$ (c) $\dfrac{5}{2a^4b^6c}$ (e) $\dfrac{c^8}{a^4b^3}$ (g) $\dfrac{7b^4y^3}{a^6c^2x^5}$ (i) $-\dfrac{5v^3z}{w^2x^4y^5}$

4. (a) $\dfrac{1}{x^9}$ (c) $\dfrac{6}{t^{13}}$ (e) $x^{12}y^{20}$ (g) $\dfrac{x^{18}y^{15}}{8}$ (i) $\dfrac{a^6b^{14}c^{10}}{25}$

EXERCISES 19.3 (page 292)

1. (a) x^{30} (c) y^3 (e) $\dfrac{1}{m^3}$ (g) x^4y^3 (i) x^4y^3 (k) $\dfrac{1}{a^4b^{11}}$

(m) $\dfrac{1}{p^5q^3r}$

2. (a) $\dfrac{4x^7}{y^7}$ (c) $\dfrac{4x^8}{5yz}$ (e) $\dfrac{4x^7}{5w^7}$ (g) $\dfrac{2x^7z^4}{y^4}$ (i) $\dfrac{m^2}{np^5}$ (k) $\dfrac{1}{7w^4x^6y^9z^5}$

EXERCISES 19.4 (page 294)

1. (a) 1   (b) 3   (c) 1   (e) 5   (g) 6   (h) 2   (i) 7   (k) -3
(m) 1   (n) $5x-1$ (o) 0

EXERCISES 19.5 (page 298)

1. (a) 5   (c) 10   (e) 2   (g) 3   (i) 5   (j) 10   (k) 63   (m) 15
(o) 1   (q) 3/5   (s) 2/3   (u) 6
2. (a) 8   (c) 64   (e) 25   (g) 20   (h) 133   (i) 10   (j) 10
(k) 270   (l) 264
3. (a) 1/7   (c) 1/3   (e) 1/32   (g) $2\frac{1}{2}$ (i) 28
4. (a) 32   (c) 243   (e) 1   (g) 1/27   (i) 1
5. (a) x^4y^2 (c) $7ab^3$ (d) $-2x$ (g) x^2 (i) $25r^2s^6$ (k) $1331x^{15}y^{24}$
(l) $5a^3y^5$ (n) $2xy^2z^3$ (o) $12a^5b^3$ (q) $6m$ (r) $-4xy^2$ (s) $2x^2$
(t) $-2y$ (w) $3a^3b^7$
6. (a) $x^{\frac{3}{2}}$ (c) $x^{\frac{5}{3}}$ (e) $y^{\frac{7}{4}}$
7. (a) $\sqrt{x^5}$ (c) $\sqrt[3]{n^7}$ (e) $\sqrt[5]{t^4}$

EXERCISES 19.6 (page 302)

1. (a) 4.7×10^6 (c) 3.5×10^{11} (e) 4.53×10^{17} (g) 2.3×10^{-4}
(i) 7.0×10^{-7} (k) 1.3×10^{-12} (m) 1.2×10^{13}
2. (a) 340,000,000   (c) 942,000,000,000   (e) .0000000013
(g) .00000000050   (i) 75.63
3. (a) 4.5×10^9 (b) 1.2×10^{19} (c) 6.2×10^{22} (d) 7.31×10^{12}
(e) 4.73×10^{-10} (f) 8.41×10^{-14} (g) 6.21×10^{18} (h) 4.56×10^{-22}
(i) 3.0×10^{24} (j) 5.2×10^{-22}
4. (a) 5.0×10^9 (b) 2.0×10^3 (c) 2.0×10^{-11} (d) 10^{11}

Chapter 19. REVIEW EXERCISES (page 303)

1. (a) x^{12} (c) x^{48} (e) $15x^{16}$ (g) $8m^{12}y^{15}$ (i) $45x^{15}y^{23}$ (k) $\dfrac{81x^8}{y^{20}}$
(m) t^{21} (o) $-32x^{10}y^5$ (q) $-35a^4b^{11}c^4$

2. (a) $\dfrac{1}{x^7}$ (c) $\dfrac{4}{m^2}$ (e) $-\dfrac{7}{x^3}$ (g) $\dfrac{1}{3y^6}$

3. (a) $x^3 y^7$ (c) $\dfrac{y}{x^3}$ (e) $\dfrac{1}{16x^2}$ (g) $\dfrac{5y^4}{x^3 z}$ (i) $-\dfrac{3}{x^7 y^8}$ (k) $\dfrac{4s^{12}u^5}{r^6 t^3}$

(m) $\dfrac{1}{x^{10}}$ (o) $\dfrac{3}{x^9}$ (q) $b^{12}c^{15}$ (s) $\dfrac{m^{13}}{n^{10}p^4 q^3}$

4. (a) 24 (b) 1 (c) 8 (d) -2 (e) 10 (f) 2 (g) 2 (h) 7/8
(i) 12 (j) 32 (k) 9 (l) 24 (m) 1/16 (n) 5/9 (o) 1
(p) 128

5. (a) $7m^5 n^{12}$ (c) $16x^2 y^8 z^{20}$

EXERCISES 20.1 (page 311)

1. (a) $r = \dfrac{d}{2}$ (b) $r = \dfrac{d}{t}$ (c) $d = \dfrac{C}{\pi}$ (d) $r = \dfrac{C}{2\pi}$ (e) $l = \dfrac{A}{w}$

(f) $h = \dfrac{V}{lw}$ (g) $r = \dfrac{i}{pt}$ (h) $h = \dfrac{V}{\pi r^2}$ (i) $h = \dfrac{2A}{b}$ (j) $b = \dfrac{2A}{h}$

(k) $m = \dfrac{2E}{v^2}$ (l) $a = \dfrac{2s}{t^2}$

2. (a) $x = \dfrac{y}{6}$ (b) $x = -\dfrac{y}{5}$ (c) $x = \dfrac{y - b}{a}$ (e) $a = \dfrac{y - b}{x}$

(g) $w = \dfrac{P - 2l}{2}$ (i) $h = \dfrac{3V}{\pi r^2}$ (k) $x = \dfrac{5y + 2}{4}$

3. (a) $x = \dfrac{8 - 7y}{5}$ (c) $b = -ax$ (d) $x = -\dfrac{b}{a}$ (e) $x = \dfrac{-by - c}{a}$

(g) $E = IR$ (h) $R = \dfrac{E}{I}$ (k) $m_1 = \dfrac{Fd^2}{m_2}$ (l) $h = \dfrac{2A}{b_1 + b_2}$

(m) $d = \dfrac{l - a}{n - 1}$ (n) $l = \dfrac{2S - na}{n}$ (o) $p = \dfrac{R}{1 + rt}$ (q) $p = \dfrac{A - m}{m}$

4. (a) $E = \dfrac{C(R + nr)}{n}$ (b) $R = \dfrac{nE - Cnr}{C}$ (c) $r = \dfrac{nE - CR}{Cn}$

(d) $n = \dfrac{CR}{E - Cr}$ (e) $b_2 = \dfrac{2A - hb_1}{h}$ (g) $p = \dfrac{fq}{q - f}$

(i) $l = \dfrac{Sr - S + a}{r}$ (j) $r = \dfrac{S - a}{S - l}$ (k) $t = \dfrac{a - p}{pr}$ (m) $x = \dfrac{by}{y - a}$

Chapter 20. REVIEW EXERCISES (page 313)

1. (a) $e = \dfrac{p - 3d}{4}$ (b) $w = \dfrac{t - uv}{u}$ (c) $a = cm - b$ (d) $b = 2a - c$

2. (a) $v = \dfrac{T - 4n}{nr}$ (b) $a = \dfrac{cx + dx + 2b}{3}$ (c) $x = \dfrac{wy}{y - w}$

(d) $c = \dfrac{a + b - dx}{x}$

EXERCISES 21.2 (page 318)

1. (a) $2\sqrt{2}$ (b) $2\sqrt{3}$ (c) $3\sqrt{3}$ (d) 6   (e) $5\sqrt{2}$ (f) $2\sqrt{15}$
(g) $2\sqrt{7}$ (h) $2\sqrt{11}$ (i) $3\sqrt{5}$ (j) 10   (k) $5\sqrt{5}$ (l) $3\sqrt{7}$
(m) $4\sqrt{2}$ (n) $4\sqrt{3}$ (o) $2\sqrt{13}$ (p) $2\sqrt{14}$ (q) $10\sqrt{2}$ (r) 30
2. (a) $2\sqrt{17}$ (c) $3\sqrt{10}$ (e) cannot be simplified   (g) $4\sqrt{7}$ (i) $8\sqrt{2}$
(k) 50
3. (a) 7   (c) $5\sqrt{5}$ (e) 9   (g) $2\sqrt{5}$ (i) 10
4. (a) $10\sqrt{2}$ (b) 15   (e) $20\sqrt{17}$ (g) $6\sqrt{5}$ (i) $50\sqrt{3}$
5. (a) $\sqrt{2}$ (c) $\sqrt{5}/6$ (e) $\sqrt{5}$
6. (a) $3\sqrt{13}$ (c) $7\sqrt{2}$ (e) $2\sqrt{26}$ (g) $9\sqrt{2}$
(i) cannot be simplified   (k) $7\sqrt{3}$ (m) $3\sqrt{17}$ (o) cannot be simplified
(q) $3\sqrt{31}$

EXERCISES 21.3 (page 320)

1. (a) $2\sqrt{7}$ (c) $14\sqrt{2}$ (e) $11\sqrt{2}$ (f) 6   (g) $10\sqrt{2}$ (i) $5\sqrt{2}$
(k) cannot combine   (l) 104   (m) $32\sqrt{3}$ (o) $11\sqrt{5}$
(p) cannot combine   (q) $5\sqrt{10}$ (r) $12\sqrt{3}$ (s) $3\sqrt{5} + \sqrt{10}$

EXERCISES 21.4 (page 324)

1. (a) $\dfrac{3\sqrt{2}}{2}$ (c) $\sqrt{5}$ (e) $\dfrac{3\sqrt{10}}{2}$ (g) 15   (i) $\dfrac{2\sqrt{3}}{5}$
2. (a) $\dfrac{2}{5\sqrt{2}}$ (c) $\dfrac{4}{3\sqrt{2}}$ (e) $\dfrac{1}{\sqrt{6}}$
3. (a) $\dfrac{\sqrt{3}}{3}$ (c) $\dfrac{2\sqrt{14}}{7}$ (e) $\dfrac{1}{3}$
4. (a) $1 + \sqrt{5}$ (c) $2 + \sqrt{7}$ (e) $\dfrac{1 + \sqrt{6}}{3}$
5. (a) $\sqrt{30}$ (c) $3 + 2\sqrt{2}$ (e) 18
(g) $48 - 6\sqrt{3} - 8\sqrt{13} + \sqrt{39}$
6. (a) $\sqrt{5} - 2$ (b) $3\sqrt{2} - 3$ (c) $2\sqrt{7} + 4$ (d) $\dfrac{28 - 7\sqrt{10}}{6}$

EXERCISES 21.5 (page 328)

1. (a) 81   (c) 17   (e) 17/20   (g) 4   (i) 10/11
2. (a) 7   (c) no solution   (e) 23/2   (g) 3/4   (h) 11   (i) -1
(j) 447/7

Chapter 21. REVIEW EXERCISES   (page 329)

1. (a) $2\sqrt{6}$ (c) $10\sqrt{5}$ (e) $3\sqrt{6}$ (f) $2\sqrt{19}$ (g) $4\sqrt{10}$
(i) cannot be simplified   (j) $3\sqrt{19}$ (k) $9\sqrt{3}$
2. (a) $26\sqrt{2}$
3. (a) $7\sqrt{5}/5$
4. (a) $\sqrt{11}/11$
5. (a) $119/3$

EXERCISES 22.1   (page 332)

1. (a) ± 8 (b) ± 1 (c) ± 6 (d) ± 10 (e) $\pm\sqrt{7}$ (f) $\pm\sqrt{11}$
(g) $\pm\sqrt{15}$ (h) $\pm\sqrt{47}$
2. (a) $-1 \pm \sqrt{11}$ (c) $3 \pm \sqrt{19}$ (e) $4 \pm 2\sqrt{2}$ (g) $-1 \pm 2\sqrt{7}$
(i) $2 \pm \sqrt{3}$ (k) $-1 \pm \sqrt{7}$

EXERCISES 22.2   (page 336)

1. (a) $-4 \pm \sqrt{15}$ (c) $-3 \pm \sqrt{2}$ (e) $-6 \pm \sqrt{21}$ (g) $4 \pm 2\sqrt{2}$
(i) $-5 \pm \sqrt{29}$
2. (a) $\dfrac{3 \pm \sqrt{5}}{2}$ (c) $\dfrac{-5 \pm \sqrt{21}}{2}$ (e) $\dfrac{-3 \pm \sqrt{41}}{2}$ (g) $\dfrac{-3 \pm \sqrt{11}}{2}$

(i) $\dfrac{-3 \pm \sqrt{29}}{10}$

EXERCISES 22.3   (page 339)

1. (a) $\dfrac{-3 \pm \sqrt{5}}{2}$ (b) $\dfrac{-5 \pm \sqrt{17}}{2}$ (c) $\dfrac{5 \pm \sqrt{37}}{2}$ (d) $\dfrac{3 \pm \sqrt{5}}{2}$

(e) $\dfrac{-5 \pm \sqrt{97}}{6}$ (f) $\dfrac{-3 \pm \sqrt{33}}{4}$

2. (a) $2 \pm \sqrt{3}$ (c) $-3 \pm 2\sqrt{2}$ (e) $\dfrac{1 \pm \sqrt{22}}{3}$ (g) $\dfrac{2 \pm \sqrt{3}}{2}$

3. (a) $\dfrac{-7 \pm \sqrt{69}}{2}$ (c) $\dfrac{3 \pm \sqrt{41}}{2}$ (e) $\dfrac{-9 \pm \sqrt{89}}{2}$ (g) $-3/5, 2$

(i) $-6, 0$ (k) $\pm\sqrt{3}$

EXERCISES 22.5   (page 343)

1. (a) $4i$ (c) $i\sqrt{3}$ (e) $2i\sqrt{6}$ (g) $3i\sqrt{3}$ (i) $4i\sqrt{5}$
2. (a) $\pm 2i$ (c) $1 + 2i$ (d) $\dfrac{-1 \pm i\sqrt{3}}{2}$ (e) $\dfrac{-1 \pm i\sqrt{7}}{2}$

(g) $\dfrac{3 \pm i\sqrt{23}}{4}$ (i) $1 \pm 2i\sqrt{2}$

EXERCISES 22.6 (page 345)

1. (a) real and unequal (c) imaginary and unequal (e) real and equal
(g) real and unequal (h) imaginary and unequal (i) real and unequal
(l) real and unequal (m) real and unequal (n) real and unequal

EXERCISES 22.7 (page 348)

1. (a)

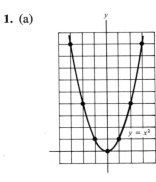

(b)

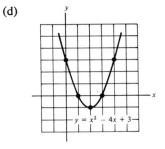

(c)

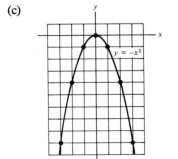

(d)

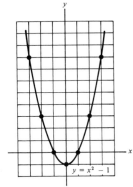

(e)

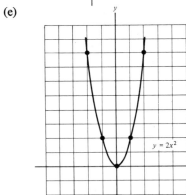

(f)

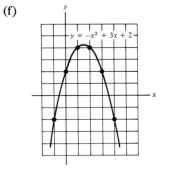

2. (a)

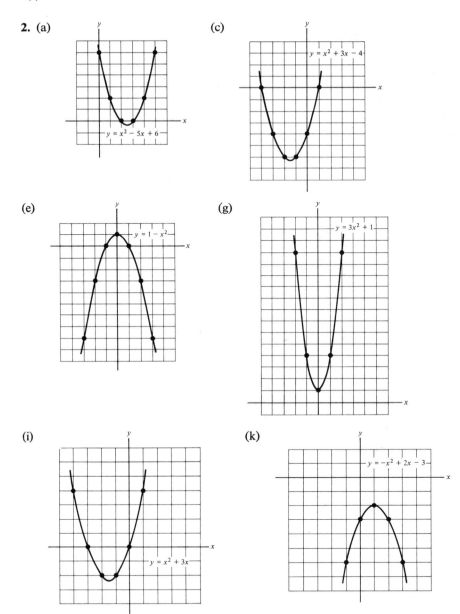

(c)

$y = x^2 + 3x - 4$

(e)

$y = 1 - x^2$

(g)

$y = 3x^2 + 1$

(i)

$y = x^2 + 3x$

(k)

$y = -x^2 + 2x - 3$

Chapter 22. REVIEW EXERCISES (page 348)

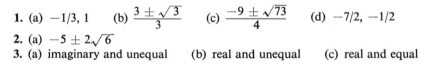

1. (a) $-1/3, 1$ (b) $\dfrac{3 \pm \sqrt{3}}{3}$ (c) $\dfrac{-9 \pm \sqrt{73}}{4}$ (d) $-7/2, -1/2$

2. (a) $-5 \pm 2\sqrt{6}$

3. (a) imaginary and unequal   (b) real and unequal   (c) real and equal

4.

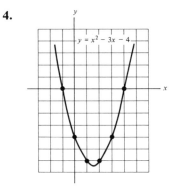

$y = x^2 - 3x - 4$

EXERCISES 23.1 (page 350)

1. (a) 5   (c) −3   (e) −1   (f) 1/3   (g) 4   (i) 1   (k) 1/3
(m) 1/2   (o) −5/4   (q) −2/3

EXERCISES 23.2 (page 357)

1. (a) 1/2   (c) 3   (d) −1   (e) 5   (g) −4   (h) 5/4   (i) −1
(k) −7/9

2. (a)

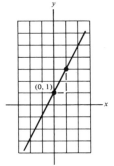

(0, 1)

(c)

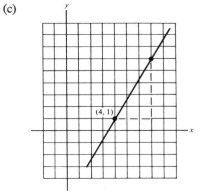

(4, 1)

(e)

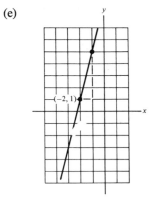

(−2, 1)

(g)

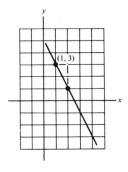

(1, 3)

(i)

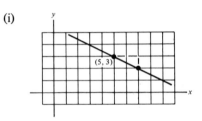

EXERCISES 23.3   (page 360)

1. (a) $m = 2, b = -1$ (c) $m = 1, b = 2$ (e) $m = 3, b = -3$
(g) $m = 2, b = -6$ (i) $m = 1/3, b = 4/3$ (j) $m = 2, b = 0$
(k) $m = 2, b = -3$ (m) $m = 2/3, b = -6$ (o) $m = -2/3, b = 1/3$
(p) $m = -1, b = 2$

2. (a)

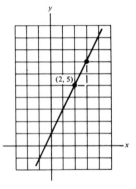

(c)

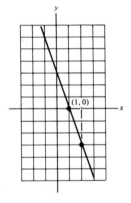

(e)

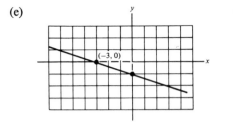

(g)

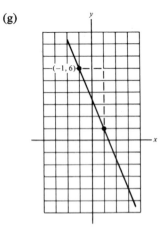

3. (a)

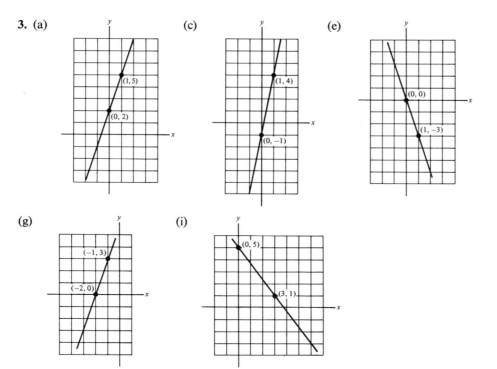

Chapter 23. **REVIEW EXERCISES (page 361)**

1. (a) 3   (b) −1   (c) 5/2   (d) 1/3
2. (a) 6   (b) −3
3. (a) $m = -6, b = 3$ (b) $m = 2, b = -9$ (c) $m = 2/9, b = -16/9$
(d) $m = 3/7, b = 3$

EXERCISES 24.1 (page 364)

1. (a) {2, 4, 6, 8, 10, 12, 14, 16, 18, 20, 22, 24}   (c) {7, 8, 9, 10, 11, 12, 13, 14, 15}
(e) {March, May}
2. (a) $5 \in \{1, 3, 5, 7, 9, 11, 13, 15\}$ (b) $7 \notin \{18, 16, 14, 12, \ldots\}$
(c) $x \notin \{a, e, i, o, u\}$
3. (a) true   (b) false   (c) false   (d) false   (e) false   (f) false
(g) false

EXERCISES 24.2 (page 365)

1. (a) $\{x \mid x > 13\}$ (b) $\{x \mid x < 1/2\}$ (d) $\{x \mid x \le 2\}$ (e) $\{x \mid x \ge 0\}$
(g) $\{x \mid -5 \le x \le 16\}$ (i) $\{x \mid x \text{ even}, 3 < x < 95\}$
(k) $\{x \mid -25 \le x \le 25, x \ne 0\}$

2. (a) all the positive numbers   (b) all numbers except zero
(d) 6, 8, 10, 12, 14, . . .   (f) 0, 1, 2, 3, 4, 5, . . .
(i) all numbers between 3 and 9 exclusive
(k) all numbers between 3 and 9, including 3 but excluding 9

EXERCISES 24.3   (page 366)

1. (a) {5, 6, 8, 12, 19, 35}   (c) {3, 4, 7, 8, 12, 14}   (e) {1, 2, 3, 4, 5, 6}
2. (a) {19}   (c) $\varnothing$ (e) {1, 3, 4}
3. (a) {x, y, z}   (c) {a, b, c, d, e, f}

EXERCISES 24.4   (page 368)

1. (a) false   (b) false   (d) true   (f) true   (i) true   (j) true
2. (a) {a, b}, {a}, {b}, { }   (b) {x, y, z}, {x, y}, {x, z}, {y, z}, {x}, {y}, {z}, { }

EXERCISES 24.5   (page 369)

1. (a) {7, 9, 11, 13, 15, 17}   (b) {1, 3, 5, 9, 11, 13, 17}   (c) $\mathcal{U}$ (d) { }
(e) {1, 3, 5, 7, 9, 11, 15, 17}

EXERCISES 24.6   (page 374)

1. (a) 1, 2   (b) 2, 3   (c) 1, 2, 3   (d) 2   (e) 1, 2, 3, 4   (f) 3, 4
(g) 1, 4   (h) 4   (i) 1, 3, 4
2. (a) 1, 2, 3, 4   (b) 3   (f) 5, 6   (g) 1, 2, 4, 5, 6
3. (a) 1, 2, 4, 5   (d) 1, 2, 3, 4, 5, 6, 7, 8   (e) 2, 5   (h) 5
(i) 1, 2, 3, 4, 5, 6   (l) 1, 2, 3, 4, 5, 6, 7

4. (a)   (c)

     (e)   (g)

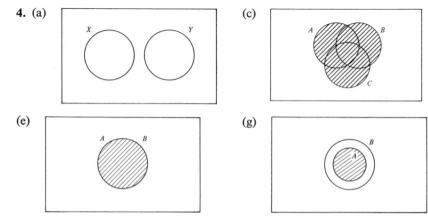

(j)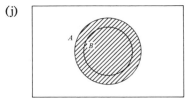

EXERCISES 24.7 (page 378)

1. (a) yes   (b) yes   (c) no   (d) yes   (e) yes   (f) no
2. (a) yes   (b) yes   (c) no   (d) yes   (e) yes   (f) yes   (g) no
(h) no

EXERCISES 24.8 (page 380)

1. (a) -2 (b) 1   (c) -5 (d) -20 (e) $3a - 5$ (f) -4
(g) -6 (h) $9m - 5$ (i) $3x - 2$ (j) $3x - 20$
2. (a) 19   (b) 32   (c) 179   (d) -1 (e) -8 (f) -16
(g) $m^2 + 8m - 1$ (h) 13/4   (i) $-47/16$ (j) $x^2 + 12x + 19$

Chapter 24. REVIEW EXERCISES (page 380)

1. All numbers between 4 and 9, including 4 but excluding 9.
2. (a) $\{1, 2, 3, 4, 5, 6\}$ (b) $\varnothing$ (c) $\{1, 2, 3, 5, 7\}$ (d) $\{5\}$
(e) $\{1, 2, 3, 7\}$ (g) $\varnothing$
(h) Yes, all elements of A are also elements of the universe.
4. (a) no   (b) yes
5. (a) 33   (b) 5

REVIEW PROBLEMS FOR PART 3 (page 382)

1. (a) 20/7   (c) $\dfrac{20 + 9y^2}{24xy}$

3. (a) $64x^{30}y^{18}$ (c) $-\dfrac{5y^6}{x^{19}}$

4. (a) $1\frac{1}{2}$ (c) 27

5. (a) $m = \dfrac{y - b}{x}$ (c) $a = \dfrac{x(c + d)}{b}$ (e) $d = \dfrac{a - b + ce}{c}$

6. (a) $2\sqrt{13}$
8. $-63/2$
9. (a) 3/2, 7   (b) $-2 \pm i$
11. (a) 7
12. (a) $\{3, 7, 11\}$ (c) $\{3, 5, 9\}$
14. (a) 2   (b) 2

EXERCISES C.1   (page 405)

1. (a) $x = 16$ (c) $x = 10$ (e) $m = 12\frac{1}{2}$ (g) $x = 12/7$ (i) $x = 75/14$
2. (a) 80   (b) 200   (c) 120   (d) 60   (e) $17\frac{1}{2}$
3. (a) 70 feet and 30 feet   (c) 25,000   (e) \$1.45   (f) 1050 square feet
(g) 4 pounds   (h) 8 gallons   (i) 20 inches   (k) 5 teaspoons

EXERCISES C.2   (page 409)

1. (a) 14   (b) 7   (c) 133   (d) 2   (e) 10/7
2. (a) 10   (b) 3   (c) 30   (d) 1/2   (e) 60
3. (a) 22   (c) 126
4. (a) 112   (c) 8
5. (a) $c = 6$; \$192   (b) 17 hours   (d) $c = 2800$; 175; 350

EXERCISES D.1   (page 416)

1. (a) 8/10 or 4/5   (b) 6/10 or 3/5   (c) 8/6 or 4/3
2. (a) $\dfrac{3}{\sqrt{13}}$ (b) $\dfrac{2}{\sqrt{13}}$ (c) $\dfrac{3}{2}$ (d) $\dfrac{2}{\sqrt{13}}$ (e) $\dfrac{3}{\sqrt{13}}$ (f) $\dfrac{2}{3}$
3. (a) $\dfrac{1}{\sqrt{2}}$ (b) $\dfrac{1}{\sqrt{2}}$ (c) 1   (d) $\dfrac{1}{\sqrt{2}}$ (e) $\dfrac{1}{\sqrt{2}}$ (f) 1
5. 86 feet   **6.** 848 feet   **9.** 70 feet

EXERCISES E.1   (page 425)

1. (a) 12   (b) 19   (c) 26   (d) 1111   (e) 100001   (f) 1010111
(g) 15   (h) 44   (i) 63   (j) 35   (k) 62   (l) 143   (m) 171
(n) 249   (o) 316   (p) 26   (q) 60   (r) D0   (s) 10001011   (t) 213
(u) 46   (v) 231   (w) 1502
2. Base 3: 1, 2, 10, 11, 12, 20, 21, 22, 100, 101, 102, 110
Base 4: 1, 2, 3, 10, 11, 12, 13, 20, 21, 22, 23, 30
Base 5: 1, 2, 3, 4, 10, 11, 12, 13, 14, 20, 21, 22
3. unacceptable in bases less than 6

EXERCISES E.2   (page 429)

1. (a) 1100   (b) 10111   (c) 10010   (d) 110000   (e) 1111   (f) 11010
2. (a) 101   (b) 1017   (c) 133   (d) 1450   (e) 135
3. (a) AE   (b) 178   (c) 1AAB   (d) 1663   (e) CB3

Index

A

Abscissa, 201
Absolute value, 136
Addend, 24
Addition:
    fractions, 62-65, 72-74, 268-72
    involving zero, 115
    like terms, 141
    radicals, 319-20
    signed numbers, 110-15, 126-29
    table, 6
    unsigned numbers, 5-8
Addition method, 212-14, 220-22
Algebraic expressions, 140-48
Approximately ($\doteq$), 83
Area, 32
Area of a circle, 36
Area of a rectangle, 32-33
Area of a trapezoid, 33-34
Associative property, 46
Axes, 200

B

Base, 25
Base 2, 419-22
Base 8, 422-24
Base 10, 418-19
Base 16, 424-25
Binary number system, 419-22
Binary operation, 45
Binomial, 141
Braces (set), 363

C

Cartesian coordinates, 198
Celsius, 102, 305, 307
Centi-, 98
Centimeter, 99

471

Checking:
    equations with fractions, 279
    equations with radicals, 326-28
    linear equations, 174-76
    quadratic equations, 246-47
    systems of equations, 218-19
Circumference of a circle, 34
Closure property, 46
Coefficient, 141
Combining like terms, 142-43
Common denominator, 63
Commutative property, 45
Complement of a set, 369
Completing the square, 332-36
Complex fractions, 273-76
Composite number, 69
Constant, 140
Coordinates, 198, 201
Cosine, 412
Counting number, 109
Cube of a number, 25
Cube root, 295

D

Deca-, 104
Deci-, 104
Decimal number system, 418-19
Decimals, 74-75, 78-86
Denominator, 52, 257
Descartes, René, 198
Difference, 24
Difference of two squares, 241-42
Digit, 4
Directed number, 110
Direct variation, 406-8
Discriminant, 343
Disjoint sets, 366
Distributive property, 47-49, 142,
    146-51, 234
Dividend, 24

Divisibility tests, 56-58
Divisible, 56
Division:
    algebra, 152-56
    fractions, 60-62, 265-68,
      273-76
    involving zero, 22-24
    signed numbers, 130-33
    unsigned numbers, 15-20
Divisor, 24, 56

E

Element of a set, 362
Elimination method, 212-14,
    220-22
Empty set, 366
Equation, 159
Equations:
    check, 174-76, 218-19, 246-47
    fractional, 277-79
    linear, 159-78
    manipulation, 305-13
    quadratic, 246-50, 330-43
    radical, 325-28
    systems, 211-22
Evaluation of expressions, 128-35
Evaluation of formulas, 32-38
Even number, 123
Exponents:
    fractional, 294-99
    negative, 287-92
    positive integral, 24-25, 124-25,
      144-45, 154, 282-86
    zero, 293-94
Expression, 140
Extraneous root, 328
Extremes of a proportion, 308,
    402
Euler, Leonhard, 379
Euler's circles, 369

F

Factor, 25, 56
Factoring, 234-50, 259-68
Fahrenheit, 102, 305, 307
Finite set, 364
Formula manipulation, 305-13
Fractional equations, 277-80
Fractional exponents, 294-99
Fractions:
    addition, 62-65, 72-74, 268-72
    common denominator, 63
    complex, 273-76
    conversion to decimals, 79-80
    denominator, 52, 257
    division, 60-62, 265-68, 273-76
    equations, 277-79
    exponents, 294-99
    improper, 52, 67-68
    LCM, 70-73, 269-72, 277-79
    meaning, 51-52
    multiplication, 52-56, 257-58,
        264-65
    proper, 52, 67
    reduction, 54-56, 58-59, 259-68
    subtraction, 65-66, 270
Function, 376
Function notation, 379-80

G

Geometric formulas, 32-36
Gram, 101
Graphs, 196-210, 228-32, 345-48
Greater than, 136, 223

H

Hecto-, 104
Hexadecimal number system,
    424-25

Hipparchus, 411
Hypotenuse, 40-42, 411

I

i, 342
Imaginary number, 342
Improper fraction, 52, 67-68
Indeterminate, 23
Inequalities, 136-37, 223-33
Infinite set, 364
Integers, 110
Intercepts, 358-60
Interest, 182
Intersection of lines, 207-8
Intersection of sets, 366
Interval notation, 224-25
Inverse variation, 408-9
Irrational number, 315

K

Kilo-, 98
Kilogram, 101
Kilometer, 98

L

LCM, 70-73, 269-72, 277-79
Least common multiple, 70-73,
    269-72, 277-79
Less than, 137, 223
Like terms, 141
Linear equations, 159-78
Lines, 201-10, 211-12, 217,
    228-32, 349-61

Liter, 100
Long division, 15-19, 154-56
Lowest common denominator, 72, 268-69

M

Magnitude, 110, 136
Means of a proportion, 308, 402
Meter, 97
Metric system, 97-106
Milli-, 98
Milliliter, 100
Millimeter, 99
Minuend, 24
Mixed number, 67-69
Monomial, 141, 145
Multiplicand, 24
Multiplication:
    algebraic expressions, 144-151
    fractions, 52-56, 257-58,
        264-65
    radicals, 323-24
    signed numbers, 120-25
    table, 12
    unsigned numbers, 11-14
    by zero, 13
Multiplier, 24

N

Natural number, 109
Negative exponent, 287-92
Negative integers, 110
Negative number, 109-10
Not equal to, 137
Null set, 366

Number:
    counting, 109
    directed, 110
    imaginary, 342
    integer, 110
    irrational, 315
    natural, 109
    negative, 109-10
    prime, 69
    signed, 109-39
    whole, 3, 109
Number lines, 110
Number structure, 3-5
Number systems, 418-29
    binary number system, 419-22
    decimal number system, 418-19
    hexadecimal number system,
        424-25
    octal number system, 422-24
Numerator, 52, 257

O

Octal number system, 422-24
Odd number, 123
Opposite, 114, 160, 262
Ordered pair, 198, 376
Order of operations, 27-30,
    128-29
Ordinate, 201
Origin, 197

P

Parabola, 345-47
Parallel lines, 208, 212
Percent, 80-89
Perfect square, 40, 314
Perimeter, 32

Perimeter of a triangle, 34
π, 34, 306
Place value, 4
Plotting points, 196-201
Plus or minus, 330
Points, 196-201
Positive integers, 110
Powers of directed numbers,
   124-25
Powers of whole numbers, 24-25
Prime factorization, 69-70
Prime numbers, 69-71
Product, 24
Properties of whole numbers,
   45-48
Proportions, 308, 402-5
Ptolemy, 411
Pythagorean Theorem, 41, 413

multiplication of, 323-24
rationalizing, 320-22, 324
simplifying, 315-18
Ratio, 401
Rationalization, 320-22, 324
Rational number, 315
Real number, 341
Reciprocals, 164-66
Rectangular coordinates, 198
Reducing fractions, 54-56, 58-59,
   259-68
Relation, 376
Remainder, 17, 24
Right triangle, 40-44, 411-17
Root of a linear equation, 159
Roots of a quadratic equation,
   246-50, 330-47
Rounding, 20-22

Q

Quadrant, 199
Quadratic equations, 246-50,
   330-43
   solution by completing square,
     332-36
   solution by factoring, 246-50
   solution by formula, 336-43
Quadratic expression, 237
Quadratic formula, 336-44
Quotient, 24

R

Radical, 40, 314
Radicals, 294-98, 314-28, 331,
   334-44
   combining, 319-20
   equations involving, 325-28

S

Scientific notation, 300-302
Set builder notation, 364-65
Sets, 362-75
Short division, 152-54
Signed numbers, 109-39
   addition, 110-15, 126-29
   division, 130-33
   expressions, 128-35
   multiplication, 120-25
   powers, 124
   subtraction, 116-20
Sine, 412
Slope, 349-57
Smallest common denominator,
   72, 277
Solution:
   equations with fractions, 277-79
   equations with radicals, 325-28
   inequalities, 225-31
   linear equations, 159, 176-77

Solution *(cont.)*
  quadratic equations, 246-50,
    330-39, 342-43
  systems of equations, 212
Square, of a number, 25, 39
Square root, 39-40, 294, 314-28
  definition, 39, 314
  simplification, 315-18
Squaring both sides of an
    equation, 325
Standard notation, 300-302
Straight lines, 200-210, 211-12,
    217, 228-32, 349-61
Subscripts, 310
Subset, 367
Substitution, 215-17
Subtraction:
  fractions, 65-66
  signed numbers, 116-20
  unsigned numbers, 9-11
Subtrahend, 24
Such that, 364
Sum, 24
Systems of equations, 211-22

T

Tangent, 412
Term, 141
Trigonometry, 411-17
Trinomial, 141

U

Unknown, 140
Union of sets, 366
Units of measure, 90-96
Universe, 368

V

Variable, 140
Variation, 406-10
Venn diagrams, 369-75
Volume, 32
Volume of a box, 33
Volume of a cone, 36
Volume of a sphere, 36

W

Whole number, 3, 109
Word phrases, 179-84
Word problems, 185-95

X

x-intercept, 360

Y

y-intercept, 358

Z

Zero:
  in addition, 6, 115, 160
  in division, 22-24
  in multiplication, 13